SOCIETY AND TECHNOLOGICAL CHANGE

THIRD EDITION

RUDI VOLTI
Pitzer College

ST. MARTIN'S PRESS
New York

Editor: Sabra Scribner
Manager, publishing services: Emily Berleth
Publishing services associate: Kalea Chapman
Project management: Omega Publishing Services, Inc.
Cover design: Rod Hernandez
Cover art: Isamu Noguchi, "One Thousand Horsepower Heart." F. S. Lincoln, Courtesy the Isamu Noguchi Foundation, Inc.

Library of Congress Catalog Card Number: 94-65242

Manufactured in the United States of America.
9876
fedc

For information, write:
St. Martin's Press, Inc.
175 Fifth Avenue
New York, NY 10010

ISBN: 0-312-09642-9

Preface

Few human creations have as much influence over our lives as the technologies we use. Yet technology has only recently become a topic of concentrated scholarly attention. While law, religious beliefs, and political doctrines have long been the subject of teaching and research, technology generally has been pursued only as a practical subject in schools of engineering. But just as war is too important to be left to the generals, technology is too important to be the exclusive province of its practitioners.

The use of technology is a basic feature of all human societies, and our technologies strongly influence the way we live. Equally important but less obvious, technology itself is a product of social, economic, political, and cultural patterns. The kinds of technologies found in a particular society reveal a great deal about the nature of that society. The study of technology is therefore important not only for its own sake, but also for what it tells us about the kinds of societies we make for ourselves.

This book is intended to be used in the growing number of courses on technology and society, as well as in other courses that touch upon technology's role in human affairs. It presents perspectives, theories, and facts that will help the reader understand the consequences of technological change, as well as the forces that produced it. Many specific examples of the interaction between technological change and other changes will be introduced, for general processes are often best understood through references to particular instances.

Part One provides a general introduction to the nature of technology, and delineates some basic themes that will appear in subsequent pages. Chapter 1 traces in broad outline the key characteristics of technology in conjunction with a description of the cultural and organizational patterns that affect its development and application. Particular attention is paid to the connection between technological advance and rational modes of thought, for technological advance is often taken as the exemplification of the uniquely human ability to solve problems through rational thought and action. Still, there is a difference between technological advance and the more elusive concept of "progress," and this distinction is presented toward the end of the chapter. Chapter 2 takes up an issue that will be prominent throughout the book: that technological change is rarely neutral. It affects groups and individuals in different ways, often leaving some in a better position while others lose ground. This chapter narrates a number of historical examples that illustrate this process. It also considers the appropriateness of technological solutions to general social problems, noting when they are likely to be successful, and when they are doomed to failure.

The chapters in Part Two examine how new technologies originate and diffuse. Chapter 3 describes the contributions to technological advance of both epochal breakthroughs and slow, cumulative changes. It looks at the research and development process through the use of a number of case studies, and at the same time provides some conceptual tools for understanding the derivation of new technologies. This chapter concludes by analyzing the differences between market and planned economies in their capacity to engender sustained technological advance.

It is often thought that scientific knowledge is the prime source of technological advance. Yet this is only a partial truth, and it certainly did not hold until quite recent times. Chapter 4 traces the connections between science and technology, noting how they are different and how they are similar. This chapter also describes and analyzes the social processes that link the two.

Chapter 5 considers the diffusion of technology, again using a number of examples and brief case studies. In so doing, it covers two general topics: the transfer of technology from one nation to another, and from one organization to another. In both instances, the diffusion of technology is not treated as a one-dimensional process. One section of this chapter describes how technologies are often altered as they pass from one place to another, and another considers the problems encountered by poor countries when they adopt technologies that originate in rich ones. This chapter also includes a section that looks into the feasibility of restricting the export of technology. The final section presents some advantages and disadvantages of the patent system.

Chapter 6 is devoted to two of the key problems surrounding the application of technology: the depletion of natural resources and the destruction of the natural environment. It documents some unfortunate consequences of modern technology, but at the same time it shows that the degradation of the environment is not a phenomenon unique to recent times. On a more positive note, this chapter presents a number of technological solutions to problems of pollution and depletion, while noting that these solutions often require changes in individual, governmental, and corporate actions.

Chapter 7 introduces a topic new to the third edition: medical and genetic technologies. More than any other set of technologies, medicine reveals the dilemmas that often inhere in technological advance. Through a consideration of kidney dialysis, heart transplants, diagnostic technologies, and the effort to create an artificial heart, this chapter presents the difficult trade-offs and hard choices that often accompany medical advances. The chapter concludes with a brief discussion of the promise and perils of genetic engineering, the application of a set of technologies that may eventually dwarf all past technological advances in scope and significance.

The next eight chapters examine the interactions between technological and social change in three crucial areas: work, communication, and warfare. These areas have been selected because of their central importance to our lives. They also provide an abundance of illustrative material concerning the interaction of technological change with other sources of change.

Chapter 8 examines technology and work in preindustrial societies. This chapter describes how people have earned their livings, as well as how the technologies they use have meshed with particular kinds of social arrangements. This chapter also presents the paradoxical notion that some technological advances required people to work longer and harder than they had done before.

Chapter 9 takes up an issue that often comes up when technological advance is discussed: the apparent danger that growing numbers of people will lose their jobs to increasingly sophisticated machines. This issue is covered through a presentation of the historical record, a discussion of the job-reducing capacities of present and future technologies, and an analysis of the changing structure of the economy. The chapter concludes on a cautiously optimistic note, but at the same time stresses that some segments of the workforce will inevitably suffer job losses. Accordingly, some programs for alleviating their plight are described.

Chapter 10 shifts the focus from the quantity of jobs to their quality. It provides a brief historical narrative of the history of industrialization, and describes how specific technologies interacted with new organizational forms to create a distinctive work environment, with attendant costs and benefits. Echoing a theme introduced earlier, this chapter also considers who gains and who loses when new technologies appear in the workplace. The chapter concludes with a consideration of the effects of technologies on workers' skills.

The next two chapters are concerned with communications technologies. Chapter 11 looks at the communications revolution that took place when printing with movable type was introduced in the fifteenth century. It describes the basic technology involved, and traces the economic, social, political, religious, psychological, and cultural consequences of this new medium. The chapter then moves forward several centuries and looks into the rise of the mass-circulation newspaper, considering both the technological advances and the social changes that made this new form of communication possible.

Chapter 12 analyzes the dramatic changes in communication that followed the invention of radio and television. It provides a brief outline of the technological developments that produced these media, and then shifts to a discussion of the social context of radio and television broadcasting. It describes how commercial interests and federal regulation produced the system we have today. The narrative then turns to one of the most important and controversial issues surrounding television: the extent to which televised violence stimulates real-life violence. The chapter then considers how television news has shaped our perceptions of the world, and how TV has influenced the political process. It then moves to some speculations regarding the effect of television on the way we perceive our world, and concludes with a general assessment of radio and television's place in modern society.

The next two chapters deal with the most unpleasant aspect of technological development: the constantly expanding capacity of military technologies to kill and destroy. Chapter 13 traces the evolution of weapons systems from antiquity to modern times. Special attention is paid to how new military technologies required different methods of military organization, and how they affected social organization in general. Chapter 14 turns things around and looks at how the development of military technologies has been shaped by social forces. This chapter examines a variety of military technologies, ranging from the longbow to the battleship, in order to discern how social structure, cultural orientations, and organizational interests have influenced the development of weaponry. This chapter also considers how major historical changes, specifically industrialization and the rise of political democracy, accelerated the tempo of technological advance in the military sphere.

This chapter then takes up a topic that outweighs all other technological issues: nuclear weapons. It notes some past efforts to restrict the spread of new weapons, notes which of them were successful, and why. The discussion then shifts to the control of nuclear weapons. Nuclear strategies are briefly described, accompanied by a consideration of why modern weapons present special problems for the creation and administration of arms-control treaties. This chapter also considers nuclear proliferation and what can be done to arrest it.

The control of nuclear weapons, crucial as it is, is part of a larger problem: the control of technology in general. The chapters in Part Six address this topic. Chapter 15 begins with a consideration of the possibility that technology is out of control, and that it develops according to its own inner dynamic. This issue is explored in conjunction with a related matter: that the application of modern technology necessarily results in the homogenization of the world as all societies converge toward a common culture. Although there is some merit in these ideas, they have their deficiencies, and these are considered next. The chapter then turns to an examination of the influence wielded by specific groups, and how this affects the course of technological change. The first topic is the role played by technical experts in shaping technology. The spotlight then shifts to engineers, who occupy a potentially pivotal position in the development of technology. This section narrates how their role has been analyzed by some influential social theorists. It then proceeds to an examination of the work environment of engineers, and in so doing assesses the actual ability of engineers to influence the course of technological change.

Chapter 16 considers how organizations and the roles played by organizational actors can strongly influence technology. It also explains how relations between different organizations can significantly influence the course of technological development. The chapter concludes with a discussion of the tensions between entrepreneurial activities and organizational routines, and their consequences for technological innovation.

Chapter 17, the final chapter, takes up the very large issue of controlling technology through the institutions of representative government. It notes the reasons for government involvement in technological change, and describes some of the key governmental institutions that are involved in the direction and control of technology. The following section demonstrates that key decisions are not guided by technical criteria, but by a variety of political factors, including the power and self-interest of particular governmental agencies and their clients. The chapter concludes with some thoughts on the democratic control of technology. It indicates some of the ways in which citizens have been able to influence technological decisions, while at the same time noting some obstacles to democratic participation. It also considers how particular technologies have themselves affected the democratic process for both good and ill.

No pretense is made that these chapters provide an all-encompassing view of the topics they explore. Much has been left out because of space limitations, and my own limitations of time and energy. At the same time, systematic study of the interactions between technology and society is a recent endeavor, and many gaps remain to be filled. It can only be hoped that this book will be a useful foundation for thought and further study. The questions that follow each chapter and the bibliography at the end should be of some use in extending the material presented in these pages. If irritation at the inadequacy of the coverage leads the reader to engage in further study, then this book will have served its purpose.

ACKNOWLEDGMENTS

Writing can be lonely activity. While putting this book together some of my loneliness was alleviated by being able to call on a number of colleagues for assistance. I would like to thank the following people for reading portions of the manuscript and making invaluable suggestions: Hugh G. J. Aitken, Stephen Cutcliffe, Newton Copp, David Cressy, Barbara Gutek, Margaret Hamilton, Lamont Hempel, Sue Mansfield, Richard Olsen, Mark Rose, John Truxal, and Andrew W. Zanella. I would also like to thank those who have reviewed this and previous editions for St. Martin's Press: Elazar Barnette, North Carolina A&T University; R. Valentine Dusek, University of New Hampshire; Martin Friedman, SUNY Binghamton; Gary Gappert, The University of Akron; James Gerhardt, Southern Methodist University; Kenneth Gould, Northwestern University; James P. Hamilton, Pennsylvania State University; Charles Jaret, Georgia State University; Felix Kaufmann, Eastern Michigan University; Diane N. Long, California Polytechnic University; Carol MacLennan, Michigan Technological University; Marilyn Mertens, Midwestern State University; Dretha M. Phillips, Roanoke College; Terry Richardson, Northern State College; James Steele, James Madison University; L. E. Trachtman, Purdue University; and Rollin Williams III, East Tennessee State University. Their knowledge and expertise exceed my ability to make complete use of the help they have given me, and they are not responsible for any errors of fact or interpretation that may be found in these pages. I would also like to thank Ross Huggins for his research assistance; Andrea Guidoboni, Randi Israelow, Huntley Funsten, Louise H. Waller, Elizabeth Bast, and Sabra Scribner of St. Martin's Press for their help and guidance; and Richard R. Wright of Omega Publishing Services for his thoughtful editorial insights. In addition to reading parts of the manuscript, my wife, Ann Stromberg, was an unfailing source of encouragement and support. I am grateful to Pitzer College for naming me Scholar in Residence for one semester, during which time I was able to focus my attention on this book. Finally, in acknowledgments of this sort it is customary for an author to thank the typists who prepared the original manuscript. I cannot do so here, for I typed all of the drafts and revisions by myself, a job made much easier by word processing technology. Accordingly I would like to thank my personal computers and printers for the tireless service they have given me.

Rudi Volti

Contents

SOCIETY AND TECHNOLOGICAL CHANGE

THIRD EDITION

PART ONE

ORIENTATIONS

Chapter 1 The Nature of Technology

Today's technology leaves us both exhilarated and terrified. Recent technological developments have presented us with such marvels as the exploration of the farthest reaches of the solar system, personal computers capable of storing an encyclopedia on a small optical disc, and the transplantation of human organs. At the same time, however, the seemingly inexorable march of technology has produced global pollution, overpopulation, and the threat of nuclear annihilation. On many occasions technological change has also produced social disruptions, as when automation destroys jobs in a particular industry or a new weapon upsets the balance of power between nations. And when technologies fail, some of them do so in a big way, as exemplified by the explosion of the *Challenger* space shuttle, the disastrous venting of poison gas at the Bhopal chemical plant in India, and the massive release of radiation at the Chernobyl nuclear plant in the Soviet Union.

Despite all of the crises, disruptions, and disasters produced by modern technology, the general public still supports it, although perhaps at a diminished level compared to a generation ago.[1] Although specific technologies such as nuclear power or abortion elicit considerable hostility, the positive achievements of technology as a whole seem to outweigh the negative ones. But this support of technology is based more on faith than on understanding. When confronting technology, most of us are poorly informed spectators, seemingly incapable of understanding an esoteric realm of lasers, microprocessors, gene splicing, and cruise missiles. Although it is embedded in the fabric of our lives, technology with its associated triumphs and tragedies remains poorly understood. This inability to understand technology and perceive its effects on our society and on ourselves is one of the greatest, if most subtle, problems of an age that has been so heavily influenced by technological change.[2] But ignorance need not be a permanent condition. Although no one can hope to comprehend the inner workings of even a small number of the most significant technologies, it is still possible to come to a better understanding of the major causes and consequences of technological change. All technologies, be they stereo systems or reinforced concrete bridges, have some basic features in common. It will be the task of this chapter to show what they are.

DEFINING TECHNOLOGY

Gaining an understanding of the meaning of words is often the beginning of knowledge. Before plunging into a discussion of the nature of technology, it is necessary to provide a more precise definition of what is meant when we use the term. "Technology" is a fairly new word, having been coined by a Harvard professor,

Jacob Bigelow, in the late 1820s. The roots of the word are, however, much older. At the heart of the word is the ancient Greek word *techne*, which can be variously translated as "art," "craft," or "skill." This word is itself derived from an even more ancient Indo-European root, *teks-*, which meant to weave or fabricate.[3] (It serves as the source of our modern word "textile.")

Technologies are developed and applied so that we can do things not otherwise possible, or so that we can do them cheaper, faster, and easier. The capacity of human beings to employ technologies sets us apart from other creatures. To be sure, beavers build dams, otters crack open shellfish with rocks, and chimpanzees use sticks to extract termites from their nests. But no other animal comes close to humans in the ability to create tools and techniques, and no other creature is so dependent on them. The development of technology is in large measure responsible for the survival and expansion of a species that lacks many of the innate abilities of other animals. Left with only their innate physical capabilities, humans cannot match the speed of a cheetah, the strength of an elephant, or the leaping ability of a kangaroo. They do not possess the eyesight of an eagle, or the defensive armament of a porcupine, and they are among the 25 percent of all species that are incapable of flying. All in all, humankind is a physically puny bunch. But compensating for this physical weakness is an intelligence that is the ultimate source of technology. Humans stand apart from all other animals in their ability to gain and transmit knowledge, and to use this knowledge to develop tools and techniques. Without this capacity to invent and use a great variety of technologies, *Homo sapiens* would have never been able to establish themselves on virtually every part of the globe.

Reliance on technology is as old as the human race. Whatever evils have accompanied the use of particular technologies, it is pointless to indict technology as being somehow "unnatural." Our past as well as our future as a species is inextricably linked to our capacity to shape our existence through the invention and application of implements and techniques that allow us to transcend our meager physical endowments. It is certainly true, as Jacob Bronowski observed, that "to quarrel with technology is to quarrel with the nature of man—just as if we were to quarrel with his upright gait, his symbolic imagination, his faculty for speech, or his unusual sexual posture and appetite."[4]

Tools and techniques have been of unquestioned importance in allowing the physical survival of the human species. Still, they are not the whole story. It is necessary to add some elements to our definition of technology that go beyond the usual identification of technology with pieces of hardware and ways of manipulating them. The first of these is *organization*. This follows from the fact that the development, production, and employment of particular technologies require a group effort. Even a relatively simple technology, such as one centering on the use of earthenware pots, requires a complex network of material suppliers, potters, tool makers, marketing agents, and consumers capable of making good use of the pots. Of course, one person can learn to do all these things adequately if not expertly, but the day is not long enough for him or her to do them all on a scale that produces a reasonable degree of efficiency. In the case of a complex technology like a computerized manufacturing system, there is no possibility of a single individual developing even a tiny fraction of the requisite skills. For a technology to be developed and used, the energies and skills of many individuals have to be combined, and this is done through some system of organization.

For these reasons, the development and operation of any technology requires an organizational framework so that individual activities can be coordinated and focused. Organization may be likened to the software that controls and guides a computer. Without an operating system and application programs, a computer is a useless arrangement of capacitors, transistors, resistors, and other bits of hardware. In similar fashion, an organizational system allows the integration of diffuse human and material inputs for the attainment of particular tasks. From this standpoint, there is considerable merit in Lewis Mumford's assertion that the first "machine" was not a physical object, but the organizational structures that the Egyptian pharoahs employed to build the pyramids.[5]

When technology is seen as a combination of devices, skills, and organizational structures, it becomes natural to think of it as a *system*. For an individual technology to operate effectively more is required than the invention of a particular piece of hardware; it has to be supported by other elements that are systematically interconnected. When Thomas Edison began to work on electrical illumination, he realized that this technology would require the development of a complete system. The invention of a practical, long-lasting light bulb rested on the development of a serviceable filament and the use of a recently invented vacuum pump that evacuated the interior of the bulb, thereby preventing the combustion of the filament. But by itself, a light bulb was useless. An effective electrical generator was needed to supply the current that produced the incandescence of the filament. A network of electrical lines had to be strung up between the generator and individual homes, shops, and factories. And metering devices were necessary so that users could be accurately billed for the electricity they used. Edison and his associates worked out all of these problems, and in so doing brought large-scale electrical illumination to the world.[6]

The development of all the elements of a technological system can be an uneven process, for technological advance often entails the resolution of tensions that are generated by changes in one part of the technological system. This process is exemplified by the development of the modern airplane. Early biplanes with their drag-inducing wires and struts could not make effective use of more powerful engines. The possibility of using these engines became a strong inducement to the design of aerodynamically cleaner aircraft. The faster aircraft that resulted from the marriage of streamlined airframes and powerful engines produced a new problem: dangerously high landing speeds. This in turn stimulated the invention of wing flaps and slots. By the 1940s it had become apparent that improved airframes could achieve still higher speeds if provided with more powerful engines; this possibility gave a strong stimulus to the development of the turbojet.[7]

The basic components of a technological system are not just material artifacts; human skills, organizational patterns, and attitudes are of equal importance. An example of how these fit together is provided by Lewis Mumford's analysis of the technology of handwriting.[8] Two hundred years ago, the standard writing instrument was a gooose-quill pen. Based on an organic product and sharpened by the user, it represented the handicraft technologies typical of its time. Cheap and crude, it called for a fair degree of skill if it was to be used effectively. In contrast, the steel-nib pen of the nineteenth century was a typical artifact of the industrial age, the product of a complex manufacturing process. Less adaptable than the quill, it was mass-produced in many different forms in order to meet specialized needs. Although Mumford's ideas were formulated before the invention of the ballpoint pen, his analysis fits this implement perfectly. Made from a variety of artificial materials and built to close tolerances, the ballpoint pen could only be produced through sophisticated industrial processes. It is completely divorced from the organic world and requires very little skill from its user. Indeed, the technological artistry embodied in the pen itself stands in sharp contrast to the poor quality of the writing that so often comes from the hand that wields it.

A technological system does not emerge all at once with every one of its components neatly fitting together. In addition to changes in tools, techniques, and organizational structures, many social, psychological, economic, and political adjustments may be required for the support of a technological system. Technological change is not always a smooth process, and many of the necessary changes may entail considerable pain and disruption. Seeing technology as a system should help us to understand that technological change is closely connected with a variety of associated changes, and that the creation of a technological system may be fraught with tension and discomfort.

It is time to put all of these ideas together into a basic definition of technology: **a system based on the application of knowledge, manifested in physical objects and organizational forms, for the attainment of specific goals.**

While other definitions of technology are possible, this one should serve our purposes reasonably well, for it highlights a number of themes that will appear continuously in this book: the systematic interconnection of technology with other human endeavors, its close ties to human knowledge, and its intimate connection with the organized structuring of human activities.

Useful as it may be, the above definition of technology is incomplete and possibly misleading in one important respect. The last part of the definition implies that technological change comes about as a response to existing needs: its purpose is "the attainment of specific goals." In the first place, one could legitimately ask *whose* goals are to be attained. This is an important issue, but it is best left for the next chapter. For now, we should note that technology does not always respond to existing needs; a new technology may in fact create its own needs. The development of technology on occasion exemplifies a phenomenon that has been dubbed "the law of the hammer": give a six-year-old a hammer, and to the child everything starts looking like a nail.

The history of technology is replete with examples of inventions looking for problems to solve. One example that illustrates this point is found in almost every medicine chest: a bottle of aspirin. One of the most common uses of aspirin is to suppress fevers that accompany various illnesses. But recent medical research (as well as some ancient practices) has demonstrated that running a fever is a therapeutic process that aids in a patient's recovery; it is the body's way of naturally combating infection. Yet since the introduction of aspirin in the early 1900s fever has been seen as a problem requiring intervention. As one medical researcher has noted, "It's no surprise that society's deep worries about fever closely followed the synthesis of aspirin, the first drug that could safely reduce it."[9] In short, a new technology created its own need.

It is also important to note that the goals achieved through the use of a technology do not have to be "practical" ones. Some technologies have been developed so that we can grow more food or construct more comfortable buildings, but others have been developed simply for the challenge and enjoyment of solving technological problems.[10] The prodigious efforts that went into the Daedalus Project, a successful attempt to build a human-powered aircraft capable of flying forty miles across the open sea, were certainly not motivated by an effort to produce a new form of transportation. Nor were economic motivations paramount, for the financial rewards were miniscule in comparison with the time and effort put into the project. The major reason for creating the aircraft was that its construction posed an intriguing technological challenge to those who designed, built, and flew it.

Flight seems to be a particularly attractive object for this kind of spirit. Immensely expensive technological endeavors such as the supersonic Concorde airliner and manned space exploration programs are hard to justify on practical grounds, although their supporters have made valiant efforts to do so. Their primary purpose seems to be the elevation of national prestige by demonstrating a nation's collective ability to solve daunting technological problems. At the same time, many other technologies have a dual nature; they serve a practical purpose, but they are not valued only for this reason. An outstanding example is the automobile. It would be hard to justify the enormous resources employed for the building and operation of cars if transportation were the only goal. For many people (the author included), cars are objects of inherent fascination. Technological features like turbocharging and active suspension systems are hard to justify on utilitarian grounds. The appeal is at least as much in the sophisticated technologies themselves as in the purposes that they serve.

TECHNOLOGICAL ADVANCE AND THE IMAGE OF PROGRESS

The development of technology is an inherently dynamic and cumulative process. It is dynamic because a technology is never perfect; there is always room for improvement. As Henry Ford said of his firm, "If we have a tradition it is this: Everything can always be done faster and better."[11] It is cumulative, for one advance paves the way for another. The lessons learned in working with an existing technology very often provide materials, tools, and most importantly a knowledge base for the next stage of development.

The dynamic and cumulative nature of technological change sets it apart from many other human endeavors. Ignoring for the moment the social consequences of technology, the process of technological change is usually one of continuous improvement in the internal workings of a particular technology: As they evolve, engines develop more power and are more efficient; integrated electronic circuits pack more components on a single chip; aircraft fly higher and faster.

Although about as different as two airplanes can be, the Concorde SST and the Gossamer Condor both exemplify the desire to advance beyond existing technological frontiers. (Air France and AeroVironment, Inc.)

The process of technological advance can be graphically portrayed according to the following diagram, in which the horizontal axis represents time and the vertical axis represents just about any aspect of technological advance: the speed of commercial airliners, the production of synthetic materials, or the number of articles in engineering journals. Although there are inevitable fits and starts over time, the general trend can be depicted as a sigmoid or S–shaped curve:

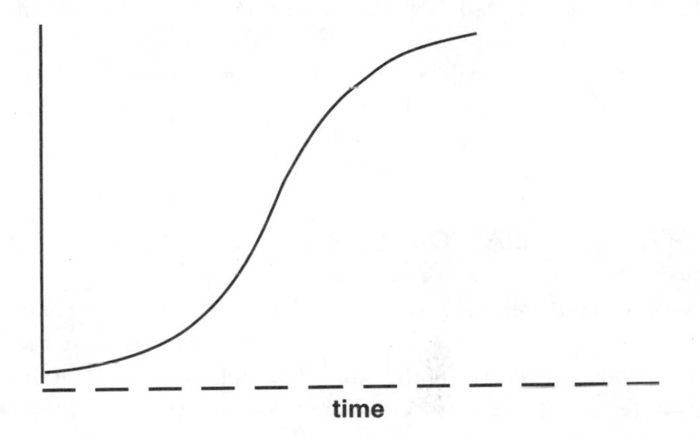

Note that at first the curve rises rather slowly, inclines steeply in the middle, and then begins to slow down. That is, after an initial period of slow growth, the rate of advance accelerates, reaches a maximum, and then begins to proceed at a slower pace, but never completely levels off. Although the rate of increase is smaller as the curve moves toward the right, this rate is applied to an increasingly larger base, so the actual addition is still substantial.

Not all human endeavors can be fitted to this sort of curve. While technology tends to be dynamic and cumulative, the same cannot always be said of other manifestations of human creativity. Although there is ample room for debate, a good case can be made that succeeding generations of writers, composers, and painters have not produced works superior to the ones created by Shakespeare, Beethoven, and Vermeer. And while we continue to take great pleasure in the artistic creations of eras long past, few of us would be satisfied with the technologies that were prevalent in those times.

We also see few indications that people are more humane than they were centuries ago. Our own century certainly provides a multitude of horrifying examples of human cruelty, many of them augmented by enlisting technology in the service of slaughter and destruction. Still, when judged solely according to its own criteria, technology is perhaps the best example of humankind's largely unrealized dream of continual progress.

Technological progress, however, is not the same thing as progress in general. The fact that a society is able to develop and make use of advanced technologies does not guarantee that it will be equally advanced in other areas. Nazi Germany produced many technological triumphs such as the all-conquering Mercedes and Auto Union grand prix racing cars of the late 1930s and the V–2 rocket used during World War II, but in its ideology and treatment of people it can only be described as barbaric. Conversely, many technologically primitive peoples have exhibited a high level of sophistication in their artistic creations, religious beliefs, and social relationships. The term "progress" can be used with some precision when applied to the development of technology per se, although even here problems can crop up because different standards of evaluation may lead to conflicting conclusions. Is it really "progress" when a new medical technology maintains an individual's life, but does so only at enormous expense, while preserving nothing but the maintenance of organic functions? Does maintaining a "life" of this sort justify expenditures that otherwise might be used for expanded prenatal care or other preventative measures? Given all of the value judgments, ambiguities, and complexities surrounding the word "progress," its use is avoided here unless a clear meaning is evident.

TECHNOLOGY AS A METAPHOR

Despite the above qualifications, it is evident that beginning in the late eighteenth century and continuing today, technology's stunning advances have fueled a belief in generalized human progress. In this way, technology has operated as a metaphor—the transference of an idea from one area to another. Technology has provided many other metaphors that have affected our way of looking at ourselves and the world, as when human thought is made analogous to the operation of a digital computer.

A further example of the power of a technology to shape our way of thinking comes from the late eighteenth century. At that time the designers of windmills and steam engines discovered the important principle of *feedback*, which the great twentieth-century mathematician Norbert Wiener defined as "a method of controlling a system by reinserting in it the results of its past performance."[12] When a steam engine begins to rotate too rapidly, a feedback device such as a flyball governor closes the valve that admits the steam, thereby bringing the engine back into its proper operating range. When it slows down, the reverse happens, and the governor opens the valve to admit more steam.

During the late eighteenth century the feedback principle offered a suggestive metaphor for the workings of the economic system: instead of being guided by a centralized authority, an economy might best be organized through the operation of a self-regulating market, with the actions of independent buyers and sellers providing the feedback. Thus, when buyers wanted a particular commodity, its price would be high, motivating sellers to produce more of it. If the price were low, less would be produced. In similar fashion, an increase in production would cause the price of a commodity to fall, so more of it would be purchased, while a drop in production would cause the price to rise, leading to a reduction of purchases. In this way, the actions of buyers and sellers in the market provide a feedback mechanism through which supply and demand are supposedly brought into equilibrium. It is probably no coincidence that the Scottish economist Adam Smith developed this basic concept at the same time that the steam engine was being put into service. Today, the widespread use of the feedback principle makes its apparent applicability to the economic system even more appealing, even though the real-world economy is hardly a neat closed system like a steam engine. Laws and regulations, as well as a host of other extraneous elements may strongly affect individual feedback loops, thereby preventing a complex economy from operating solely on the basis of supply-and-demand signals. Technological development has supplied a useful metaphor in the feedback principle, but like all metaphors it cannot be taken as a literal depiction of reality.

TECHNOLOGY AND RATIONALITY

The development of technology has stimulated a belief that progress is a natural part of human life. At the same time, the progressive development of technology has itself been the product of a distinctive set of cultural values and mental processes that are characterized by a rational approach to the world and how it is to be controlled. Technological development is more than the random accumulation of tools, techniques, and organizational forms. Underlying the process is a set of attitudes and orientations that are collectively described as "rational."

What makes a technologically progressive society different from others is that its methods of problem-solving are oriented toward an objective scrutiny of the problem at hand, coupled with a systematic examination of possible solutions and a logical selection of the most appropriate ones. Beyond this approach to the solution of problems lies another cultural attribute: a basic belief that solutions are *possible* and that constant changes are necessary in order to realize them. A society imbued with a rational ethos is dynamic and essentially optimistic, and it exhibits the confidence necessary to alter existing ways of doing things in order to gain particular benefits.

These abstract concepts may be illustrated through a simple example. All societies are faced with the problem of coping with the capriciousness of the weather. A great deal of human suffering has been the result of the vagaries of rainfall, and history provides many examples of the tragic consequences of drought. A number of responses are possible when people are confronted with this problem. The simplest is to succumb to despair, and perhaps try to find meaning in it by attributing the drought to fate or God's will. A more active approach might be to offer prayers, perform a special ceremony, or sacrifice a member of the community. But these latter activities are not likely to meet with success. There is no logical or empirically verifiable connection between them and the circumstances that produced the drought, a fact that could be demonstrated by a systematic inquiry into the long-term connection between prayers, ceremonies, or human sacrifices and the incidence of rainfall.

These attitudes and behaviors stand in sharp contrast with rational ones. Through the use of logic and empirical observation it is possible to develop ways of dealing with problems like drought that are both more effective and more closely connected to the way the world actually works. A systematic and empirical observation of weather patterns might allow the prediction of a drought so that necessary steps can be taken to alter farming practices and conserve water. Other solutions could be the development of drought-resistant crops, methods of conserving water, or possibly the distillation of sea water. It might also be possible to artificially stimulate rainfall through cloud seeding. In short, a rational approach to problem-solving is continuously concerned with the identification and development of appropriate means for the achievement of particular ends.

The above remarks are not meant to convey the ethnocentric belief that modern, Western culture is superior to all others. The intention is not to ridicule the beliefs and practices of people and societies not imbued with rational thought patterns. No assumption is being made here that rationality has been and always will be the special attribute of a particular group of people. Moreover, modern societies often manifest behaviors and patterns of thought that are anything but rational, as when large numbers of people continue to believe in astrology, numerology, and the predictions of supposed psychics.

It is also important to recognize that rational ways of thinking do not confer moral superiority. To the contrary, the rigorous development and use of rational procedures can be accompanied by major moral and ethical transgressions. The rational method of problem solving, with its overarching concern for devising appropriate means for attaining particular ends, makes no distinction concerning the ends being pursued. There is nothing in the rational approach to the world that prevents the use of logically and empirically derived means in the service of goals that are neither rational nor ethically justifiable. We can take note of the words of Captain Ahab, the main figure in Herman Melville's novel *Moby Dick:* "All my means are sane, my motive and subject mad." Nazi Germany provides many ghastly historical examples of human destruction ensuing from rational thinking and its resultant technologies. As Albert Speer, Hitler's Minister of Armaments ruefully noted, "The criminal events of these years were not only an outgrowth of Hitler's personality. The extent of the crimes was also due to the fact that Hitler was the first to be able to employ the implements of technology to multiply crime".[13]

Even when rationality is not used for manifestly immoral purposes, it can still leave a dubious spiritual legacy. The very strength of rationality and the scientific and technological accomplishments that flow from it lie in their matter-of-fact approach to the world. A rational approach to things is often accompanied by a reluctance to admit there are any forces incapable of withstanding logical and empirical scrutiny. As the great German sociologist Max Weber put it, the world defined by rational thought processes had become "disenchanted," for it was bereft of the gods, genies, and spiritual forces that people not imbued with the spirit of rationality used to explain their world. But "disenchantment" is a two-edged sword, as the everyday meaning of the word makes clear. To be disenchanted is to lose the sense of awe, commitment, and loyalty that is a necessary part of a meaningful existence. Weber's melancholy analysis of a world that has lost its enchantment is summarized by the French sociologist Julian Freund:[14]

> With the progress of science and technology, man has stopped believing in magic powers, in spirits and demons; he has lost his sense of prophecy and, above all, his sense of the sacred. Reality has become dreary, flat and utilitarian, leaving a great void in the souls of men which they seek to fill by furious activity and through various devices and substitutes.

Similar misgivings were voiced by the eighteenth-century political philosopher Edmund Burke. Burke's primary concern was the destruction of traditional authority by modern mass movements, as exemplified by the French Revolution. Burke attributed much of the demonic energy of that movement to the spread of rational modes of thought that left no room for the traditional attitudes, values, and political structures that had long sustained European civilization. Burke's comment on the downfall of the queen of France, Marie Antoinette, thus contains a sharp indictment of the bearers of rational values that were leading Europe to its doom:[15]

> . . . little did I dream that I should have lived to see such disasters fallen upon her in a nation of gallant men, in a nation of men of honor and of cavaliers. I thought ten thousand swords must have leaped from their scabbards to avenge even a look that threatened her with insult. But the age of chivalry is gone. That of sophisters, economists, and calculators, has succeeded; and the glory of Europe is extinguished for ever.

Rationality also implies objectivity; coolness and detachment are part of the rational approach to understanding and changing the world. Guided by a rational outlook, scientific inquiry and technological application are usually based on the abstraction or isolation of the part of the natural world that is being studied or manipulated. This isn't always a good thing, for it can produce a sharp separation between the individual and the rest of the world. The scientist or technologist stands apart from the system that is being studied and manipulated, resulting in a kind of tunnel vision that all too often ignores the larger consequences of gaining and applying knowledge.[16]

The tendency of science and technology to take nature as an object to be understood and then conquered is part of the general tendency of rationally oriented people to dominate their world. This attitude is described in the legend of Faust, whose pact with the devil gave him universal knowledge in return for the loss of his soul. For a society with no greater goal than gaining knowledge and power, there lies the danger of succumbing to what Lionel Rubinoff has called "the law of the possible": "What is possible for science to know science must know. What is possible for technology to do technology is obliged to do. Whatever is possible is obligatory."[17] Knowledge-based power can thus become an end in itself. In the absence of any clear vision of the proper purposes of technology, it is all too easy to let technology's agenda be set by little more than a self-interested search for novelty. Should this happen, life becomes, in the phrase of the philosopher Alfred North Whitehead, "a welter of minor excitements completely devoid of any sense of plan or purpose."[18]

The development and application of a vision that determines the proper purposes of technology requires the participation of a wide range of people. Unfortunately, the very nature of modern technology places severe limits on popular understanding. The sophistication and complexity of contemporary technology preclude direct involvement by all but those immediately concerned with them. The rest of us are passive consumers, content to reap the benefits of rationally derived knowledge, but woefully ignorant of it. This creates the fundamental paradox of modern society: technology has generated massive powers available to human society, while as individuals we exert very little of that power. We have access to a wide range of powerful technologies, yet our inability to understand them often leaves us with feelings of impotence and frustration.[19]

As was noted above, the application of rationality for the solution of human problems is both the consequence and the cause of optimism and a willingness to accept constant change. Yet one cannot help but wonder if these characteristics can be sustained in an environment that sharply limits participation and inculcates widespread feelings of having no power over the process of technological change.

The mixed consequences of a technologically advanced society are well illustrated by the enthusiasm evoked in certain quarters for Laetrile, an alleged cancer cure that achieved notoriety in the 1960s. The medical profession has almost universally denied the efficacy of Laetrile, yet hundreds of cancer patients have gone to clinics in Mexico for Laetrile treatments. Underlying this faith in a bogus medication is the belief, common in technologically advanced societies, that modern medicine is capable of providing miraculous cures. After all, since "wonder drugs" have cured a wide variety of illnesses, why can't the same thing be done with cancer? Yet at the same time, a belief in the curative powers of Laetrile is accompanied by a distrust in the keepers of medical technology, the health-care "establishment." Laetrile enthusiasts seem to be of the opinion that conventional doctors are intent on keeping tight control over medical technologies, zealously guarding against the use of curative practices that do not meet with their approval. From this standpoint, health-care professionals are engaged in a conspiracy against the public, using their monopolization of medical practice to reinforce their privileged positions. There is certainly an element of truth in this accusation; control over technology can be a source of power. A belief in the efficacy of Laetrile thus combines an excessive confidence in technological solutions with a distrust of those who are entrusted with their dispensation.

The example of Laetrile illustrates how the lack of participation in technological development and application can produce a mixture of naive hope and paranoid reaction. Technology is an immensely powerful force in our world, and its material benefits cannot be separated from its consequences for people and the societies in which they live. The next chapter therefore will take a closer look at some of the social and political issues surrounding the development and use of technology.

Questions for Discussion

1. In your opinion, which recent technology has produced the greatest benefit? Which has produced the most harm? Are there any harmful elements to the beneficial technology, and has anything good come from the harmful one?
2. Do all technologies require tools of some sort? Does it make any sense to speak of *bureaucracy* as a kind of technology?
3. What sort of things went into the technological system that allowed the landing of people on the moon? How were all of these elements put together?

4. Can you think of any technologies that were developed simply because of the technical challenges involved? How can these "impractical" technologies be justified?
5. How do you feel when a technological device upon which you depend malfunctions? What do these feelings tell you about your attitude toward technology in general?

Notes

1. Eliot Marshall, "Public Attitudes to Technological Progress," *Science* 205, 4403 (July 1979): 284–285; "Technology: Uses, Promises, Fears," *The Public Perspective*, 5, 2 (January 1993).
2. James D. Carroll, "Participatory Technology," in Thomas J. Kuehn and Alan L. Porter (eds.), *Science, Technology, and National Policy* (Ithaca, N.Y.: Cornell University Press, 1981), p. 416.
3. Langdon Winner, *Autonomous Technology: Technics-Out-of-Control as a Theme in Political Thought* (Cambridge, Mass.: The MIT Press, 1977), pp. 11–12.
4. J. Bronowski, "Technology and Culture in Evolution," *Philosophy of the Social Sciences* 1, 3 (1971): 199.
5. Lewis Mumford, "Technics and the Nature of Man," *Technology and Culture* 7, 3 (July 1966): 303–317.
6. Matthew Josephson, "The Invention of the Electric Light," *Scientific American* 201, 5 (November 1959): 98–114.
7. John B. Rae, *Climb to Greatness: The American Aircraft Industry, 1920–1960* (Cambridge, Mass.: The MIT Press, 1968), p. 74; Edward Constant, *Origins of the Turbojet Revolution* (Baltimore: The Johns Hopkins University Press, 1980).
8. Lewis Mumford, *Technics and Civilization* (New York: Harcourt, Brace and World, 1934), p. 110.
9. Edwin Kiester, Jr., "A Little Fever is Good for You," *Science 84* 5, 9 (November 1984): 172.
10. Daedalus of New Scientist, "Pure Techhnology," *Technology Review* 72, 7 (June 1970): 38–45.
11. Quoted in Edward Constant, op. cit., p. 12.
12. Otto Mayr, "The Origins of Feedback Control," *Scientific American* 223, 4 (October 1970): 110–118.
13. Albert Speer, *Inside the Third Reich* (New York: Macmillan, 1970), p. 212.
14. Julian Freund, *The Sociology of Max Weber* (New York: Pantheon, 1968), p. 24.
15. Edmund Burke, *Reflections on the Revolution in France* (New York: Holt, Rinehart and Winston, 1959), p. 91.
16. Richard Schlegel, "Why Can Science Lead to a Malevolent Technology?" *The Centennial Review* 21, 1 (Winter 1977): 14.
17. Lionel Rubinoff, "Technology and the Crisis of Rationality: Reflections on the Death and Rebirth of Dialogue," *Philosophy Forum* 15, 3–4 (1977): 273, 278.
18. Quoted in Elting E. Morison, *Men, Machines, and Modern Times* (Cambridge, Mass.: The MIT Press, 1966), p. 14.
19. N. Bruce Hannay and Robert E. McGinn, "The Anatomy of Modern Technology: Prolegomenon to an Improved Public Policy for the Social Management of Technology," *Daedalus* 109, 1 (Winter 1980): 30.

Chapter 2 Winners and Losers: The Differential Effects of Technological Change

The last chapter may have seemed a bit negative in its assessment of technology and the culture that supports it. In one regard, however, there is no denying technology's positive consequences: technological advance has been the greatest single source of economic growth. If our material lives are better than those of our grandparents, it is largely because technological development has boosted the production of goods and services. Equally important, it has created entirely new products while at the same time improving the quality of existing ones.

Curiously, economists were slow to grasp this seemingly obvious fact. Conventional economic analysis identifies three basic "factors of production": land (which includes natural resources), labor, and capital. Any increase in production is therefore taken to be the result of an increase of these factors. This view began to change in the 1950s when the historical course of economic development in the United States was analyzed through the use of sophisticated statistical techniques. It then became apparent that increases in the traditional factors of production did not adequately explain the actual record of economic growth. The amount of land had remained constant, and capital accumulation and increases in the labor force accounted for only 10 to 20 percent of economic growth during the first half of the twentieth century.[1] Accordingly, the major source of economic growth was a "residual" factor of overwhelming importance. Most economists agree that technological advance is the main element of this residual, although organizational development and improved worker skills, along with economies of scale, are also key components. Still, as we have already seen, organization and skill are integral parts of technology, so it is reasonable to consider technological change as the major source of economic advance.

TECHNOLOGY AS A SUBVERSIVE FORCE

While technological development has been responsible for a vast amount of economic advance, it has not been without cost. One of the most pleasant myths about technology is that it can work its wonders without altering existing social arrangements. Americans in particular have often seen technological progress as the surest basis for progress in general, and have tended to believe that technological solutions to problems are less painful than solutions that require political or social changes.[2] These beliefs are not easily sustained after an examination of the actual pattern of technological advance.

It has become a cliché that a particular technology can be used for either good or evil purposes; a construction team employs dynamite to build a road, while a terrorist uses it to blow up an airplane. But there is less appreciation for a more subtle point: technological change is often a subversive process that results in the modification or destruction of established social roles, relationships, and values. Even a technology that is used exclusively for benign purposes will cause disruptions by altering existing social structures and relationships. There are many technological changes that are small in scope, the effects of which are felt by only a few. A few technological changes are massive, and they lead to vast social restructuring. In either case, technology does not yield its benefits without exacting a cost.

The disruptive effects of technological change can readily be seen in the economic realm, where new technologies can lead to the destruction of obsolete firms, as when the fabled Pony Express rapidly lost its customers after telegraph wires had been strung across the West. Of course, sometimes the disruption is less apparent when technological innovation results in the creation of entirely new industries that are not in direct competition with existing ones. Many new industries and individual firms owe their existence to the emergence of a new technology. Witness for example the explosive growth of microcomputer manufacturers, software publishers, and producers of peripheral equipment that followed the invention of the integrated circuit. Even so, lurking behind these successes were a number of failures, most notably the manufacturers of vacuum tubes and transistors, who faced a diminished market for their products.

Concerns about the disruptive effects of technological change are not new, as can be seen in an English magazine editor's fulminations against the first railroads in 1835: "Railroads, if they succeed, will give an unnatural impetus to society, destroy all the relations that exist between man and man, overthrow all mercantile regulations, and create, at the peril of life, all sorts of confusion and distress."[3] Anyone convinced of the virtues of technological change could easily criticize this reactionary view by noting how the railroad stimulated economic development and produced many social benefits. Even so, there is more than a grain of truth in the concerns expressed by the agitated magazine editor. Technological changes, both major and minor, often lead to a restructuring of power relations, the redistribution of wealth and income, and an alteration of human relationships.

The experiences of the Yir Yoront, a group of Australian aboriginals, gives us an excellent, albeit sad, example of the disruptive effects of a new technology.[4] The Yir Yoront were a truly paleolithic people whose highest technological achievement was the stone axe. These axes were simple implements, but a considerable amount of skill went into their production. Several different materials had to be gathered—wood for the handle, bark for binding, and gum for fixing the head to the handle. The stone itself was obtained through an elaborate trading network that involved only adult males. The actual possession of the axes was also an exclusively male prerogative. Women and children could only borrow an axe, and even then only from close relatives. The axe also had an important symbolic value, for it was a totemic symbol that was used in certain religious ceremonies performed by men only. Thus the production and use of the axes reflected and reinforced traditional social relationships based on age, sex, and kinship.

All this changed when steel axes began to be introduced into Yir Yoront society during the early twentieth century. These axes were dispensed as gifts by missionaries, and they were given to all "worthy" members of the society, including women, young men, and even children. As a result, mature men lost an important indicator of their distinctive status. At the same time, the trading networks between men of different tribes were bypassed. In their place new trading relationships emerged, with some men even prostituting their wives in return for the axes. The possession and distribution of axes no longer symbolized traditional relationships; a certain kind of freedom was achieved, but only at the expense of confusion and insecurity. A more general malaise spread through the entire tribe, for the steel axes had no clear links with the religiously based explanations of how the world came to be as it was; they were alien objects whose origin could not be explained. Symbolically, steel axes represented a new world that the Yir Yoront could not comprehend. The result was rapid cultural disintegration and a bewildered and apathetic populace.

To be sure, it wasn't the axes themselves that produced these disruptions. Steel axes were part of an outside world that was impinging on the traditional aboriginal order. Stone axes were an integral part of the indigenous technological system, while steel axes were alien intrusions that represented both a new technology and a new pattern of social relationships. For the Yir Yoront, the two were so closely intertwined that the introduction of a new artifact produced a social and cultural crisis that could not be surmounted.

"Primitive" people are not the only ones subject to the unpleasant consequences of technological change. On occasion, technological advance has fatally disrupted modern communities and the people living in them. One such place was Caliente, Nevada.[5] Caliente was a small town with a variety of civic amenities—schools, churches, a hospital, a theater, a park, and many prosperous small retail businesses. Many of its inhabitants were proud members of civic organizations such as the Chamber of Commerce, the Rotary, the Masons, and the American Legion. It was a typical American small town, with typical American small-town values.

The life of the town was supported by a single industry: the servicing of steam locomotives. Caliente was an important division point on a transcontinental railroad, and many of the town's people worked as machinists, boilermakers, and repairmen. Their incomes in turn supported Caliente's commercial and civic establishments. Then, in the late 1940s, the diesel-electric locomotive rapidly replaced the steam locomotive. Diesels had many advantages; they were more fuel-efficient, hauled longer trains, and did less damage to the rails and roadbed. They also required less frequent servicing. When servicing was required, it took place in large centralized shops. As a result, servicing was eliminated at many division points, and Caliente was one of them. The town lost its economic base, and within a few years it had become a shell of its former self. People moved out, homes were abandoned, and shops were boarded up. The local newspaper sadly noted that "Employees who have given the best years of their lives to this railroad are cut off without anything to which they can turn, many of them with homes in which they have taken much pride; while others, similarly with nice homes, are told to move elsewhere. . . ."

The tragedy of this small town has been repeated in many other communities affected by technological change. Many places of employment have closed down as new products and processes have replaced old ones, leaving communities and their inhabitants in desperate straits. The technological advances that produced these dislocations may have benefited society as a whole, but at great cost to the people who were immediately affected.

Technological changes do not always result in the destruction or modification of an existing social order; sometimes they may help to preserve it, as happened when pneumatic molding machines were adopted by the McCormick reaper manufacturing plant in the 1880s.[6] These machines were not installed, as conventional analysis would lead us to think, in order to reduce costs or to produce a better product; in fact, they were deficient on both counts. They were installed for the sole purpose of eliminating the skilled workers who formed the backbone of the National Union of Iron Molders, an organization that was challenging the entrenched authority of McCormick's management. The molding machines allowed the replacement of skilled workers by unskilled ones, and three years later, having served their purpose, they were discarded by McCormick's management.

Groups that are threatened by a technological innovation are not always as helpless as the iron molders apparently were. Many affected parties have been able to defend themselves against changes in the way of doing things. To take one example, prefabricated buildings have been vigorously resisted by many local construction workers' unions because they threaten their members' jobs. One sad tale is narrated by Peter Blake:[7]

> Shortly after the end of World War II, an enterprising manufacturer decided to mass-produce a so-called service core: a complete "package" containing kitchen, bathroom, and utility room, with all fixtures, pipes, ducts, and wires in place, ready to be plonked down in any typical suburban house.
>
> The first twenty of these beautifully designed and beautifully made "packages" arrived on a site near Detroit; local union plumbers and electricians promptly refused to install them. Finally, after nine months of heated debate (during which the units, parked on a sidewalk, were exposed to weather and vandalism), the local unions agreed to handle the "packages"—by disassembling them on the sidewalk and then reassembling them, piece by piece, in each of the houses. The manufacturer, needless to say, thereupon went out of business.

Nineteenth-century China provides another example of the efforts of a group of people defending their interests in the face of a potentially disruptive technological change.[8] For centuries, the Chinese had produced silk thread by manually unwinding silkworm cocoons. The technology employed, although unsophisticated, was adequate to serve a substantial domestic and export market. Then, in 1859, a representative of the British Jardine Matheson trading company arrived in Shanghai with the intention of building a modern factory that would use steam-powered machinery to reel the silk. The machinery required skilled labor for its operation, and many problems were encountered in mustering an adequate labor force. This obstacle was eventually overcome, and the factory enjoyed an adequate measure of technical success. Unfortunately, it was not an economic success, for the high price of its basic raw material, silkworm cocoons, was not offset by increased productivity, and the enterprise suffered chronic losses until it closed down less than ten years after its founding. The significant point here is that the factory could not obtain cocoons at reasonable prices due to the opposition of an entrenched silk-makers' guild. Accustomed to monopolizing silk manufacture, the guild prevented most individual cocoon producers from having any dealings with the foreign operation, while the few who did were able to charge high prices for their wares. As happened with the disgruntled construction workers, the Chinese guild members effectively undermined a technology that threatened their established ways of doing things.

THE LUDDITES

There have been many other occasions when individuals and groups have recognized that certain technological changes were not working to their advantage. In some cases, their reactions have taken a violent turn. The most famous of these are the outbreaks of machine-smashing that occurred in early nineteenth-century England.[9] These attacks were the work of different groups who were collectively known as Luddites, a name that was derived from one Ned Ludlum, an apprentice stocking maker who, as legend had it, answered his master's reprimand by smashing his stocking frames with a hammer. There was really nothing new about these attacks; the breaking of machines by disgrunted workers had a long history in England, the earliest recorded episode taking place in 1663. But the Luddite disturbances that began in 1811 did represent a substantial increase in the scale of these attacks; by the following year the government had to deploy 12,000 troops in order to restore order to the parts of England affected by the movement.

Since these attacks coincided with an era of rapid technological change, it is easy to draw the conclusion that they were motivated by the fear of many workers that their jobs would be lost to new machinery. The actual story is a bit more complicated. Luddite attacks occurred in a number of separate branches of the textile industry, and each was characterized by a distinctive set of motivations and responses. The Luddite movement began in the hosiery trades, where there long had been opposition to the use of wider stocking frames that allowed the employment of poorly paid unskilled labor for the manufacture of an inferior product. The situation might have been resolved in a peaceful manner had it not been for the dire conditions encountered by many of England's working people at the time. The Napoleonic wars had resulted in the closure of many export markets, leading to a general trade depression. To make matters worse, a series of bad harvests led to sharp increases in the cost of food, and many workers found that their wages were insufficient to meet their basic needs. These conditions produced a fertile ground for the spread of "collective bargaining by riot," and Luddite attacks were soon fomented by handloom weavers and shearers in the textile industry. The weavers viewed the advance of steam-powered weaving machinery with understandable apprehension, and, following the example of workers in the hosiery trade, some of them attacked the factories housing the steam looms as well as the houses of their owners. Only in a few instances was the machinery itself directly attacked.

Luddite disturbances were expressly oriented toward the prevention of technological change in the cropping trade. Wool cloth was traditionally finished by raising the nap and then leveling the surface through the use of a heavy set of shears. The growing use of the gig mill, a device for raising the nap, along with the employment of a crude device for the mechanized cropping of cloth threatened the livelihood of the traditional hand workers. They responded with some of the most severe attacks of the Luddite epoch. Although the machinery had been used for many years in many textile establishments, the severe economic conditions of the time brought matters to a head. More than the other instances of Luddite revolt, the attacks on cropping equipment were motivated by a deep fear of unemployment induced by technological change.

Within a few years the Luddite assaults came to an end due to the deployment of government troops; the execution, imprisonment, and transportation to Australia of a number of the participants; and the general improvement in living conditions after the defeat of Napoleon. The succeeding decades of the nineteenth century also saw the replacement of the small manufacturing establishment by the large factory. Machine-smashing by riotous crowds was an understandable form of labor protest when workers were scattered and lacking in permanent organizational linkages. In contrast, the large factory served as a fertile ground for the development of labor unions and other organizational vehicles for pressing the interests of workers. Industrial sabotage did not come to an end, but it was generally superseded by unionization and more effective forms of worker protest.

NEO-LUDDISM

These early episodes of machine-smashing have led to the application of the "Luddite" label to anyone opposed to modern technology and its extension. But it is perhaps unfair to impute to the original Luddites a hostility to technology per se. As we have seen, most instances of Luddism were not motivated by a fear and hatred of new machinery; their grievances were those of people suffering from the low wages and unemployment caused by a generally depressed economy. The machines were seen as convenient targets of their ire rather than the sources of it.

This is not to say that attacks on new technologies are always motivated by concerns that transcend the technology in question. As the pace of technological change has quickened and people have become more aware of its consequences, numerous efforts have been made to prevent or restrict the spread of technologies that are perceived as threats. For example, in recent years computers have threatened many established occupational roles and procedures, leading to significant resistance to their installation and use. In one case that received a good deal of national publicity during the mid-1970s, newspaper linotype operators in Washington, D.C., demonstrated their opposition to computerized typesetting equipment by engaging in large-scale industrial sabotage.

Often, however, the resistance is more subtle. Consider the case of a group of county probation officers that viewed the introduction of a computerized information system as a threat to their professional autonomy. The probation officers were able to blunt this threat by making sure that they designed the forms to be used by the computer and by controlling the input of information and limiting its distribution. These limitations left the computerized information system as little more than a sophisticated recordkeeping system, and prevented its being used as a basis for managerial decision-making. The result was an underutilized system, the costs of which could not be justified by the county government. The system was scrapped, and the probation department went back to a manual system of entering and keeping records.[10]

WHOSE TECHNOLOGY?

We have just seen how specific technologies were used and resisted by particular groups in accordance with their own needs and concerns. These examples should help us to realize that technology does not proceed solely through its own momentum; its development is strongly influenced by existing social and political arrangements. Technological changes may take place because they advance the interests of a particular group. Conversely, some technologies may meet with stiff resistance because they threaten a group's interests. Technologies do not stand or fall on their intrinsic merits. The decision to develop and deploy a new technology is often shaped by the distribution of power in a society.

Social and political arrangements affect the course of technological change by influencing the kinds of investments that are made, the research projects that are funded, and the general priorities that are established.[11] Large organizations, such as corporations and government agencies, often wield disproportionate influence over the process of technological change. This can be seen in the role played by the federal government, which funds nearly half of the research and development done in the United States. During most of the post-World War II era, 90 percent of nonmedical research expenditures have been administered by only three agencies: the Department of Defense, the National Aeronautics and Space Administration (NASA), and the Atomic Energy Commission.[12] Although alternative outcomes cannot be predicted with certainty, it seems likely that American technology would have diverged markedly from its present path if financial support for research and development had been entrusted to a different set of agencies.

Perhaps with a different set of sponsors, technological development might have made greater contributions to the solution of a number of pressing social problems, such as poverty and crime. At the same time, however, it can be argued that certain kinds of problems are simply not amenable to technological solutions. Even with significant changes in the funding of research, technological solutions to many social problems will not be forthcoming. This is an important objection, and we will examine it in the next section.

What Technology Can Do— And What It Cannot Do

The growth of technology has brought dazzling changes to our lives. At the same time, we seem to be mired in problems for which there seems to be no solution. The continued existence of these problems is all the more frustrating when contrasted with the rapid progress of technology. For example, we can use all kinds of sophisticated medical equipment and techniques to preserve the lives of sickly infants who have been born many weeks premature, but we can't seem to conquer the poverty that often produces these sick infants. Why, it is often asked, is there such a gulf between technological progress and social progress? Why can't technology be applied as a solution for more, if not all, of our problems? If we can put a man on the moon, why can't we . . . ?

THE TECHNOLOGICAL FIX

These are troubling paradoxes, and in recent years we have searched for ways of finding technological solutions to a host of problems. In response to widespread heroin addiction, the U.S. government has engaged in a large-scale program to eliminate the addict's craving for the drug through the administration of Methadone. As highway accidents continue to result in tens of thousands of deaths and hundreds of thousands of injuries each year, efforts have been mounted to develop and manufacture cars capable of protecting their occupants from the consequences of incompetent driving. Cities befouled by graffiti have turned to the use of new paints and cleaning solutions that resist the endeavors of spray-can artists.

The list of technologies that have been or could be applied to the alleviation of social problems is an extensive one, and examples could be supplied almost indefinitely. What they have in common is that they are "technological fixes," for they seek to use the power of technology in order to solve problems that are nontechnical in nature. In this section we will briefly examine a few of these technologies and consider the extent to which technology can alleviate these pressing problems.

One study of a number of technologies directed at the solution of social problems bears the significant title "Technological Shortcuts to Social Change."[13] The authors examined a number of case studies, ranging from instructional television to intrauterine devices for birth control. As might be expected, the application of different technologies for the solution of social problems resulted in varying degrees of success, but a few generalizations can be made about the efficacy of technological solutions to social problems. First, even if a technology "works" by producing the desired result, the actual mechanisms through which the technology produces a change are often poorly understood. This is particularly evident when the technology is used in conjunction with other interventions, such as the coupling of Methadone maintenance with individual counseling. Technological shortcuts also produce uneven results; they work when applied to some segments of the targeted population but do nothing for the rest. Above all, technological solutions only eliminate the surface manifestations of the problem and do not get at its roots. A Methadone program does not address the social and psychological causes of drug addiction, while improved methods of removing graffiti do nothing to mitigate the anger and alienation that may motivate the defacement of public spaces. These criticisms aside, technological shortcuts may be effective in alleviating a range of problems, and even though these problems may not be eliminated, their alleviation may at least come at a lower price than would be the case if nontechnological efforts at solutions were employed.

Many other technological fixes have been employed over time, although not always with the conscious understanding that technology was being used in lieu of some other method of achieving a desired end. To take one example, at the beginning of the twentieth century the United States was undergoing severe growing pains; the urban population was expanding at a rapid rate, accompanied by congestion, pollution, and a host of other urban ills. In a nation steeped in the Jeffersonian belief that cities were inherently evil and that the countryside was the best location for virtuous living, the conversion of the American populace into a race of unhealthy and disaffected city dwellers was viewed with alarm. A number of technologies did make urban life more tolerable, most notably those concerned with public health and sanitation, but these only served to ameliorate living conditions without addressing the real issue: the desire of many Americans to escape the city and return to a vaguely perceived rural idyll.

The pursuit of this goal gave a great impetus to the development of transportation technologies that would allow the solution of urban problems by eliminating the need for cities, at least as places of residence. Instead of comprehensively addressing urban ills through planning and the development of social programs, Americans pinned their hopes on new transportation technologies. The first of these was the electric trolley. Through the construction of extensive networks of interurban electric lines, it was hoped, America's urban problems could be literally left behind as a new generation of workers could commute from their places of work to their rural or suburban homes.[14]

The trolley was soon rendered obsolete by the automobile, yet a great deal of automobile ownership was motivated by similar sentiments. Widespread automobile ownership promised an escape from the harsh realities of America's cities through individual commuting. As Henry Ford neatly summed things up, "We shall solve the city problem by leaving the city."[15] Ford's sentiments were taken to rhapsodical levels by one early twentieth-century journalist:

> Imagine a healthier race of workingmen, toiling in cheerful and sanitary factories, with mechanical skill and tradecraft developed to the highest, as the machinery grows more delicate and perfect, who, in late afternoon, glide away in their own comfortable vehicles to their little farms or houses in the country or by the sea twenty or thirty miles distant! They will be healthier, happier, more intelligent and self-respecting citizens because of the chance to live among the meadows and flowers of the country instead of in crowded city streets.[16]

It is hardly necessary to note that these hopes were not realized. The mushrooming growth of suburbias spawned by trolleys and automobiles did not create a harmonious social order based on rural values. All too often the legacy has been suburban sprawl, the deterioration of city centers, visual blight, pollution, traffic fatalities, and many other social costs. This is not to say that the automobile has

The trolley held out the promise of an escape from the noise, dirt, and congestion of the early twentieth-century city. (The Chicago Historical Society.)

been an unmixed curse; the benefits of personal mobility, privacy, and a sense of power have been too eagerly accepted to allow such a judgment. But the automobile, just like its predecessor the trolley, was hardly the technological panacea that was envisioned. The examples of the trolley and the automobile remind us that while some specific problems may be amenable to technological solutions, larger issues rarely admit of easy solutions through the application of technological fixes.

WHY TECHNOLOGY CAN'T ALWAYS FIX IT

The main difficulty underlying the use of technology to solve social problems is that these problems are fundamentally different from technical problems. In the first place, social and technical problems differ in their specificity. If you intend to design an air conditioner, you at least know what your goal is: to keep a space cool. In many ways this problem is similar to the far more grandiose objective of landing a man on the moon; although there are enormous technical problems to overcome, at least the goal is clear and unambiguous. But what if your goal is to "stop crime"? Crime, unlike air temperature, is a very diffuse concept, encompassing everything from forgery to murder. Even when a particular crime is singled out for treatment, its causes are likely to be manifold, and not easily addressed by a single technique.

To make matters even more difficult, social problems are directly concerned with human motivations and behaviors. It is one thing to change the temperature of the air by inventing and installing an air conditioning system; it is quite another to attempt to change human behavior through some kind of technological intervention. Human beings are wonderously intricate creatures whose actions are governed by extremely complex motivations. Trying to understand, let alone change, human actions is an exceedingly difficult task. And humans are likely to resist when attempts are made to change their behavior.

It is also apparent that technological solutions work best when they operate within closed systems, that is, when the issue to be addressed is sealed off from outside influences. Of course, no technology exists in isolation from the surrounding society. A transportation system based on private automobiles, for example, is the result of choices exercised within the economic and political realm, such as a government's decision to build a highway network. But within a given technology there are many specific matters that can be treated as purely technical problems. In these cases, it is possible to approach the problem directly, and not worry about the influence of other factors. If your car fails to start one morning, you can be sure that the problem lies only with its components; you need not concern yourself with sunspot activity or a recent presidential election in Peru. When a problem is not so easily isolated, a technological solution is much less likely. Attempts to decrease heroin addiction through the administration of Methadone to addicts fall into this category. Heroin addiction is a multifaceted problem, affected by such diverse matters as the unemployment rate, racial discrimination, constitutional checks on police activities, and the drug-control policies of other countries. Although Methadone maintenance may be effective in treating some individual addicts, the systemic nature of the heroin problem works against the possibility of a single technologically based solution.

As a final point, it should be noted that no problem, technical or otherwise, is ever really "solved." Not only are most solutions incomplete, they also generate new (and sometimes very different) problems. These "residue problems" may be considerably more intractable than the original problem.[17] This process has been dramatically illustrated by the rapid development of modern medical technologies, a topic that will be explored in greater depth in Chapter 7. Technical solutions such as the development of an artificial heart, organ transplants, and sophisticated diagnostic techniques have proliferated, but at the same time they have created a host of new dilemmas. Given the expense of many of these new technologies, it may be necessary either to spend more on medical care or to attempt to ration it. If these technologies are to be rationed, will this take place through the price mechanism, or will it be done according to some formalized procedure? In either case, serious ethical issues will have to be faced. Life-extending technologies have also raised vexing questions about the morality of prolonging a life under conditions that seem dismal indeed. Moreover, a longer individual life span leads to an aging population and the necessity for a wide range of adjustments to the society, the economy, and even the culture. Without belaboring the point, it should be apparent that no set of technologies will make our lives better without requiring the enactment of other changes.

THE APPEAL OF TECHNOCRACY

These inherent limitations have not deterred a number of individuals and groups from trying to convert social problems into technical problems. There have been numerous flirtations with technocracy—the governance of society by engineers and other people with technical expertise, who attempt to develop policies based on technical and "scientific" principles. There is no denying that the technocratic vision is at first glance an appealing one. In a world too often governed by venal and incompetent politicians, there is something very attractive about a system of governance that supposedly bases itself on logic and the use of expertise. Moreover, where conventional political systems of all types seem endlessly involved with apportioning pieces of a small pie, adherents of some form of technocracy often promise a social and economic order that produces an ever-expanding pie through the application of the methods that have served technological development so well.

The promises and pitfalls of a technocratic approach to the solution of social problems are well illustrated by the theories of Scientific Management, as developed by Frederick W. Taylor and his followers during the early decades of the twentieth century. Scientific Management arose in an era marked by a profound paradox: industrial production was increasing at a rapid pace, but at the same time American society was racked by large-scale and potentially explosive conflicts between workers and management. Many cures for labor unrest had been proposed, but for Taylor all of them missed the mark. Taylor had earned an international reputation as a metallurgical engineer, and his systematic studies on the cutting tools used for machining metal had resulted in major technological advances. If obdurate metals could be better controlled and shaped through the application of new technologies guided by scientific principles, why couldn't the same thing be done with workers?

To achieve this goal, Taylor and his colleagues developed a "scientific" regimen for studying work. The main technique used for this task was the time-and-motion study through which workers were systematically observed and their work motions precisely timed. Through an analysis of these observations and measurements Taylor came up with a supposedly optimum set of motions for a given job, all of them subject to rigid time constraints. Equally important, the development and administration of these motions were the business of management exclusively, and any attempt by workers to go about their tasks independently would necessarily result in wasted motions and general inefficiency. A basic tenet of Scientific Management was that the planning and organization of work had to be separated from its actual execution. Only specially trained managers had the time and expertise necessary for the devising of optimal methods of production. The prime obligation of the workers was that they did what they were told to do.[18]

Although they had no power to plan and manage their own work, workers were supposed to benefit from the system. Because their work activities were now optimized, production would supposedly increase significantly. Workers would necessarily share in these higher returns, for Taylor also advocated that workers be paid according to piece rates rather than straight wages; the more they produced, the more they earned.

The technocratic spirit of Scientific Management is thus evident: the tasks and prerogatives of management rested not upon the exercise of raw power but on management's technical superiority in guiding the production process. At the same time, Scientific Management promised relief from continual squabbling over relative shares of the fruits of production; an optimal system of organization would result in more of everything for everybody. Taylor was not content with using Scientific Management as a solution for the problems of the workplace; its principles, he claimed, "can be applied with equal force to all social activities: to the management of our homes; the management of our farms; the management of the business of our tradesmen large and small; of our churches, our philanthropic organizations, our universities; and our governmental departments."[19]

The appeal of Scientific Management was not confined to the United States, or even to the capitalist world. In the years immediately following the Russian revolution, the Soviet leadership expressed a deep admiration for American technology and American forms of industrial organization. In particular, Lenin was a strong believer in the efficacy of Taylor's principles. Although he duly noted that Scientific Management embodied "the refined cruelty of bourgeois exploitation," he made it clear that its basic principles and procedures could contribute to the realization of Soviet economic goals: "The possibility of building Socialism will be determined precisely by our success in combining Soviet government and the Soviet organization of administration with the modern achievements of capitalism. We must organize in Russia the study and teaching of the Taylor System and systematically try it out and adopt it to our purposes."[20]

THE TECHNOCRAT'S DELUSION

Although some of its elements, such as the use of time-and-motion studies, can still be found in contemporary managerial practices, Scientific Management in its pure form never took hold in the United States, the Soviet Union, or anywhere else. A number of technical problems impeded its use. Considerable skill was required for the administration of time-and-motion studies, and they were especially difficult to conduct in work settings not characterized by repetitious actions. But of equal or greater importance, both management and labor realized that the implementation of Taylor's system posed fundamental threats to their own interests. Most managers were highly reluctant to delegate their authority to the dictates of "scientific" procedures.[21] Workers, on the other hand, resented the loss of what little autonomy they had, and they widely believed—with considerable justification—that higher levels of productivity would result in the downward adjustment of piece rates, leaving them no better off than before the program had been enacted.

Scientific Management, like all technocratically inspired systems, ignored the distinction between technical and sociopolitical problems. Even if Scientific Management had generated the productive increases it promised—which is unlikely—it would still have been strongly resisted by those who had to submit to it. Scientific Management promised a conflict-free method of administration where no such thing was possible. Workers and managers had their separate interests, and each group was unwilling to entrust its fate to Taylor and his disciples.

The basic fallacy of Scientific Management, one shared by all other variants of technocracy, is that *administration* can replace *politics*. Administration is based on the application of rules that allow the realization of given ends. It is thus a manifestation of the rational spirit of applying the best means for the achievement of a particular goal. It does not, however, determine these ends. The Internal Revenue Service officials who administer the tax system are not the authors of the tax code. Around April 15 we may get angry about the perceived unfairness of the tax code, but it is pointless to blame the officials at the local IRS office.

Tax codes and other policies are formulated through choices made in the political arena. Neither technology nor administration can supply the values that form the basis of these choices. They cannot tell us what we should do with our lives, nor can they help us to resolve the fundamental issue that all societies confront: how to distribute fairly life's necessities and luxuries. The resolution of these issues will always be marked by sizeable differences of opinion and a good deal of conflict. The technocrat's hope that society can be run on the basis of engineering principles will always remain an illusion.

This chapter has shown how technological changes inevitably produce social changes. These changes, in turn, do not affect everyone equally. Although many technologies produce widespread benefits, not everyone benefits to the same degree, and there are instances where particular individuals and groups lose out completely. A choice of technology is often a determination of who wins and who loses; it is therefore only proper that such decisions be made in a political arena where these choices can be made explicit.

This issue will be taken up in greater depth in the last three chapters. At this point it can at least be hoped that without deflating the very real achievements of technology, some sense of its inherent limitations has been conveyed. Technology and the procedures underlying its development have been immensely powerful in their own realm; outside this realm, however, they are less likely to be effective. Equally important, the methods that have been so successful in developing and applying new technologies cannot be transferred to the governance of society. Technological development may make some aspects of our lives better, but it can never substitute for a just and effective political and social system.

Questions for Discussion

1. Can you think of some established industries that have been undermined by technological advance? What industries might be threatened in the near future?
2. Were the Luddites justified in mounting their attacks on machinery? How else might they have expressed their grievances? Would other kinds of actions have been more successful?
3. What examples of technological "fixes" can you think of? Have they been successful or not? What are your criteria for judging success and failure?

4. Political leaders at home and abroad are occasionally described as "technocrats." What are the implications of this description? Would you be more or less likely to vote for somebody who was described in this way?

Notes

1. Moses Abramowitz, "Resource and Output Trends in the United States Since 1870," *American Economic Review, Papers and Proceedings* 56 (May 1956): 5–23; John W. Kendrick, "Productivity Trends, Capital, and Labor," *Review of Economics and Statistics* 37 (August 1956): 248–257; R.M. Solo, "Technical Change and the Aggregate Production Function," *Review of Economics and Statistics* 39 (August 1957): 312–320.
2. See Howard P. Segal, *Technological Utopianism in American Culture 1830–1940*, (Chicago: University of Chicago Press, 1985).
3. Quoted in Herbert J. Muller, "Human Values and Modern Technology," in Edwin T. Layton, Jr. (ed.), *Technology and Social Change in America* (New York: Harper & Row, 1973), p. 159.
4. Lauriston Sharp, "Steel Axes for Stone Age Australians," in Edward H. Spicer (ed.), *Human Problems in Technological Change: A Casebook* (New York: John Wiley & Sons, 1967).
5. W. F. Cottrell, "Death by Dieselization: A Case Study in the Reaction to Technological Change," *American Sociological Review* 16 (June 1951): 358–365.
6. Langdon Winner, "Do Artifacts Have Politics?" *Daedalus* 109, 1 (Winter 1980): 123–125.
7. Peter Blake, *Form Follows Fiasco: Why Modern Architecture Hasn't Worked* (Boston: Little, Brown, 1974), p. 59.
8. Shannon R. Brown, "The Ewo Filature: A Study in the Transfer of Technology to China in the 19th Century," *Technology and Culture* 20, 3 (July 1979).
9. George Rude, *The Crowd in History: A Study of Popular Disturbances in France and England, 1730–1848* (New York: John Wiley & Sons, 1965), pp. 66–92; Malcom I. Thomis, *The Luddites: Machine-Breaking in Regency England* (New York: Schocken Books, 1972).
10. Gary L. Albrecht, "Defusing Technological Change in Juvenile Courts: The Probation Officer's Struggle for Professional Autonomy," *Sociology of Work and Occupations* 6, 3 (August 1979): 259–282.
11. Reinhard Rürup, "Reflections on the Development and Current Problems of the History of Technology," *Technology and Culture* 15, 2 (April 1974): 165.
12. Nathan Rosenberg, *Perspectives on Technology* (Armonk, N.Y.: M.E. Sharpe, 1985), p. 223.
13. Amatai Etzioni and Richard Remp, "Technological 'Shortcuts' to Social Change," *Science* 175, 4017 (7 January 1972): 31–38.
14. James C. Williams, "The Trolley: Technology and Values in Retrospect," *San Jose Studies* 3, 3 (November 1977): 74–90.
15. James J. Flink, *The Car Culture* (Cambridge, Mass.: The MIT Press), p. 39.
16. William F. Dix, "The Automobile as a Vacation Agent," *Independent* 56 (2 June 1904): 1259–1260, quoted in Ibid., pp. 39–40.
17. See Kan Chen, et al., *Growth Policy: Population, Environment, and Beyond* (Ann Arbor: University of Michigan Press, 1973) pp. 105–112.
18. Harry Braverman, *Labor and Monopoly Capital: The Degradation of Work in the Twentieth Century* (New York: Monthly Review Press, 1974), pp. 85–138.

19. Quoted in Samuel Florman, *The Existential Pleasures of Engineering* (New York: St. Martin's Press, 1976), p. 8.
20. V. I. Lenin, "The Immediate Tasks of the Soviet Government," *Izvestia*, 28 April 1918, translated in V. I. Lenin, *Selected Works*, vol. 2 (Moscow: Foreign Languages Publishing House, 1947), p. 327.
21. Daniel Nelson, *Managers and Workers: Origins of the New Factory System in the United States, 1880–1920* (Madison: University of Wisconsin Press, 1975), pp. 75–76.

PART TWO

THE PROCESS OF TECHNOLOGICAL CHANGE

Chapter 3 The Sources of Technological Change

What accounts for the emergence of particular technologies? Why do they appear when they do? What sort of forces generate them? How is the choice of technology exercised? To put it more concretely, why were digital computers developed only during the second half of the twentieth century even though their basic principles were understood more than a hundred years earlier? Why did photography undergo rapid development during the nineteenth century? What were the inventors of radio trying to accomplish, and how did their intentions differ from those of subsequent developers? These are the questions that this chapter will address as it considers some of the most basic issues in the study of technology.

THE GREAT BREAKTHROUGH

When we think about the history of technology our thoughts usually turn to solitary inventions, the dramatic breakthroughs that by themselves fundamentally change the way things are done. The electric light bulb ushered in an era of artifical illumination that turned night into day. The discovery of penicillin, the first antibiotic, allowed the effective treatment of infectious diseases, thereby producing vast changes in medical practice. The invention of the transistor ushered in a new era of solid-state electronics that was manifest in everything from better automobile ignition systems to the rapid development of digital computers.

Inventions and discoveries such as these truly merit the term "revolutionary." They are not simply improvements or extensions of existing technologies, but are completely new approaches. And perhaps because of their novelty, they are often the work of talented individuals. Many histories of technology therefore fasten upon individual genius as the chief explanation for technological change. We have all heard or read stories of how the inspired labors of Thomas Edison or the Wright Brothers resulted in epochal inventions that transformed the world. Histories of technology written from this perspective tend to become a roster of the names of great inventors whose brilliance is taken to be the sole source of technological advance.

Other histories of technological advance have remained within this framework, but have looked to a different kind of genius as the source of advance. The key players here are not the inventors of new technologies, but the entrepreneurs who make inventions into commercial successes by taking risks, moving into uncharted territory, and in general doing what hadn't been done before.[1] There have been some individuals who have been both inventor and successful entrepreneur—for example, Edwin Land, the inventor and moving force behind the

Polaroid camera—but they have been rare. Entrepreneurs generally take other people's inventions and make them into commercial successes. From this perspective, the key figure in the development of the steel industry is not Henry Bessemer, the co-inventor of the iron-refining furnace that bears his name, but Andrew Carnegie, who laid the commercial and organizational foundations of the industry.

Sometimes the hero is neither an inventor nor an entrepreneur, but a scientist. Although usually less well-known than the inventors and entrepreneurs of legend, scientists have produced the knowledge that forms the basis of many technological applications. However, the connections between science and technology are complex. Scientific discovery has been an important source of technological advance, especially in our own times, but technology cannot be described simply as applied science. The subject merits extended consideration, and for this reason the following chapter will be devoted exclusively to it.

IT ALL ADDS UP

Although we tend to think of technological change as the result of the sudden emergence of epochal inventions, the process is not always this dramatic. A great amount of technological advance is the result of slow but steady incremental changes. And in sharp contrast to the "heroic" approach to the history of technology, a narrative of these changes would introduce us to a cast of largely unknown engineers, mechanics, and technicians whose individual accomplishments may seem modest, but in aggregate have been an extremely important source of technological advance.

These incremental changes are the result of a learning process that occurs as a technology is used. Problems are identified and overcome, bugs are worked out, and improvements are made. In many cases, the cumulative results of these efforts are technological advances at least as important as those that stem from fundamental breakthroughs. In industries as different as petroleum refining and building construction, the major source of productivity improvements has been a multitude of small technological improvements that have resulted in large cumulative gains.[2]

This process is nicely illustrated by Louis Hunter's narrative of how the impressive development of nineteenth-century steamboats was the result

> of plodding progress in which invention in the formal sense counted far less than a multitude of minor improvements, adjustments and adaptations. . . . The story of the evolution of steamboat machinery in the end resolves itself in a large part into such seemingly small matters as, for instance, machining a shaft to hundredths instead of sixteenths of an inch, or devising a cylinder packing which would increase the effective pressure a few pounds, or altering the design of a boiler so that cleaning could be accomplished in three hours instead of six and would be necessary only every other instead of every trip. Matter such as these do not get into the historical record, yet they are the stuff of which mechanical progress is made. . . .[3]

One can also witness the far-reaching consequences of numerous small improvements in the development of railroad technology. According to one calculation, if the traffic loads borne in 1910 had been carried by railroads employing the technologies of 1870, the additional costs would have amounted to $1.3 billion. Fortunately, by 1910 American railroads had benefited from steady improvements in the size of cars and the power of locomotives, which in turn were the result of slow but steady evolution.[4] And so it goes today. Even though the railroad may be described as a "mature" industry, this pattern of incremental yet substantial technological development continues. American railroads have significantly lowered their costs through the implementation of a series of small improvements: better insulation for electrical components (thereby allowing higher power loads), improved turbochargers and fuel injection systems, higher compression ratios, more efficient motors in cooling radiators, two-speed cooling fans, redesigned air ducts, lower idling speeds, and the substitution of alternators for direct-current generators. By themselves, none of these innovations is terribly significant. But when they are all put together in a modern locomotive, the result is a 10 percent savings in fuel costs, and a gain of 24 percent in the ton-miles of freight carried per gallon of fuel consumed. When it is noted that fuel costs for railroads were $3.1 billion in 1982, the consequences of these improvements are all the more impressive.[5]

In a world where the ability to produce successful technological innovations is increasingly a requirement for a firm's success, if not its survival, each firm needs to tap every source of technological advance, no matter how modest it may seem. This is a point stressed by Anthony Athos and Richard Pascale in their book about American and Japanese management: "careful scrutiny reveals that despite the exalted status of 'strategy' in the lexicon of American management, few great successes stem from one bold-stroke strategic thrust. More often, they result from one half-good idea that is improved upon incrementally. These improvements are invariably the result of a lot of 'little people' paying attention to the product, the customer, and the marketplace."[6]

THE "D" IN R&D

The process of making a technology work is often summarized by the abbreviation R&D, which stands for research and development. "Research" calls to mind images of cutting-edge work in well-equipped laboratories, where great breakthroughs produce dramatically new technologies. Research can be an exciting, even glamorous activity, and we naturally look to it as the basis of technological progress. It is the source of fundamental change in technology, like the invention of integrated circuits, monoclonal antibodies, and graphite composite materials. Still, this sort of research rarely results in useable products. The realization of the potentialities created by research breakthroughs usually requires a lengthy process of development. Numerous problems have to be resolved, and, equally important, the new material or device has to be put into a form that allows it to be produced at a reasonable cost.

Here again we can see the importance of the slow, unspectacular improvements that turn a good idea into a working product or process. And here too we can often see a substantial outpouring of money. If basic research is an expensive process, development is often even more so. The development work that goes into preparing a new technology for actual production can entail massive expenditures for equipment, material, manpower, pilot plants, and the like.

A great deal of development work is oriented toward "scaling up," that is, making the transition from a successful research result to large-scale production. It is one thing to invent a device or process that works under laboratory conditions, and quite another to make it happen in an industrial setting where quantity production is the goal. The development of penicillin provides an excellent illustration of the many facets of the scaling-up process.[7] Although the discovery of penicillin was used at the beginning of this chapter as an example of a major technological breakthrough, the development phase was certainly no less important.

Penicillin, a fermentation product of the mold *Penicillium notatum* was discovered in 1928 by Alexander Fleming, who observed that bacteria were destroyed in a culture that had been accidentally contaminated by the mold. *Penicillium* cultures grown in small quantities were the basis of laboratory and clinical research, but this process could not yield the large quantities of the drug needed for widespread therapeutic use. Large-scale production of penicillin was eventually done in huge fermentation vats, a process that required the solution of many technical problems. The key step was the development of a submerged fermentation process that allowed the mold to be grown directly in the nutrient medium. The success of this process in turn required a number of other improvements, such as new tank designs with special cooling systems and turbine mixers, which also had to be developed. The use of corn steep liquor (a by-product of corn starch production) as a culture medium increased yields tenfold, but it created a new problem. Penicillin requires air in order to grow, but severe foaming occurred when the culture was aerated. Anti-foaming products therefore had to be developed to alleviate this problem. The extraction of penicillin from the moldy brew in the vats also created problems that were ultimately solved by the use of freeze drying, which was itself scaled up from a process first used to preserve blood plasma. As a result of all of this development work, production of penicillin had reached 650 billion units by the end of World War II, and the cost of a dose was 55 cents instead of the $20 it had been three years earlier.

ALL TOGETHER NOW

The first chapter stressed the importance of thinking of technologies as systems. This point is reinforced by a consideration of how particular technologies develop. Quite often, a technological leap forward takes place because of the availability of complementary technological developments that allow the resolution of fundamental problems. The history of the computer provides a good example of how complementary changes are essential for the translation of an idea into a workable technology. Back in the 1820s Charles Babbage began to develop an "analytical engine" that contained a set of input devices, a processor, a control unit, a memory storage, and an output mechanism—the esssential elements of today's computer. But Babbage's computer was operated by an exceedingly complex set of gears, rods,

Scaling up: fermentation tanks for the large-scale production of penicillin. (Courtesy of Bristol-Meyers Squibb Company.)

and other mechanical linkages. Unfortunately, even with the best machining techniques of the time, it was impossible to achieve levels of speed and accuracy required for the effective operation of the device. Babbage's ambitions were not realized for more than a century, when solid-state electronics, the cathode ray tube, and magnetic storage devices allowed the development of practical computers.

Another illustration of the importance of complementary technological changes can be drawn from the history of one of the twentieth century's most important devices, the internal combustion engine. When these engines were first produced during the late nineteenth century, the spark that ignited the air-fuel mixture was timed to occur at the top of the compression stroke. This did not allow a sufficient time for the mixture to be completely ignited, and efficiency consequently suffered. When, after much experimentation, the spark was timed to occur before the piston reached the top of its stroke, the combustion process was

Ada, Countess of Lovelace worked with Charles Babbage on his Analytical Engine. She is often described as the world's first computer programmer. (Culver Pictures, Inc.)

A portion of Charles Babbage's computer. (IBM Corporation.)

greatly improved, and much more power was consequently delivered. Still, early engines were deficient in power because compression ratios were kept low in order to prevent the sudden detonation of the air-fuel mixture and resultant damage to the engine. This problem was solved in part by conducting careful inquiries into the combustion process and by re-shaping the combustion chamber and piston crown in ways suggested by this research. But this was not enough. In order to employ higher compression ratios successfully, it was necessary to modify not just the engine, but also the fuel it used. In particular, the research of Thomas Midgley and Charles Kettering demonstrated that the addition of tetraethyl lead to gasoline allowed higher compression ratios and a subsequent improvement in both power and economy. It was thus through a combination of cumulative improvements in both the engine and the fuel it burned that the internal combustion engine reached an acceptably high level of efficiency.

It is important to note that these changes in engine technology did not take place when the internal combustion engine was first invented, but during a period when it was already being sold to customers. This illustrates a point made by Nathan Rosenberg: "The idea that an invention reaches a stage of commercial profitability *first* and is then 'introduced' is, as a matter of fact, simple minded. It is during a (frequently protracted) shakedown period in its early introduction that it becomes obviously worthwhile to *bother* making the improvements."[8]

The commercial viability of a new technology may thus stimulate the development of complementary technologies. A bottleneck that restricts the continued development of a particular technology creates strong economic incentives to find new technologies that clear the bottleneck.[9] Many obstacles are surmounted through the use of technologies that have been developed for use by a different industry or in a different sector. This can be seen today in the automobile industry, where emissions requirements have necessitated the use of microprocessors and computers for the regulation of spark advance and fuel mixtures. Thus, a nineteenth-century technology, the four-stroke internal combustion engine, owes its survival to its marriage to a late twentieth-century technology.

Technological advance in one area is often stimulated by the emergence of new technologies in different, but related, areas. This process can be seen in the development of nineteenth-century metal-working industries. Although their products differed substantially, the processes employed by these industries were basically the same: turning, boring, drilling, milling, and planing. At the same time, they all confronted similar technical problems of transmitting power, reducing friction, and controlling the rate of feed. This meant that a technological solution arrived at by one industry was often directly applicable to the problems of another industry.[10] The early automobile industry made abundant use of the products and manufacturing techniques that had been developed by the bicycle industry during the 1880s and 1890s. Ball bearings, drive chains, and the use of electrical resistance welding had been extensively employed by the bicycle industry during its boom years. One of the most novel and significant technologies entailed the use of stamped components to take the place of forgings. Their use eliminated a great deal of machining, with a consequent lowering of production costs. The cheap, mass-produced automobile thus owed much to technologies initially developed to make a product that it completely eclipsed.[11]

This is hardly the only example of technologies developed by an established industry paving the way for a radically new one. Although the turbojet engine was a novel method of propulsion when it first appeared in the 1940s, it drew heavily on designs, components, and processes that had already been developed for steam turbines. In Edward Constant's summary, "All the work done on blade design, gas flow, shaft and bearing loads, temperature distribution, lubrication systems, governors, blade-cutting machines, test procedures and instruments, and countless other facets of design and production could be applied to gas turbine development."[12]

Inter-industry transfers of technology do not happen automatically. The effective transfer of hardware, information, or simply (but importantly) the belief that a problem is *solvable* requires individuals and organizations that are capable of functioning in both worlds and have the incentives to do so. It also requires that these individuals and organizations are acceptable to the other individuals and organizations with which they interact, and that they are capable of speaking the same technical language. Technological innovation is, in Christopher Freeman's phrase, a "coupling process" that occurs at the interfaces between science, technology, and the market. This does not take place solely through intuitive flashes: "It is a continuous creative dialogue over a long period of research, experimental design, and development."[13]

Finally, it should be noted that sometimes the complementary changes necessary for the success of a new technology are not technological. A successful technological change may require changes in basic habits and attitudes. This can be seen in the failure of agricultural extension agents in New Mexico to get farmers to adopt hybrid corn in the late 1940s. There was no question about the technical superiority of the corn: demonstrations showed that its use resulted in a potential trebling of yields. Impressed by this improvement, half of the farmers planted the new variety, thereby immediately doubling their output. But after two years virtually all of the farmers had abandoned hybrid corn and reverted to their traditional low-yielding crop. The problem was that the cornmeal made from the hybrid variety could not be made into good tortillas; it did not taste right and it couldn't be easily shaped. In the absence of a change in culinary patterns, a technically superior product could make no lasting impact.[14]

THE SOCIAL CONTEXT OF TECHNOLOGICAL CHANGE

The nonuse of hybrid corn in New Mexico demonstrates the perils of viewing technological change largely as a self-contained process. Unfortunately, quite a few histories of technological change seem content to analyze the process of change largely in terms of developments within the technology itself. As with a kind of history that has been characterized as "one darn thing after another," this "internalist" approach narrates the series of changes that unfold as a technology develops, with little attention paid to social, political, and economic forces. Because it often emphasizes specific devices and their evolution, this sort of history is appealing to those with an antiquarian turn of mind. But these narratives are insufficient if we want to find out why these technologies appeared in the first place, and why they were readily assimilated after they made their appearance.

As was noted earlier, there has been a tendency to write technological history as a series of biographies of great scientists, inventors, and entrepreneurs. There can be no denying that many technologies owe their existence to the inspiration and hard work of individuals, including the unsung heroes who were responsible for the myriad improvements necessary for the realization of a new technology. But surely more is involved than their efforts. After all, human ability is presumably spread evenly throughout cultures and historical epochs, yet significant technological changes are not equally distributed over time and place. Thomas Edison's genius produced 1,093 patents, and some of his inventions transformed the world. Had he been born in ancient Rome or dynastic China, he might have helped to design aqueducts or sections of the Great Wall, but it is unlikely that his talents would have changed the course of Roman or Chinese history. Geniuses require appropriate social settings for the realization of their talents.

What kind of social system is required if inventive ability is to flourish? Why does technological innovation occur in some places and times and not in others? In beginning to answer these questions, it is helpful to apply to technological change the concepts that have been so useful to economists: supply and demand. We have already considered some of the agents and processes that "supply" technology. Everything from fundamental scientific breakthroughs to minor refinements serve to "push" new technologies into the world. Still, simply having an available supply of new devices and techniques does not guarantee that they will be used. Many examples of technologies that languished because they were "ahead of their time" can be cited. The pneumatic tire was patented in 1845 and then forgotten until it was reinvented by John Dunlop in 1888. DDT was first synthesized in 1874, but it was not put to use until 1941. Several decades elapsed before the laser passed from being a laboratory curiosity to a practical device used for everything from supermarket scanners to instruments for microsurgery.

For a technology to make the transition from the potential to the actual requires not just that it exist; there must also be a desire for it, coupled with the ability to pay for it. Economists call this "effective demand." Seen in this light, technology is like any other good or service; it will not be produced unless some person, group, or organization wants it and is willing to buy it. Technology is "pushed" by the forces described at the beginning of the chapter, but it also has to be "pulled" by effective demand. To understand why certain technologies have flourished while others have languished it is therefore necessary to consider the configuration of a society and the way in which it determines the effective demand for particular technologies.

The most influential research on the importance of effective demand inducing technological change was done by Jacob Schmookler.[15] By examining a long series of patents in various industries, Schmookler found that their emergence was closely related to the level of demand for the products of these industries. To take one rather obvious example, inventors' interest in improving the horseshoe was strong when the horse was a primary means of transportation, but interest evaporated when the steam engine and the internal combustion engine began to displace it.[16]

Another illustration of the importance of demand inducing technological development can be extracted from the history of photography in the nineteenth century.[17] For centuries painters and scientists had made use of the *camera obscura*, a darkened room into which light was admitted through a pinhole, resulting in the projection of an inverted image of an outdoor scene on the opposite wall. Later

The interior of a nineteenth-century photographer's portrait studio. (Southern Oregon Historical Society Photograph No. 4977.)

developments substituted optical lenses for the pinhole, which made for a sharper image. Photographs were first produced during the 1820s and 1830s when pioneers such as Niepce, Daguerre, and Fox Talbot devised chemical emulsions that preserved the image on paper or a metal plate. But the rapid growth of photography cannot be attributed simply to the supply of these inventions. At this time, social changes were sweeping across Europe, resulting in the ascendence of a new social elite, the property-owning commercial and industrial bourgeoisie. The members of this group had a keen desire to flaunt their affluence by taking on characteristics of the old aristocracy. For the latter, a key artifact of their status was the painted portrait; we have all seen renditions of persons such as the Eighth Duke of Puddleswallop hanging in our local museums. But many of the rising bourgeoisie lacked the money or the time for such symbols of their importance, and, in any event, there were not enough skilled portrait painters to serve the needs of this growing group of people. Their aspirations were therefore met by the photographic portrait studio, where the subject posed with the trappings of upper-class status, such as rich draperies and elegant furniture. In the later decades of the century new and cheaper methods of photography emerged to meet the demands of poorer yet increasingly affluent people, such as American immigrants who wanted portraits that could be sent back home. Today, the effective demand produced by a great mass of consumers has stimulated the development of a huge variety of photographic apparatus, ranging from simple Instamatics to sophisticated 35-millimeter cameras.

Good business practice is often implictly based on the realization that successful technological development requires the presence of effective demand. One English study found that the key determinant of a firm's innovative success was an understanding of customer requirements. This meant that from its very inception, a new product or process had to be developed with an eye toward meeting the needs of actual or potential customers.[18] Similarly, a Canadian study found that the commercial failure of many inventions was due to an inability to evaluate the true extent of demand. A great deal of product development was done with little consideration of market potential. It was often the case that entrepreneurs were so infatuated with their product innovation that they were incapable of realistically assessing opportunities and the nature of the market.

BELATED DEMAND

At the same time, however, gauging the potential demand for a new product can be a tricky task. Many of today's "essential" technologies were not at first recognized as such. When in the late 1930s Chester Carlson attempted to interest established business machine manufacturers in his photocopying device—the first Xerox machine—they were of the uniform opinion that there was no point in employing complicated apparatus and chemicals simply to replace carbon paper. Even inventors can badly misjudge the ultimate cconsequences of their own creations. Alexander Graham Bell initially thought that the telephone he invented would primarily be used to transmit operas and other musical performances from the concert hall to the home.[19] Edison at first believed that one of the main applications of his phonograph would be to record the last words of dying men. We can also take note of the judgment of Howard Aiken, the director of the team that built one of the world's first computers, who in the early 1950s prophesied that in the foreseeable future, the total need for computers in the United States could be met by no more than a half-dozen machines. In similar fashion, before 1950 Thomas J. Watson, the president of IBM, was of the opinion that there would be no commercial market for computers.[20]

The history of radio also shows how a technology may be put to uses not envisaged by its inventors. When equipment for sending and receiving radio waves was first developed during the late nineteenth century, no one imagined that it would be used for commercial broadcasts. The first transmitters and receivers were devised for purely intellectual purposes—in order to test the validity of James Clerk Maxwell's theories about the nature of electromagnetic waves. Only after the passage of more than a decade did some visionaries perceive a commercial use for radio apparatus, and then their imaginations were limited to the use of the radio for ship-to-shore communications. Decades passed before the idea of broadcasting to a mass audience emerged.[21]

What are we to make of these examples? They seem to refute the theory that technologies are primarily the result of "demand-pull," for all of the technological developments just described emerged in the absence of apparent demand for them. Perhaps we can salvage the demand-pull theory by distinguishing two different

An early IBM computer. (IBM Corporation)

kinds of technological advances. The first kind consists of refinements and improvements to an existing way of doing things, while the second (and far less frequent) is the truly revolutionary breakthrough—the digital computer, radio, the telephone, and the like. In cases such as these, the very novelty of a revolutionary breakthrough makes it difficult to determine what its ultimate uses will be, and who, if anyone, will want it. By contrast, advances of the first kind occur within a known context; the basic technology is already in use, and there are likely to be people and business firms that want, and are willing to pay for, the new wrinkles that promise to improve an existing technology. Improvements of this sort therefore have a predictable market. Conversely, radically new technologies confront a great deal of uncertainty. They may satisfy a latent need or they may create a new one. They are flights into the unknown, and it is hazardous to guess what sort of a reception they will meet. If nothing else, they confirm the old Chinese saying that it is dangerous to make predictions—especially about the future.

MARKET ECONOMIES AND TECHNOLOGICAL ADVANCE

The forces that "push" and "pull" technological advance do not exist everywhere or in equal measure. In many places and at many times, the distribution of wealth and power retarded these forces, resulting in a slow pace of technological advance. Moreover, the particular technological advances that do occur usually reflect a society's general configuration of wealth and power. In the European Middle Ages, the landowning aristocracy and Church officials controlled most of the wealth and exerted great power. The monastic orders often played an important role in land-clearing, farming, and the construction of mechanical devices, but for the most part the religious and secular establishment showed little interest in such matters. While the era gave rise to significant technological advances in water power, mechanical clocks, and weaponry, its most evident technological triumph was the great symbol of the traditional order: the Gothic cathedral.

As European history unfolded, the interests and demands of a growing merchant class led to the development of technologies that eventually surpassed even the soaring cathedrals. The great technological innovations that began in the mid-fifteenth century with improvements in shipbuilding and ocean navigation were closely associated with the rise of capitalism and the emergence of a market system. A market system organized around the principle of private property was of crucial importance for the stimulation and guidance of inventive and innovative abilities, and their application to production.[22]

One of the strongest accolades to the technological dynamism of capitalist society can be found, of all places, in *The Communist Manifesto*. With unfeigned admiration Karl Marx and Friedrich Engels note that: [23]

> The bourgeoisie, during its rule of scarce one hundred years, has created more massive and colossal productive forces than have all preceding generations together. Subjection of Nature's forces to man, machinery, application of chemistry to industry and agriculture, steam-navigation, railways, electric telegraphs, clearing of whole continents for cultivation, canalisation of rivers, whole populations conjured out of the ground—what earlier century had even a presentiment that such productive forces slumbered in the lap of social labour?

In the time of Marx and Engels, and in our own time, a market economy driven by the activities of self-interested businessmen has produced the most receptive environment for technological innovation. There are several reasons for this. A market economy will stimulate inventive efforts, for it promises financial rewards to those able to meet the needs of consumers. For example, somebody invents a better mousetrap in the hope of selling it in the market. If the demand is there, eager customers will buy it. Everybody is better off: consumers have a better mousetrap, while the inventor gets rich and retires to Palm Springs, and nobody worries about how technological change has lowered the quality of life for mice. Second, a market economy is characterized by the presence of numerous competitors. Under these circumstances, a producer is strongly motivated to develop and apply new technologies in order to make better products and to reduce production costs. Conversely, failure to do so may result in the eventual collapse of the enterprise, as the history of many once-successful firms demonstrates. Finally, a market system is particularly effective in eliciting the production of the auxiliary items necesssary for technological innovation. A new technology will require special materials, components, and services. Because of its responsiveness to new sources of demand, a market economy is well-suited to meet these requirements.

PROBLEMS OF TECHNOLOGICAL DEVELOPMENT IN PLANNED ECONOMIES

The advantages of a market economy in stimulating technological advance are further demonstrated by an examination of centrally planned economies. For decades the economies of the former Soviet Union and the People's Republic of China were organized through the mechanisms of central planning, but during the 1980s it became painfully evident that these mechanisms were fatally flawed. One of the chief manifestations of that failure has been a retarded technology. It cannot be denied that the Soviet Union produced some impressive technological achievements, most notably in its space and military programs, while China made significant progress in industrializing an impoverished country, but taken as a whole their level of technological development remained stagnant while the United States, Western Europe, and Japan moved rapidly forward.

The backwardness of centrally planned economies has had significant political repercussions. In China, dissatisfaction with the pace of technological change has produced some economic decentralization, as well as a greater openness to contacts with the outside world. The occupation of Tiananmen Square in Beijing by workers and students was a consequence of these external influences, as demands for a more democratic order were stimulated by exposure to foreign ways. In the Soviet Union, the fear of falling even further behind the West motivated the Gorbachev regime to introduce some elements of a market economy and to attenuate the role of central planning. The subsequent collapse of communism left the former Soviet Union in an uncertain state, with little prospect of a rapid economic turnaround. As with China the future course of reform cannot be determined, but it does seem evident that economic and technological modernization will require the displacement of a centrally planned economy by something closer to a market economy.

Many of the difficulties experienced by centrally planned economies in achieving technological advance are the result of a basic tension between their system of economic management and the requirements of technological innovation. Centrally planned economies rest on the assumption that economic activities can be reduced to predictable routines. But the course of technological innovation is notoriously difficult to predict. The bureaucratic procedures that work tolerably well for the administration of routine productive tasks usually fail when they are applied to technological innovation. A planning agency can set goals and quotas for the production of established goods, and various ministries can oversee the actual operation of individual enterprises through routine bureaucratic administration. But these procedures work much less well when innovation is the goal. Innovation is an activity full of risk and unpredictability, and it cannot easily be accommodated to pre-programmed structures and activities.

To make matters worse, centrally planned economies attempt to motivate workers and managers through the allocation of rewards that create disincentives for technological innovation. A factory manager typically receives bonuses for the fulfillment and overfulfillment of quotas for established products, as given by the central plan. The production of an innovative product is not rewarded, for it has not been stipulated by the plan. The uncertainties and unpredictabilities that surround technological innovations create risks for those who seek to develop and use them, but these risks are not matched by commensurate rewards for those who take them.

There are also disincentives for the development and use of new process technologies that would allow the cheaper production of an existing product. In the first place, the possibilities of using new production technologies are constrained by the inability of the central planning apparatus to provide complementary inputs. A new product or process usually needs a variety of new materials and components, but the provision of these requires substantial modifications to the existing economic plan. It isn't hard to imagine the immense readjustments that would have to be made if hundreds or thousands of innovations were to be accommodated. Second, even if it is possible to successfully install a new technology that promised higher productive levels, a factory manager may be reluctant to use it out of fear that his expected output quota will be revised upward. And when an enterprise develops a new way of doing things, it is likely to remain bottled up within the enterprise; the manager will keep it a secret out of fear that higher quotas will be assigned by the planning authorities. As a result, the diffusion of innovations is severely impeded.

NONECONOMIC SOURCES OF TECHNOLOGICAL ADVANCE

It is not the intention here to convey the impression that only market forces can produce technological innovation; as we shall see, government institutions have become increasingly important sources of technological advance. And no inference should be made that the historical superiority of a market economy in promoting technological advance makes it a superior system in general. There is more to life than technological advance, and, as we have already seen and shall see some more, both capitalism and the market have produced technologies that have been detrimental to large numbers of people.

Furthermore, technological innovation cannot always be traced to economic motives or even to the desire to address practical problems. To be sure, we tend to think of technology as the result of efforts to solve problems of this sort; after all, technology has already been defined as the product of knowledge that is used in order to get something done. The very word "technology" conjures up images of useful devices, and technology's practitioners—engineers, managers, and skilled workers—are often viewed as a serious bunch, sitting rigidly in front of computer terminals, making precise measurements, and above all, applying their talents to the solution of practical problems that are usually tied to economic concerns.

In fact, even the most practical of inventions may owe its origin to a spirit that seems more closely connected to play than to "productive" work. When Willis Whitney served as the first director of the research laboratory of the General Electric Company, he often asked his scientists and technicians there if they were "having fun." For Whitney, "fun" was working on problems that had stumped everyone. Pursuing these problems was nothing less than the most exiting thing that a person could do.[24]

Consider, too, one of America's most famous inventors, Benjamin Franklin. With typical American pragmatism he wrote, "Utility is in my opinion the test of value in matters of invention, and that a discovery which can be applied to no use, or is not good for something is good for nothing."[25] Franklin's inquiries into the nature of electricity did result in one useful device: the lightning rod, which saved many a building from destruction. But his other inquiries had a less immediate payoff. Although Franklin devised a number of devices that helped him to learn more about the nature of electricity, none of these had any immediate payoff. Indeed, he was "chagrined a little that we have been hitherto able to produce nothing in this way of use to mankind."[26]

Later events proved him wrong, although he never shared in that knowledge. Still, the pursuit of useful innovations could not have been the prime motive for Franklin's inquiries. Franklin was an amateur in the literal sense of the word: a person who pursues an activity for the sheer love of it. For many years the leisure-time pursuits of amateur scientists such as Franklin sustained research into the nature of electricity despite the absence of direct payoffs. And these "idle" intellectual efforts were essential to the invention of a great variety of useful devices.

A century after Franklin died, a young man of seventeen climbed a cherry tree and turned his imagination to possibilities that only a few had dreamed of. The year was 1899, and the young man was Robert Goddard, who was to be the inventor of the liquid-fueled rocket. As he recalled in later years, "It was one of the quiet, colorful afternoons of sheer beauty which we have in October in New England, and as I looked toward the fields at the east, I imagined how wonderful it would be to make some device which had even the *possibility* of ascending to Mars, and how it would look on a small scale, if sent up from the meadow at my feet. . . . I was a different boy when I descended the tree from when I ascended, for existence at last seemed very purposive."[27] At that time and for many years to come Goddard could scarcely have imagined the products of that October vision: orbiting satellites for global communication links, weather prediction, remote sensing of natural resources, and the development of terrifying new weapons. Throughout his life Goddard continued to be energized by the dream of space travel for its own sake; practical consequences were at best a secondary concern.

It also should be noted that even in a predominantly market-oriented, capitalist society such as the United States, not all technologies have been generated and shaped by market forces. Chapter 17 will take up this issue by examining the role of the government in promoting technological change. At this point it will only be noted that technologies developed outside the constraints of the market system are less likely to be shaped by concerns about costs. Firms operating in a market environment know that minimizing costs is essential to success. Technologies that hold down production costs are quite appealing, as are technologies that expand sales by lowering the cost of the product itself. The Ford Motor Company during the glory years of the Model T is a striking example of a firm that owed its success to the use of new technologies in order to lower costs. This process also has been dramatically demonstrated in recent years as technological advances in the design and production of integrated circuits have led to sharp declines in the price of personal computers and a concomitant expansion of this market. At the same

The realization of a vision: Robert Goddard and his first liquid-fueled rocket. (UPI/Bettmann Newsphotos.)

time, however, there are large sectors of the economy where prices and the costs of production are secondary concerns. As a result, the pattern of technological development is quite different. This has been particularly evident in the defense industry, where the presumed dictates of national security have allowed the deployment of staggeringly expensive military technologies, such as fighter aircraft that have an initial purchase price in excess of $25 million, and also require 30 person-hours of maintenance for every hour that they are in the air. In a climate where only the presumed best will do, there are strong tendencies to "gold plate" weapons systems and to have little concern for cost constraints.

This tendency is not confined to the military sector. In recent years there has been a mounting concern about the financial costs of modern medical technologies. Some medical technologies, such as antibiotics, have undoubtedly lowered the costs of medical care, but many others have had the opposite result. When faced with a choice between controlling medical expenses or saving lives and alleviating pain through the use of sophisticated technologies, it is difficult for any individual or society concerned with the well-being of its members to put the former ahead of the latter. It is thus no surprise that, according to one government study, one-third of the cost increases borne by Medicare payments during the first half of the 1980s came from the use of new medical technologies.[28]

The case of military and medical technologies brings us back to the necessity to consider the political and social context of technological choice. Economic considerations, important as they are, are not the sole basis of decisions regarding the development, selection, and use of particular technologies. Nor does technology develop according to its own internal dynamics. Technologies are social creations, and any successful attempt at understanding why particular technologies are created, chosen, and used must take into account their social context. Having made this basic point, we will pursue the matter a bit further by looking at the processes through which technologies spread and take hold. But before we do so, we need to consider one more source of technological change: scientific advance. This will be the topic of the next chapter.

Questions for Discussion

1. Some of the most famous Americans of the past were inventors (Edison, Morse, Bell, the Wright Brothers). Can you name any contemporary inventors? Why do individual inventors appear to be less prominent today?
2. A person suffering from terminal heart disease might want to receive an artificial heart, but be unable to pay for it. What, then, is the source of effective demand for this technology? Who ultimately pays for it, and why?
3. As a would-be inventor or entrepreneur, how would you go about ascertaining whether or not a new technology is likely to find a receptive market?
4. If you were the leader of China or the Soviet Union, would sort of changes would you try to institute in order to accelerate the pace of technological innovation?
5. Does the structure of the modern American corporation help or hinder the development and application of technology? How does the long passage of time before a financial payoff affect the propensity of corporations to invest in new technologies? Should corporate structures be changed in order to generate a greater receptivity to innovation?

Notes

1. Arnold Heertje, *Economics and Technical Change* (New York: John Wiley & Sons, 1977), p. 98.
2. Nathan Rosenberg, *Inside the Black Box: Technology and Economics* (Cambridge: Cambridge University Press, 1982), pp. 62–70.
3. Louis Hunter, *Steamboats on the Western Rivers* (Cambridge, Mass.: Harvard University Press, 1949), pp. 121–122. Quoted in Rosenberg, *Inside the Black Box*, p. 64.

4. Albert Fishlow, "Productivity and Technological Change in the Railroad Sector, 1840–1910," in *Studies in Income and Wealth No. 30: Output, Employment, and Productivity in the United States After 1800* (New York: National Bureau of Economic Research, 1966), pp. 635, 641.
5. Tom Shedd, "The Little Engine That Does," *Technology Review* 87, 2 (February–March 1984): 66–67.
6. Richard Tanner Pascale and Anthony G. Athos, *The Art of Japanese Management: Applications for American Executives* (New York: Warner Books, 1981), p. 306.
7. The following is based on John A. Heitman and David J. Rhees, *Scaling Up: Science, Engineering, and the American Chemical Industry* (Philadelphia: Center for the History of Chemistry, 1984), pp. 17–21.
8. Nathan Rosenberg, *Perspectives on Technology* (Armonk, N.Y.: M.E. Sharpe, 1985), p. 167 [author's emphasis].
9. Rosenberg, *Inside the Black Box*, op. cit., pp. 60–61.
10. Peter George, *The Emergence of Industrial America: Strategic Factors in American Economic Growth Since 1870* (Albany: State University of New York Press, 1982), p. 51.
11. David A. Hounshell, *From the American System to Mass Production, 1800–1932: The Development of Manufacturing Technology in the United States* (Baltimore: The Johns Hopkins University Press, 1984) pp. 189–215.
12. Edward Constant, *The Origins of the Turbojet Revolution* (Baltimore: The Johns Hopkins University Press, 1980), p. 82.
13. Christopher Freeman, "The Determinants of Innovation: Market Demand, Technology, and the Response to Social Problems," *Futures* 11, 3 (June 1979): 211.
14. Everett M. Rogers, *Diffusion of Innovations* (New York: The Fress Press, 1962), pp. 148–149.
15. Jacob Schmookler, *Inventions and Economic Growth* (Cambridge, Mass.: Harvard University Press, 1966).
16. Ibid., p. 93.
17. W. G. L. De Haas, "Technology as a Subject of Comparative Studies: The Case of Photography," *Comparative Studies in Society and History* 21, 3 (July 1979): 367–370.
18. Christopher Freeman, *The Economics of Industrial Innovation*, 2d ed. (New York: Cambridge University Press, 1982), p. 124.
19. Ibid., p. 127.
20. Wilson Dizard, *The Coming Information Age: An Overview of Technology, Economics, and Politics* (New York: Longman, 1982), p. 33. See also Paul Ceruzzi, "An Unforeseen Revolution: Computers and Expectations, 1935–1985," in Joseph J. Corn (ed.), *Imagining Tomorrow: History, Technology of the American Future* (Cambridge, Mass., MIT Press, 1986).
21. Hugh G. J. Aitken, *Syntony and Spark: The Origins of Radio* (Princeton, N.J.: Princeton University Press, 1985).
22. Robert Heilbroner, *Between Capitalism and Socialism: Essays in Political Economics* (New York: Random House, 1970), p. 162.
23. Karl Marx and Frederick Engels, "The Communist Manifesto," in *Karl Marx and Frederick Engels: Selected Works*, vol. I (Moscow: Foreign Languages Publishing House, 1962), p. 39.
24. Elting E. Morison, *From Know-How to Nowhere: The Development of American Technology* (New York: New American Library, 1977), p. 126.
25. Quoted in Roger Burlingame, *March of the Iron Men: A Social History of Union through Invention* (New York: Grosset & Dunlap, 1938), p. 77.
26. Ibid.

27. Barton C. Hacker, "Robert H. Goddard and the Origins of Space Flight," in Carroll W. Pursell, Jr. (ed.), *Technology in America: A History of Individuals and Ideas* (Cambridge, Mass.: The MIT Press, 1981), p. 233.
28. Harvey V. Fineberg, "Irresistible Medical Technologies: Weighing the Costs and Benefits," *Technology Review* 87, 8 (November–December 1984): 17.

Chapter 4 Scientific Knowledge and Technological Advance

One of the most common beliefs about technology is that it is simply "applied science." There are certainly many examples that can be cited in support of this view. Modern medical practices have been strongly influenced by fundamental discoveries in biology. The development of the transistor depended on a thorough understanding of quantum mechanics. Synthetic materials have been made possible by research into polymer chemistry. But one should not be content to rest with these examples. When the full spectrum of technological advance is considered, it becomes evident that science does not always play the decisive role in the development of technology. Indeed, many are the times when technological advances have taken place without the benefit of scientific knowledge. Conversely, on some occasions scientific advance has depended on prior technological achievements. In this chapter we will look at the complex and shifting relationships between science and technology in order to come to a better understanding of their unique natures and the ways in which they have influenced each other.

THE HISTORICAL SEPARATION OF SCIENCE AND TECHNOLOGY

The definition of technology that was offered in the first chapter stressed that technology is based above all on the application of knowledge. But not all knowledge need be derived from scientific research. It is certainly true that today much of the knowledge required for technological advance is derived from scientific inquiries. Still, when the full history of technology is surveyed, it is apparent that most technologies have been developed and applied with little scientific input. The ancient Greeks made important contributions to many sciences—most notably astronomy, optics, and acoustics, as well as producing major advances in mathematics. Greek technology also progressed through innovations in agriculture, building construction, mining, the refining of metals, and military equipment. Yet none of these innovations drew to any significant degree on Greek science. Moreover, the Greeks' technological achievements were far less impressive than their scientific achievements, again indicating the lack of connection between the two. This lopsided pattern of development continued with the Romans, although in reverse. Roman contributions to science were minor, while Roman engineering (mainfested in such things as the construction of great aqueducts) reached a high level of development. In any event, Roman technology had little to do with science.

The European Middle Ages were a time of slow but significant technological advance. Improved agricultural practices were introduced, along with new sources of power such as windmills. An effective horse collar allowed the literal harnessing of another important source of power. Soaring cathedrals were built in many parts of Europe, where they continue to a be a source of awe and inspiration. Again, these achievements owed nothing to the scientific inquiries of the time. In fact, the designers and builders of the cathedrals apparently did not even have knowledge of multiplication tables. Then too, there was little that technology could have drawn on, for medieval science exhibited little of the dynamism of medieval technology.

This inverse relationship between scientific and technological development continued during the succeeding centuries. The sixteenth and seventeenth centuries were the scene of epochal advances in science, yet technological change occurred at a slower rate than it did during the preceding centuries when science had been largely stagnant.[1] In similar fashion, early nineteenth-century France boasted the most advanced science in Europe, yet its technology was no more advanced than that of England, its scientific inferior. In some key areas, such as mechanization and steam power, France lagged well behind England.[2] This historical record has led one historian of science, Thomas Kuhn, to speculate that for the bulk of human history, technology has flourished in societies where science has remained undeveloped, and vice versa.[3] It is possible that our era is unique in its apparent ability to simultaneously support scientific and technological advance.

STUDIES OF CONTEMPORARY SCIENCE-TECHNOLOGY RELATIONSHIPS

Even today, when the connection between science and technology is much stronger than it was in the past, a great deal of technological change takes place without substantial inputs from science. The relative unimportance of science for many technological developments was highlighted by a study that was conducted by the Defense Department in the mid-1960s. Dubbed Project Hindsight, this study assessed the extent to which pure scientific research was essential to the development of twenty major weapons systems. In conducting their study, the Hindsight researchers began with a weapon system and traced its history backwards in order to determine the key "events" that produced the knowledge that had been essential to its creation and development. The results of the study were highly unfavorable to the commonly accepted view that scientific knowledge is the primary basis of technological development. Of the 710 events surveyed, only two were the result of basic scientific research, a minuscule 0.3 percent of the total.[4] Scientific research that was specifically directed toward a particular military project was of greater importance, accounting for 6.7 percent of events, while 2 percent were the result of scientific research directed toward commercial or nondefense needs. The greatest portion of events (the remaining 92 percent) owed little to concurrent scientific research, and relied almost entirely on established concepts and principles.

Similar conclusions emerged from a research project that was conducted in England. This study examined the winners of the Queen's Award for Industry, which is given to British firms that have distinguished themselves by initiating technologically innovative products and processes. The study found that very few of these innovations were directly connected to basic scientific research. Accordingly, the authors concluded that "the great bulk of basic science bears only tenuously if at all on the operations of industry."[5] The findings of this study, as well as those of Project Hindsight, were echoed by a study of American industrial innovation that was conducted by James Utterback, who concluded that "Basic research does not seem to be significant as a direct source of innovations."[6]

Still, one should not draw sweeping generalizations from these studies. The authors of the Hindsight study, as well as a number of critics of it, were quick to note that the long-term influences of scientific research were not captured by the study's methodology. Project Hindsight considered only the effects of scientific research conducted for the most part after 1945, thereby removing from consideration the immense body of scientific knowledge that had accumulated before that time. The study's researchers found that a median delay of nine years separated the completion of a scientific research project from its application, even when research efforts targeted at specific technological missions were included. It was therefore not surprising that basic scientific research had few technological consequences during the twenty-year span covered by the study.[7]

Another study of the relationship between scientific research and technological application, Technology in Retrospect and Critical Events in Science (TRACES), contradicted the previously cited studies by determining that a number of recent innovations, ranging from oral contraceptives to videocasette recorders, depended on prior scientific research.[8] This study differed from the others in that it took a much longer chronological view of the interactions between science and technology. Consequently, a particular invention could be linked to scientific discoveries of long ago. But even here, the researchers were obliged to point out that the sequence from scientific discovery to technological innovation is not linear, and that "a better understanding needs to be achieved concerning the two-way influence between science and technology."[9]

HOW TECHNOLOGY DIFFERS FROM SCIENCE

If nothing else, these studies show that the connection between science and technology is not adequately captured by the common belief that technology is simply applied science, and that scientific discoveries are easily converted to technological applications. Some technologies draw directly on scientific research, while others make little use of it. This is rather obvious. Of greater significance is the fact that science and technology are quite different in their basic natures. This makes the translation of scientific knowledge into technological application a difficult and complex process.

Whereas science is directed at the discovery of knowledge for its own sake, technology develops and employs knowledge in order to get something done. The content of the knowledge may be rather similar, but different motivations underlie its pursuit and application. Here of course we are on slippery ground; it is often extremely difficult to discern the motivations underlying a person's activities, and it may well be the case that a particular engineer may be driven by the same desire to understand something for its own sake that animates the work of a pure scientist. Motives are often mixed and complex.[10]

Much of the prestige accorded to science is the result of its supposed purity; science is thought to be an intellectual venture free from political, organizational, and economic constraints. The insulation of scientists from the demands of their patrons confers a sense of higher ethical standards; scientists are beholden to nothing but the internal demands of science. A great deal of recent scholarship has sharply questioned this assumption by demonstrating strong political, economic, and cultural influences on science.[11] Still, the deeply ingrained belief in the purity of science gives individual scientists a claim to autonomy not enjoyed by other employees. Scientists are thus in a particularly favorable situation. The assumption that scientific progress leads to material progress confers an aura of practicality on their work, while at the same time they are in a good position to resist the overt control of their work by their sponsors.

In contrast, most engineers work under tighter constraints. Their employers expect results that have immediate applications and fall within a narrowly defined range of possibilities. A scientist may abandon a theory or an experiment in order to pursue a line of inquiry that unexpectedly arises during the course of his or her research. An engineer, however, rarely has this opportunity; there may be some room for serendipity, but the bridge has to be built within a given time frame and under definite budget constraints. For this reason, what separates scientific and technological inquiries may not be the motivations of individual practitioners, but the motivations of their employers and patrons.[12]

Technology also differs from science in the type and depth of knowledge that is required. The ultimate question asked of scientific knowledge is "is it true?" For technological knowledge, the key issue is "will it work?" Technological problems can be often solved with no understanding of what is going on. Many technologies are effective even though the basic principles underlying their operation are poorly understood, if they are understood at all. For decades iron was produced in blast furnaces even though existing levels of metallurgical knowledge provided few clues about what was actually occurring inside the furnace. Even today, successful aircraft are built in the absence of completely adequate theories of air turbulence or compressibility that would allow the design of optimal configurations.[13] It is also instructive to consider the story of the great scientist Johannes Kepler (1571–1630), who developed and employed the calculus of variation in order to derive optimum dimensions of beer kegs—only to discover that these dimensions were already being employed by the coopers who actually built the kegs![14]

Many other technological innovations seem to fall into this pattern. Although scientifically derived principles may emerge after the fact, many technologies have been guided almost exclusively by trial and error, with the successful ones informed by an intuitive sense of the right solution, and not by scientific truths. As Eugene Ferguson has observed, at the end of the nineteenth century there were no scientific principles that could be invoked during the design of the first motorcycles; the placement of the engine, fuel tank, and other major components could be determined only through the actual construction and operation of motorcycles, without the benefit of scientific principles or other forms of existing knowledge. Ferguson therefore makes the point that "there is often no a priori reason to do one thing rather than another, particularly if neither had been done before. No bell rings when the optimum design comes to mind."[15]

HOW TECHNOLOGY STIMULATES SCIENTIFIC DISCOVERY

Although we tend to think of science as the leading factor in technological advance, the reverse often occurs: scientific knowledge and discovery may be a by-product of technological achievements, as when fundamental advances in biology and chemistry were stimulated by the successful efforts of Pasteur, Lister, and Koch to solve practical medical problems.[16] There have also been cases where a technology already in operation defied accepted scientific explanations and stimulated the formulation of new theories. This process is exemplified by the story of how the steam injector contributed to the abandonment of a popular scientific theory regarding the nature of heat.[17] In the mid-nineteenth century, many scientists believed that heat was the result of the presence of a substance known as "caloric." According to this theory, when caloric combined with other materials they became hot. Also, caloric particles were supposedly self-repellent; thus, when sufficient quantities of these particles came into contact with water, their repulsive quality resulted in water turning into steam.

While this theory had its uses, it could not explain the operation of the steam injector that was patented by Henri Giffard in 1858. The injector used steam from the boiler to lift water into it, an operation that seemed to mimic perpetual motion for those who subscribed to the caloric theory. In fact, Giffard, who was well-trained in academic science, based his injector on the Bernoulli principle, which postulated that the pressure of a fluid (in this case steam) drops as its velocity increases. The operation of the injector was therefore the result of expanding steam producing a partial vacuum that sucked water into the boiler.

Giffard's injector was no perpetual motion machine; it used a quantity of heat that was equal to the quantity of work expended in raising water into the boiler, plus the losses due to radiation and contact with surrounding surfaces. Its operation therefore made sense only when the interconvertability of heat and work was understood. This idea rested on the kinetic theory of heat, and it followed the first law of thermodynamics (which stipulates that, quantitatively, energy cannot be created or destroyed). The kinetic theory of heat was formulated several years before Giffard's invention, but had been slow in winning acceptance. The rival caloric theory had many adherents in the scientific community, and it took the anomaly of the injector to convert many of them to the now universally accepted kinetic theory of heat.

The steam injector illustrates the often subtle interactions between science and technology. The operation of the injector provided a strong stimulus for the acceptance of one scientific theory and the rejection of another. At the same time, another scientific theory had been essential to the invention of the injector. But scientific theories by themselves were not enough; the design and effective use of the injector still depended on the experiments and modifications performed by practicing engineers, for no set of theories was powerful enough to guide its design. Again, we have an example of a technology that worked even though existing scientific principles did not completely explain its operation.

This example and many others that could be cited indicate that science and engineering are still separate enterprises, although there are certainly linkages between them. Scientific knowledge can result in technological advances, while at the same time many technologies create both the opportunities and the motivations for new scientific inquiries. Many technological developments reach a plateau due to a lack of scientific knowledge, thereby generating a clearly perceived need for fundamental scientific research. The knowledge obtained through technological practices and applications is thus the raw material of many scientists, whose work centers on explaining technological practices at a deeper level.[18]

The scientific discoveries so gained then become grist for the technological mill, as basic scientific research provides new sources of knowledge that have technological applications. This feedback between science and technology may be a fundamental source of their dynamism.[19] When science and technology have gone their separate ways, as has been the case for most of human history, they develop more slowly than when they interact with each other, as they have done to an increasing degree during the twentieth century.

INDIRECT EFFECTS OF TECHNOLOGY ON SCIENTIFIC ADVANCE

Technology's role in stimulating scientific advance does not end with the provision of data, problems, and research opportunities. Technological development also plays a vital role in scientific advance by supplying devices and instruments that are essential for scientific inquiry. From early telescopes and galvanometers to today's electron microscopes and computers, the products of technology have steadily increased our ability to observe and analyze the phenomena that science takes as objects of inquiry.[20] One of the essential features of science is accurate and systematic observation, and without the assistance of the technologies that have allowed a close monitoring of the world, science could hardly go beyond abstract theorizing and speculation.

There is a final and less immediately evident contribution that technology has made to scientific progress. Although an effort has been made here to demonstrate that science has not always been decisive in the development of new technologies, the opposite is widely believed. To a significant degree, this faith in the practical consequences of scientific research has given science the immense prestige and legitimacy that it enjoys today. Many areas of scientific inquiry have become increasingly expensive propositions, and without the promise of an eventual payback through technological applications, many research programs would die for lack of funding and other sources of support. This is illustrated by the great amount of sophisticated and expensive biological research that has been justified on the grounds of its potential contribution to curing cancer—a hope that has yet to be realized. Biological research is hardly unique in this aspect, for scientists have become quite proficient in writing grant applications that stress the practical value of their abstract inquiries.

Financial support, however important it is to the maintenance of scientific inquiry, is only part of the picture. The willingness of government agencies to grant money for scientific research and of citizens to have their taxes used in this manner is indicative of a widespread belief in the legitimacy of scientific research. This legitimacy is in large measure the product of the presumed ability of science to ultimately produce practical results. These ascribed powers of science have been analyzed by Langdon Winner:[21]

> [The ultimate success of science] must be accounted to its fulfillment of Baconian ambitions—the delivery of power. Other modes of knowing have been able to give an intelligible, systematic, aesthetically pleasing picture of reality. If science had only been able to accomplish this and nothing more, it is likely that it would have been supplanted by yet another philosophy of inquiry. But in the West at least, the test is not so much *what do you know?* or *how elegant is your interpretation of worldly phenomena?* but rather, *what can you actually do?* This is the conclusive factor, the reason that, for instance, social science has never fully established its credentials in the halls of science. Science succeeds over rival ways of knowing—poetry, religion, art, philosophy, and the occult—not by its ability to illuminate, not even by its ability to organize knowledge, but by its ability to produce solid results. . . . In the last analysis, the popular proof of science is technology.

This expected ability of science to "deliver the goods" is somewhat paradoxical, for science as a system unto itself responds rather poorly to economic needs. This has even been made into a virtue by many scientists who pride themselves on their autonomy and their insulation from the crass demands of the marketplace. As we have seen, the autonomy of scientists has been legitimized by the conception of science as a detached exercise in free inquiry. But it is also the case that the unpredictable nature of scientific discovery often precludes the possibility of useful discoveries being produced to order.

Scientific research, especially when directed at the discovery of basic principles, is an uncertain endeavor that cannot be guided by schedules and routinized procedures. This is illustrated by the response of one researcher who was offered more research funds by an officer of his company in the hope that the conclusion of a particular project could thereby be hastened. The researcher replied that it would be just as logical to expect that eggs could be made to hatch in half the normal time if twice as many hens were made to sit on them. Not only are the paths of scientific discovery full of twists and turns, many of them terminate in dead ends. Of course, technology may also be incapable of solving the problems presented to it. If it were otherwise, we would have a cure for cancer by now. But most technology is directed toward the solution of specific problems, which narrows its scope and makes it a more predictable enterprise than science.

Scientific knowledge often ends up being extremely useful to technology, but that is not why is was produced. Scientists typically create knowledge for other scientists. Their efforts are focused on the testing of theories and the solution of problems that have been generated by previous scientific inquiries. If scientific knowledge is used for technological applications, it is because engineers and other technologists have appropriated it for their own use. In most places where science and technology meet, engineers and technicians "pull" knowledge out of science. Only in rare instances is knowledge directly relevant to technological application "pushed" by science itself.[22]

When knowledge is "pushed" from science into technology, it often happens indirectly. The transfer of knowledge from science to technology can be a subtle process, with scientific research motivating technological change by pointing out unseen problems, while at the same time suggesting new opportunities. This happened in the 1930s when the evolving science of aerodynamics showed how the behavior of aircraft changed dramatically at high speeds. This research clearly indicated that conventional propeller-driven airplanes would encounter an insurmountable velocity barrier as they approached the speed of sound. At the same time, aerodynamic research indicated that proper streamlining could greatly increase the speed at which airplanes could fly, provided they had a different method of propulsion. In making these discoveries, aerodynamic researchers generated a powerful impetus for the development of jet engines that produced more power and did not have the inherent limitations of existing power systems.[23]

THE COMMONALITIES OF SCIENCE AND TECHNOLOGY

Up to now, this chapter has stressed the differences between science and technology. At this point some mention should be made of the characteristics that they have in common. At the most fundamental level, both are based on the gathering of knowledge, and they both advance through the cumulative development of that knowledge. Isaac Newton is reputed to have said that he could see farther because he stood on the shoulders of giants. That is, his scientific discoveries were based on knowledge produced by earlier scientists. The same holds true for modern technology. Just as the scientific achievements of an individual chemist owe a great deal to the past research efforts of other chemists, the work of an aerospace engineer draws upon the accomplishments of other aerospace engineers.

Equally important, both science and technology share the rational thought processes described in Chapter 1. Although the development of both science and technology requires intuitive and other nonrational modes of thought, rationality is an essential part of the general methodology of science and technology. In general, a rational approach includes a propensity to challenge traditional intellectual authorities; a willingness to settle questions through observation, testing, and experimentation; and a desire to develop exact methods of measurement.[24]

Some of the basic elements of this mode of inquiry are described by Robert Pirsig in *Zen and the Art of Motorcycle Maintenance*, where he explains how even a clearly technological task like determining why a motorcycle won't start is addressed through the use of procedures that have much in common with scientific inquiry.[25] As a first step, a mechanic might formulate the hypothesis that the battery is dead; he or she will then try to honk the horn to see if the battery is working. If the horn honks, the mechanic concludes that the problem doesn't lie with the battery, and proceeds to other parts of the electrical system. Should tests performed on these components show them to be in good shape, the mechanic may hypothesize that the problem lies with the fuel system, and conduct tests (experiments) to check them out. And so it goes, with the formulation of a series of hypotheses and the conducting of experiments to test them. In the end the problem is isolated and perhaps fixed; if nothing else, you know what is wrong as you push your motorcycle along the side of the road.

Of course, one shouldn't take this analysis too far. Although both science and technology make heavy use of rational modes of thought, neither can be properly characterized as the embodiment of rationality. Scientific theories must be logically consistent and rationally articulated, but their ultimate source is human creativity and imagination—qualities often at a considerable distance from rational thought processes. At the other end of the scientific enterprise, the testing of these theories, there are no perfectly rational means of determining the criteria through which theories can be validated or disproved. Even empirically derived "facts" can be subject to interpretation, while general world views can strongly affect what is acceptable as "proof."[26] In similar fashion, a great deal of technological advance is also the product of nonrational thought. And, as was noted earlier, the benefit or harm of a particular technology cannot always be adjudged according to criteria based on rationally determined principles; a great deal hinges on values and ethical standards that are derived through other means.

Other commonalities between science and technology can be noted. Mathematics is important to both as a kind of language and as an analytical tool. Also, engineers and other technological practitioners employ organized knowledge that is presented and diffused through journals, books, and professional meetings that have many similarities to those found in the realm of science, although engineers often view their findings as proprietary knowledge, and therefore may resist its dissemination.[27]

Finally, at the core of the common culture of science and technology is a sense of optimism and progress. Science and technology are dynamic enterprises that build on past successes, but also make profitable use of their failures. An inadequate scientific theory may lead to the formulation of a better one, and a collapsed bridge is likely to provide valuable lessons that help to prevent future failures.[28] Above all, science is predicated on the belief that the world is knowable, while technology is animated by a conviction that it will always be possible to do something better. Both of these beliefs contribute to the dynamic, essentially optimistic spirits of science and technology.

Although there are broad similarities between science and technology today, their coexistence is problematic, much as it has been in the past. For Melvin Kranzberg, their coexistence has been marked by the same kind of tensions and attractions that characterize the marriage of a man and a woman. In Kranzberg's words:[29]

> History suggests that science and technology, though wedded today, went through a long, indifferent courtship. They grew up independently, either oblivious to each other's existence or taking scornful note of the other's presence. When they reached

the age of puberty—the scientific revolution in the case of science and the Industrial Revolution in the case of technology—a mild flirtation ensued.

The marriage, when it came at last, was a marriage of convenience and necessity, certainly no love match. Insofar as military needs helped to bring about many a daring and secretive meeting, the ceremonies when finally reached, could be called a shotgun wedding; and the couple, predictably, has not lived happily ever after.

Each partner has retained a good deal of independence, though lately both have been having identity problems. There are constant bickerings about who is contributing more to the marriage. They quarrel over mutual responsibilities, the education of their offspring, and, as might be expected, the household budget.

It is a very modern marriage. Science and technology live independently, yet coordinately, as if they had but one joint bank account and one car. Divorce is frequently discussed. It is invariably rejected, however, because the scandal would surely deface the public image of the parties, and because, I suspect, of the indisputable pleasures and the learned frivolities of the bed.

THE TRANSLATION OF SCIENCE INTO TECHNOLOGY

Today, technology makes heavy use the products of scientific inquiry. Much of this use, however, is indirect. A great deal of scientific information finds its way into technological practice through the education of engineers.[30] The findings of basic scientific research appear in handbooks, university courses, and textbooks. Much of the scientific knowledge presented in these ways is eventually drawn on during the course of technological development.[31]

Even here there can be problems. It has been argued that a significant amount of engineering education has been distorted by overreliance on science-based instruction. This has led to a devaluation of nonverbal thought, an excessive stress on mathematics, and an unwillingness to tackle problems that do not have a single unique solution.[32] Scientific thinking converges toward a single (if temporary) set of theories, while the history of technology is replete with examples of the old saying that there's more than one way to skin a cat. An excessive focus on the principles and methods of science may therefore restrict creativity and lead to an overly rigid approach to the solving of technological problems.

There is no getting around the fact that despite all that they have in common, science and technology operate in different worlds. If the two are to share in a productive symbiosis, they must be sustained by continual efforts to span the differences that separate them. In many cases, technological development has been stimulated by the presence of individuals and organizations that simultaneously participate in scientific and technological communities. Their primary role is to serve as translators, "decoding information generated in one system and transforming it into information usable in another."[33]

This process can be seen in the events that culminated in the invention of the vacuum tube. The story began with Edison's invention of the light bulb.[34] While trying to determine why dark deposits were forming on the interior walls of the bulbs, Edison found that the needle of a galvanometer deflected when a wire probe was placed in circuit between the galvanometer and the bulb's glowing filament.

Edison did not understand what was producing the flow of electrical current through thin air, although he patented the apparatus for use as a voltage indicator. (Many years later the realization came that the current was produced by the migration of electrons from the negatively charged filament to the positively charged probe.)

Nothing practical came of Edison's discovery until John Ambrose Fleming renewed his acquaintance with these specially equipped bulbs. During the 1880s and 1890s Fleming had conducted a number of experiments using these bulbs; his sporadic efforts produced useful scientific knowledge, but no technological applications. Things began to change in 1899 when he became technical advisor to Guglielmo Marconi's Wireless Telegraphy Company. At that time the chief need of the infant radio industry was for a detector that could efficiently convert the weak oscillatory current of radio waves into direct current. After a few years' work with other devices, in 1904 Fleming came to the sudden realization that the specially equipped light bulbs with which he had previously worked might be used for this purpose. His hunch proved to be correct, and the "oscillation valve," as he named the device, began to be commercially used for the detection of radio signals a short time later.

Fleming had not been the only one to experiment with modified light bulbs, but he had been uniquely situated to act as a "translator" between science and technology. He was not an inventor like Edison or a full-fledged scientist like other experimenters. Rather, he was a scientifically trained engineer and teacher who was closely associated with the electrical industry and with engineering-training institutions. These separate but interrelated roles gave him the knowledge and the motivation to convert a scientific curiosity into a practical technology.

This chapter began with the assertion that technology is not applied science and went on to provide some evidence for this statement. It is undeniable that technology today makes extensive use of scientific knowledge. But as we have seen, scientific knowledge often makes its way into technological practice in a very roundabout way. At the same time, a considerable amount of scientific advance stems from prior achievements in technology. Science and technology have evolved along separate paths that often intersect. At these points of intersection each has often contributed to the other's development. Both science and technology seem to do best when they remain in close contact, but this should not obscure the fact that they remain very different enterprises.

Questions for Discussion

1. In 1993 the United States Congress cancelled one of the most ambitious science research projects of all time, the superconducting supercollider for high-energy physics research. The project would have cost at least $8.5 billion. How do you suppose the project was justified? Did Congress make the right decision?
2. Why have science and technology been so closely associated in popular thought? How do each of them gain from this association?

3. Monetary considerations aside, which would you find more personally satisfying: making a scientific discovery or inventing a useful technology? Why?
4. Students in engineering programs typically take a substantial number of science and mathematics courses. Should some of these courses be eliminated and replaced with different kinds of courses? If so, which courses should be taken instead?

Notes

1. Thomas S. Kuhn, "Comment on the Principle of Acceleration," *Comparative Studies in History and Society* 11, 4 (1969): 427.
2. Thomas S. Kuhn, *The Rate and Direction of Inventive Activity* (Princeton, N.J.: Princeton University Press, 1963), pp. 450ff.
3. Kuhn, "Comment," op. cit., p. 428.
4. Chalmers W. Sherwin and Raymond S. Isenson, "Project Hindsight," *Science* 156, 3782 (23 June 1967): 1571–1577.
5. Quoted in J. E. S. Parker, *The Economics of Innovation: The National and Multinational Enterprise in Technological Change*, 2d ed. (London: Longman, 1978), p. 30.
6. James M. Utterback, "Innovation in Industry and the Diffusion of Technology," *Science* 183, 4125 (15 February 1974): 622.
7. Sherwin and Isenson, op. cit., p. 1575.
8. Parker, op. cit., p. 30.
9. Edwin Layton, "Mirror-Image Twins: The Communities of Science and Technology in 19th Century America," *Technology and Culture* 12, 4 (October 1971): 564–565.
10. Otto Mayr, "The Science-Technology Relationship as a Historiographic Problem," *Technology and Culture* 17, 4 (October 1976): 668.
11. See Michael Mulkay, *Science and the Sociology of Knowledge* (London: George Allen and Unwin, 1979); John Ziman, *Reliable Knowledge* (Cambridge: Cambridge University Press, 1979); John G. Burke (ed.), *Science and Culture in the Western Tradition: Sources and Interpretations* (Scottsdale, Ariz.: Gorsuch Scarisbrick, 1987).
12. F. R. Jevons, "The Interaction of Science and Technology Today, or, Is Science the Mother of Invention?" *Technology and Culture* 17, 4 (October 1976): 731.
13. Nathan Rosenberg, *Inside the Black Box: Technology and Economics* (Cambridge: Cambridge University Press, 1982), p. 143.
14. Kuhn, "Comment," op. cit., p. 429.
15. Eugene S. Ferguson, "The Mind's Eye: Nonverbal Thought in Technology," *Science* 197, 4306 (26 August 1977): 827.
16. Joseph Ben-David, "Roles and Innovation in Medicine," *American Journal of Sociology* 65, 6 (1960): 557–568.
17. Eda Fowlks Kranakis, "The French Connection: Giffard's Injector and the Nature of Heat," *Technology and Culture* 23, 1 (January 1982): 3–38.
18. Rosenberg, *Inside the Black Box*, op. cit., pp. 146–147.
19. Edwin T. Layton, Jr., "American Ideologies of Science and Engineering," *Technology and Culture* 17, 4 (October 1976): 688.
20. John P. McKelvey, "Science and Technology: The Driven and the Driver," *Technology Review* 88, 1 (January 1985): 38–47.
21. Langdon Winner, *Autonomous Technology: Technics-Out-of-Control as a Theme in Political Thought* (Cambridge, Mass.: The MIT Press, 1977), pp. 24–25.

22. Hugh G. J. Aitken, *Syntony and Spark: The Origins of Radio* (Princeton, N.J.: Princeton University Press, 1985), p. 316.
23. Edward Constant, *The Origins of the Turbojet Revolution* (Baltimore: The Johns Hopkins University Press, 1980), pp. 15–16.
24. Peter Mathias, "Who Unbound Prometheus? Science and Technical Change, 1600–1800," in Peter Mathias (ed.), *Science and Society 1600–1900* (Cambridge: Cambridge University Press, 1972), p. 79.
25. Robert M. Pirsig, *Zen and the Art of Motorcycle Maintenance* (New York: Bantam Books, 1974), pp. 99–103.
26. John Ziman, *An Introduction to Science Studies: The Philosophical and Social Aspects of Science and Technology* (Cambridge: Cambridge University Press, 1984), p. 104.
27. Thomas Parke Hughes, "The Science-Technology Interaction: The Case of High-Voltage Power Transmission Systems," *Technology and Culture* 17, 4 (October 1976): 659.
28. See Henry Petrovski, *To Engineer Is Human: The Role of Failure in Successful Design* (New York: St. Martin's Press, 1985).
29. Melvin Kranzberg, "Let's Not Get Wrought Up about It," *Technology and Culture* 25, 4 (October 1984): 742.
30. Utterback, op. cit., p. 622.
31. Sherwin and Isenson, op. cit., p. 1577.
32. Ferguson, op. cit., p. 832.
33. Aitken, op. cit., pp. 335–336.
34. George Shiers, "The First Electron Tube," *Scientific American* 220, 3 (March 1969): 104–112.

Chapter 5 The Diffusion of Technology

A technology is of no use unless it is put to use. This is a crushingly obvious statement, but it does serve to remind us that a technology that never makes it out of a laboratory or an inventor's basement might just as well not exist. The last two chapters have examined some of the ways in which technologies have come into being; the task of this chapter is to describe how they are adopted by the individuals and organizations that actually put them to use. Our concern here is the *diffusion* of technologies, that is, the processes through which they spread from their initial sources into homes, factories, offices, and so forth. To begin, we will consider the diffusion of technologies from one country to another. In the second part the focus will narrow as we look into the process of technological diffusion at the level of the individual business firm.

The International Diffusion of Technology

There is an understandable human tendency to believe that progress, technological or otherwise, is largely the result of efforts of one's own people. We exalt native scientists, entrepreneurs, and inventors, and see in their efforts the major source of our technological advance. The Soviet Union at the time of Stalin carried this to ridiculous lengths (Soviet historians claimed that their countrymen had invented everything from the steam engine to baseball), but many other countries have exhibited strains of technological nativism. The truth of the matter is quite different. Although indigenous skills and achievements are an indispensable part of technological advance, no technologically dynamic nation has ever been isolated from the influences of other lands. Continued technological advance requires an infusion of ideas, tools, and materials from other places, coupled with an ability to make good use of them.

Consider the world in the year 1500. There can be little doubt about what country had produced the most technological innovations up to that time. It was not England, nor was it France, Italy, or Germany. The country with the greatest heritage of inventive achievements was China. By 1500, Chinese technologists had produced not only the compass, paper, and gunpowder—the familiar textbook examples of Chinese inventive ability—but also had been responsible for such inventions as the segmental arch bridge, the chain drive transmission, the spinning wheel, watertight bulkheads for ships, printing, fishing reels, paddlewheel boats, the differential gear, earthquake detection devices, and of course the fine porcelain that takes its name from its country of origin.[1] Europe too could claim some significant technological advances, most notably in mining and cathedral building, but on the whole, there could be little question where the greater degree of inventiveness had been residing.

Yet within less than two centuries China's lead had dissipated, and the European continent exhibited the greater degree of economic and technological vitality. Much of this dynamism could be attributed to the successful adoption of Chinese inventions by European countries. Paper was an essential complement to printing; together they made possible the rapid circulation of new ideas and stimulated the growth of literacy, which in turn directly contributed to major historical changes such as the Protestant Reformation and the spread of capitalism. The magnetic compass greatly aided transoceanic navigation and allowed the spread of European economic power throughout the world. And gunpowder helped to conquer and secure that world for the benefit of Europe.

The successful use of foreign technologies by early modern Europe was very much in accordance with long-standing historical patterns. Many of the most significant inventions used by the Western world up to that time did not originate there. The stirrup, which revolutionized warfare and produced profound changes in the social system, was not a Western invention, but diffused from Persia sometime after the eighth century, although it too probably originated in China.[2] Horses could not be used effectively as draft animals until an effective collar, originally designed in the Middle East for use with camels, made its way to Europe in the eighth century.[3]

Much of the economic dynamism and prosperity of the Western world therefore can be traced to the willingness of its people to accept and make good use of technologies that originated elsewhere.[4] By contrast, China was far less open to foreign inventions. Secure in the belief that they inhabited the "Middle Kingdom" (the literal translation of *Zhong Guo*, the name by which the Chinese called their country), the Chinese looked at the outside world with more than a trace of arrogance. This attitude was particularly evident when China came into contact with the expansionary nations of Europe. During the late eighteenth century English merchants and officials scoured the world in search of customers for the products of English factories, but they found no takers in China. The emperor made it very clear to one envoy of King George III that "As your Ambassador can see for himself, we possess all things. I set no value on objects strange or ingenious, and have no use for your country's manufactures."[5]

China's ruling elite persisted in this attitude. Toward the end of the nineteenth century the scholar Kang Youwei advocated the combination of Western techniques with traditional Chinese culture. Although he won the support of the emperor, the reform movement that was initiated in 1898 was aborted in a few months when the emperor was deposed and replaced by the reactionary dowager empress, Dzu Xi. The early years of the Chinese Republic (1912–1949) saw a considerable infusion of Western ideas and technologies, but these were eclipsed by Chiang Kai-shek's New Life Movement, which stressed a return to Confucian virtues as the cure for China's weakness.

Even the emergence of a Communist state in 1949 did not result in a wholehearted acceptance of foreign ideas and ways of doing things. Although a great amount of technology transfer from the Soviet Union took place during the early 1950s, the Chinese were offended by Russian arrogance. The political rift between the two countries closed off China's main source of advanced technology, while at the same time, Maoist ideology contributed to an increasing aloofness to the outside world and stressed self-reliance in technological matters. Only in recent years has China exhibited an openness to the outside world. In the words of one Chinese journalist, "No nation has a monopoly on culture. Cultural exchanges between nations are inevitable and irresistible. . . . Whatever will facilitate China's material modernization and its cultural advancement is welcomed."[6]

People and business firms in the United States have been much more inclined to borrow technologies that originated elsewhere. One could practically date the start of America's industrial revolution to the year 1790, when Samuel Slater, who had recently emigrated from England, used his accumulated knowhow to construct the first successful mechanized spinning factory in the United States. In the years immediately following, the United States was so deficient in indigenous technological capability that many implements and techniques had to be imported. American canal builders found it necessary to secure from England drawings and examples of devices as simple as a wheelbarrow in order to successfully construct early canals.[7]

These were not simply instances of a young and immature nation relying on imported technologies due to temporary incapability. Modern, quintessentially "American" innovations such as the automobile, the jet airplane, and television all owe their existence to pioneering efforts that were undertaken in other lands. And along with imported implements, materials, and processes, American technological development has been spurred on by an even more valuable import: people. Although a large reservoir of "Yankee ingenuity" cannot be denied, many of America's greatest technological achievements were produced by immigrants. It would take many pages simply to list some of the most prominent of these, but a small sample can illustrate the debt that American technology owes to immigrant scientists, inventors, and engineers. One of the first plastics, Bakelite, was the work of a man born and educated in Belgium, Leo Baekeland. The mathematical and engineering work of Charles Steinmetz, a transplanted German, was essential to the development of the American electrical industry. The science of aerodynamics owes much of its early development to an immigrant from Hungary (via Germany), Theodore von Karman. The television camera was invented by a Russian exile, Vladimir Zworykin. And just about everybody knows of Wernher von Braun's importance to the space program. Equally important, along with these famous figures there have been legions of unsung men and women who brought their skills from other lands and in so doing helped to push American technology to new heights.

A heavy reliance on foreign technologies also can be seen in the development of another great twentieth century technological colossus, Japan. The aggressive industrialization effort launched after the Meiji Restoration in 1868 was at first based on the importation of foreign equipment such as locomotives and textile equipment. Foreign technical advice was also avidly sought; during the early years

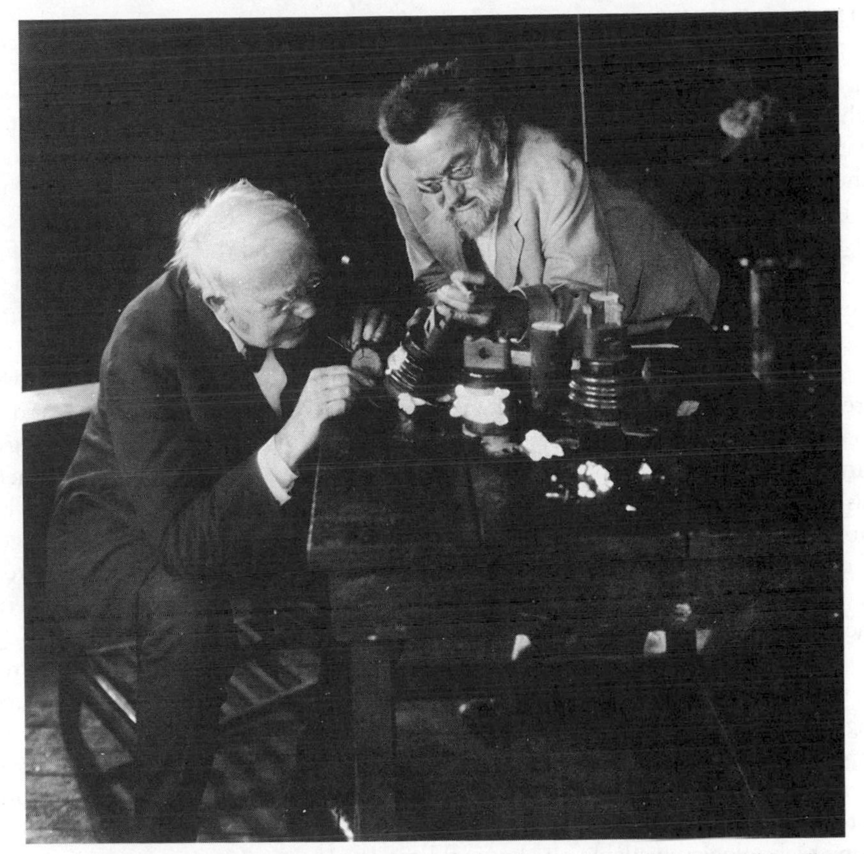

Two of the greatest geniuses of American technology, one native-born, the other an immigrant: Thomas Edison and Charles Steinmetz. (From the collections of Henry Ford Museum and Greenfield Village.)

of the new emperor's reign, the salaries of foreign technicians absorbed six percent of the central government's budget, and 40 to 50 percent of the budget of the Ministry of Industry for the duration of its existence.[8]

As Japan's industrial enterprises began to emerge, their products were often outright copies of foreign products. At times, Japanese attempts to copy foreign artifacts produced comic results, as exemplified by the factory with a sawtooth roof that faced the wrong way so no light shone through the windows, or the tailor who stitched together a Western-style suit complete with a patch on the pants. But with mounting experience Japanese workers and managers were able to effect substantial improvements until some of their products were the equal of Western manufactured goods. An early example of this was the Toyoda automatic loom, which in 1929 was exported to a textile mill in England, the birthplace of the Industrial Revolution.[9] Certainly no one laughs at Japanese technology today, although the accusation is occasionally made that the country still excels in the refinement of existing technologies, while producing few completely new ones.

CLEVER COPYISTS

In citing these examples, there is no intention to downplay the technological achievements of the U.S. and Japan, or to attribute their success to technological plagiarism. It is not an easy matter to copy a technology that was produced somewhere else. Having an example of a machine or a material in front of you helps, but it hardly guarantees success. Life would be simpler and easier if technological advance were simply a matter of importing a prototype and then "knocking it off." It does work occasionally, but all too often the process ends in frustration, as when the Chinese government reportedly spent $300 million in a failed attempt to produce a workable copy of the Boeing 707 jetliner.[10]

The very nature of technology makes copying a difficult enterprise. As was stressed in earlier chapters, technologies have to be considered as systems, and if one element is missing or deficient, even seemingly simple tasks such as copying a particular item may be impossible. A successful technology transfer requires numerous complementary inputs, many of which have to be developed and produced by the recipient. There are occasional possibilities for "leapfrogging" over whole stages of development through the importation of foreign technologies, but more often than not successful diffusion requires the preparation that comes through experience with earlier stages of technological development.

This brings us to a related point. The effective use of foreign technologies requires a labor force with a broad range of skills. It is essential that native managers, engineers, technicians, and ordinary workers are capable of making the best of opportunities that imported technologies present. As we shall soon see, this often entails significant modifications to the technology. At the very least, a fair amount of technical knowledge is required in order to make intelligent choices among the many technologies available on the world market.[11]

A successful user of imported technologies must therefore have a reasonably high level of indigenous capability. This is demonstrated by one study that found that the countries most successful in using foreign technologies tended to be the ones with high levels of indigenous research and development.[12] It is also significant that countries with the highest innovation rate make the quickest use of technologies developed elsewhere.[13] A country (or an individual firm) that draws on technologies developed elsewhere is spared the expense of "reinventing the wheel." But making effective use of imported technologies—even if only direct copying is required—often requires a stock of indigenous skills that cannot always be easily found.

ADAPTATION AND ADOPTION

Technology does not flow from one country or system to another like water through a pipe. In some cases the diffusion process may result in a technology emerging as something fundamentally different. This is exemplified by the aforementioned transfer of gunpowder from China to Europe. The Chinese at first regarded gunpowder as a medicinal substance, and only after hundreds of years of experimentation did they begin to employ it for fireworks and on occasion for military rockets, bombs, and mines. But when gunpowder diffused into Europe, it was soon used for cannon and firearms, thereby transforming the nature of warfare and ultimately society as a whole.[14]

The changes made in the course of the diffusion of gunpowder from China to Europe were monumental, for the nature of the technology changed along with the purposes to which it was put. In most cases, however, the modifications are far less dramatic. A process of "fine tuning" takes place as the technology is adapted to different conditions. In fact, many successfully diffused technologies succeed only because of modifications and improvements that were made during the process of diffusion.[15]

Successful diffusion of a technology may require adaptive changes to suit local conditions. This is illustrated by the efforts of the Japanese government during the late nineteenth century to increase agricultural yields by publicizing the successful techniques of "veteran farmers" (*rono*). Since techniques that worked in one locale might be less successful in places with different conditions, government-sponsored agricultural experiment stations tested and modified them so that they could work under a variety of conditions.[16]

Another adaptive change occurred when the steam railroad, an English invention, crossed the Atlantic. Railroad systems were built in the United States with great haste, for there were huge expanses of frontier to be opened up as quickly as possible. One consequence of these conditions was that curves were much sharper than in England. This in turn necessitated changes in locomotive design, as American engines were equipped with pilot wheels placed ahead of the driving wheels to guide them through the curves.

A JAPANESE EXAMPLE

The early development of the Japanese steel industry provides a good illustration of a successfully diffused technology that required many adaptations wrought by native workers.[17] Although the Japanese had been making small batches of high-quality steel for centuries, their indigenous industry was incapable of making iron and steel in large quantities. Japanese steelworkers therefore looked to Europe for better ways of producing steel. The first Japanese attempts to draw on foreign techniques occurred in the 1850s when they attempted to produce cannon by using methods described in a Dutch book on steel production. Success was slow in coming. Numerous trials resulted in the production of cannon that burst the first time they were fired. Yet underlying these failures was a highly useful learning process. At first, the native iron workers were incapable of melting even half of the pig iron that was to be cast into a cannon, but by the fifth attempt a complete cannon was cast, albeit not with complete success. Still, even this limited achievement required strenuous efforts to learn how to treat the raw material and to develop devices and techniques to bore the barrel of the cannon. Within about twenty years of the original attempt the foundry had successfully cast approximately 200 cannon. In accomplishing this, the cannon-founders had been able to draw on indigenous capabilities in such things as the production of fire-bricks for use in the reverberatory furnace in which pig iron was melted; the construction of water-powered boring machines; and the employment of high temperature furnaces for melting the iron.

Although their efforts represented an attempt at imitating foreign products and techniques, more was involved than mere copying. What began as an effort to copy foreign techniques ended successfully only when indigenous capabilities were applied to the solution of production and design problems. The process continued in 1880 when foreign engineers were recruited to oversee an ambitious effort to increase steel production. Initially this too was a failure, largely because the foreign experts did not take into account the special characteristics of Japanese coal and iron ore. But even this failed attempt to produce steel through the importation of foreign materials and experts ultimately met with success twelve years later when a Japanese engineer and one of his students reopened the dormant blast furnace after solving the problems created by the use of native coke and iron ore.

The point of this brief narrative is simply that Japan's successful nineteenth-century attempt to create a modern iron and steel industry was the result of importing foreign equipment, techniques, and expertise while at the same time making full use of native capabilities that were required for their successful employment. This example shows that technological diffusion is not a one-way process; without active participation by the recipients of the foreign technology, it can meet the fate of Japan's foreign-sponsored steel mill. And effective participation often is the product of previous efforts at indigenous technological transformation. The issue is not one of "imitation" versus "self-reliance," but of how to achieve an effective blending of the two.

APPROPRIATE TECHNOLOGY

Up to this point an unspoken assumption has been operating: the use of technologies developed in advanced nations is essential to the economic and technological modernization of underdeveloped countries. Modifications may have to be made to these technologies, but on the whole technologies of foreign origin are a sound basis for indigenous development.

This assumption has been strongly challenged in recent years. The countries receiving foreign technologies today are not like nineteenth-century Japan and the United States. Most are far poorer, not just in terms of wealth, but also in regard to human skills; this creates numerous problems in effectively using the sophisticated technologies that have been created for the developed nations of the world. Poor countries have large and growing populations, making the creation of jobs one of their prime needs; according to one calculation, over the next twenty years, 30 million jobs will have to be created every year just to accommodate the needs of new entrants into the poor countries' labor forces.[18]

Unfortunately, many of the technologies that have been developed in the economically advanced nations have been designed to save labor, not to maximize its use. Labor is often saved by using technologies that allow the substitution of capital for labor, yet one of the distinguishing features of poor nations is that they are critically short of capital. And finally, is it reasonable to expect that the world as a whole can sustain the global diffusion of the technologies that have been developed by the rich nations of the world? It is hard to imagine China or India consuming resources at the rate the United States does. To take one notable example, China currently has an approximate ratio of one car for every 500 people, while in the developed countries the ratio is more like 1 to 2. If China, a nation of more than a billion people, were to attempt to achieve these levels, the prospects for resource depletion and environmental pollution would be grim indeed.

Due to the drastically different circumstances of today's underdeveloped countries, technologies that have worked well in the rich nations where they originated may fail when they are taken to a different setting. Even worse, they may seriously distort the course of a nation's development, and leave it poorer than when it started out. Human and financial resources will be concentrated in a few privileged segments of the economy and society, leaving much of the population, especially those in rural areas, destitute. To take one particularly notorious example, after Western experts introduced large tractors into Pakistan, farm owners replaced farm laborers with the new machines. In one region, 40 percent of these workers lost their jobs, forcing them to migrate to the cities and lives of dire poverty. Per-acre crop yields hardly increased at all.[19]

According to critics of conventional technology transfers, there is a great need for technologies that are appropriate to the needs of these countries.[20] Sometimes they are called "alternative technologies," sometimes they are called "intermediate technologies" (because they stand in an intermediary position between modern and traditional ways of doing things), and sometimes they are simply called appropriate technologies. These technologies, it is claimed, do more than boost production; they increase employment, help to redress the maldistribution of wealth and income, empower people, contribute to economic independence, and preserve the environment.

Few could take issue with the general goals of appropriate technology. The difficulty lies in their actual development and implementation. Consider the first goal: increasing employment. As noted earlier, in poor countries there is good reason to select technologies that use relatively more labor than capital. But using labor-absorbing technologies many not result in overall increases in employment. At first glace, small-size sugar mills in India seem to be excellent examples of appropriate technology. They produce 20 percent of India's output—1.3 million tons of sugar per year—while providing jobs for more than a half a million people. Many jobs are created because the same amount of capital required to build a large modern sugar mill can be used for the construction of forty mini-mills; together these employ ten times more people than the modern mill.[21]

These apparent employment gains may be lost elsewhere in the system, however. Although the mini-mills produce two and a half times more sugar per unit of capital, they incur much higher labor costs (exactly how high depends on wage rates in the small mills compared to the large ones). This must lead to higher prices, lower profits, government subsidies, or some combination of all three. All of these prevent funds from entering other sectors of the economy. Were this not the case, there would be more money to spend on other goods and services, and a consequent stimulation of employment in these industries. Thus, the apparent creation of jobs in small sugar mills may produce no net increases in employment.

Many economists would argue that a free market system, with its realistic reckoning of capital and labor costs, will naturally result in the use of labor-intensive appropriate technologies. But the choice of technologies is not always governed by economic calculations. Technologies may be selected because they meet the requirements of powerful segments of the society, as when ultra-modern hospital technologies are installed to serve the urban upper class of a country that lacks basic medical services for most of its people.

At the same time, even practical, profit-oriented businessmen in underdeveloped countries can be charmed by the technological sophistication of foreign technologies that do not make good economic sense. According to one study of business firms in Indonesia, intermediate technologies resulted in lower costs and equal product quality, but they were often rejected in favor of advanced technologies of foreign origin. This was especially likely to happen in firms that had some monopoly control over their market. Under these circumstances, the aesthetic appeal of sophisticated technologies and the prospect of managing machines instead of people overrode the economic advantages of more appropriate technologies.[22]

Finally, a consideration of alternative technologies that is confined to economic matters is incomplete, for the appropriate technology movement has been motivated by a deep concern with how technologies are connected to larger issues. In the final analysis, a technology can be adjudged appropriate or inappropriate only by reference to particular values, and these cannot be defined exclusively in economic or technological terms. The appropriate technology movement also has directed attention to the role played by powerful groups and insitutions in the choice of technologies. Choices are likely to be seriously constrained when donors of foreign aid, international lending agencies, and multinational corporations occupy positions of dominance. It is also likely that the emergence of appropriate technologies is seriously hindered by the international distribution of R&D expenditures, only about one percent of which is spent in the poor countries of the world.[23]

Appropriate technologies, however defined, cannot be a panacea for poverty and backwardness. At the same time, the transfer of sophisticated technologies from developed countries may cause more harm than good. Choices must be made, and in poor countries as well as in rich ones, the selection of technologies reflects many larger issues. In selecting technologies there is no escaping the need to be concerned with values that transcend political, financial, and technical expendiencies.

Business Firms and Technological Diffusion

Many of the same processes that characterize the diffusion of technology from one country to another also manifest themselves when technologies diffuse to business firms. Many factors affect the speed with which new technologies diffuse into firms. According to Tanon and Rogers, these include: relative advantages over existing technology; compatibility with existing values of the firm; the ease or difficulty of understanding and applying the new technology; the ease in experimenting with the new technology or employing it on a trial basis; and the extent to which positive results are apparent.[24] In highlighting the importance of ease of use, observability, and comprehensibility, the above list carries the implication that after all is said and done, it is individual people who determine the success of a technological innovation. The diffusion of technology is still essentially a learning process, through which the recipient gains an understanding of how the technology works, while at the same time adapting it to its particular needs. In this, the activities of skilled people are usually essential. As Mira Wilkins explains, "Often the product, or the description in the patent, or mere drawings and instructions, are inadequate for transfers of technology; [people] are needed to carry, explain, and facilitate the introduction of new processes or products."[25]

When we speak of "people," we mean special kinds of people. The effective transfer of technology often requires a movement from one environment to another, such as from a laboratory to a factory. Technology is fundamentally a system of knowledge, but knowledge that is readily comprehensible in one setting may be a great mystery in another. Making a new kind of knowledge both understandable and useful often requires the services of a kind of "translator"—a person capable of functioning in both settings so that information can be transferred from one to the other.

As Hugh Aitken's study of the early days of radio indicates, the efforts of "translators" were crucial to the development of that technology. First there was the work of Heinrich Hertz, who "translated" James Clerk Maxwell's theories of electomagnetic radiation into a laboratory technology that allowed an empirical examination of the theory. Following Hertz, Oliver Lodge moved beyond the laboratory experiment through his development of an imperfect, but workable method of using radio waves for actual communication. The first stage in the development of radio was essentially completed by Marconi, who took the embryonic technology and converted it into a practical (and highly lucrative) commercial business.[26] Throughout the course of radio's history, information was put into a new form, pushing the technology to the next stage of development. In Aitken's summary, "At each stage in the process of translation, information generated in one system was converted into a form that 'made sense' in terms of another; and at each stage new information was blended with what was already known to create something essentially new."[27]

ECONOMIC MOTIVES AND TECHNOLOGY TRANSFER

Some technologies, especially those involving military matters, are sponsored by governments, but the majority of technological innovations are adopted by private businesses pursuing their economic goals. Firms may make use of "imported" technologies in order to cut their costs, improve their products, bolster their profits, penetrate new markets, or achieve some combination of any or all of these. The speed and extent of diffusion are therefore strongly influenced by expectations of costs and benefits.

The influence of economic calculations on the process of technological diffusion is illustrated by a study of the mechanical reaper conducted by Paul David.[28] Through the ages, grain had been harvested by laboriously cutting stalks with a sickle. The introduction of the cradle scythe during the eighteenth century speeded up the process and made it a bit less backbreaking, but the reaping of grain still remained a highly labor-intensive process. The invention of the mechanical reaper (a horse-drawn implement that used a large reel to gather the stalks of grain, which were then cut by a reciprocating horizontal blade) by Obed Hussey and Cyrus McCormick promised a relief from this onerous chore, and most important from an economic standpoint, a significant increase in productivity.

For two decades the reaper diffused very slowly into the farms of America. Only in the 1850s did it begin to be purchased in large and increasing numbers. According to David, this abrupt increase in the employment of reapers cannot be explained in terms of the technology itself, for the reaper did not undergo any significant improvements during this period. It is therefore necessary to look to changed economic conditions in order to explain the increased rate of diffusion that emerged in the 1850s.

In the first place, this period was characterized by rising grain prices, stimulated by urbanization and the effect of the Crimean War on the international grain market. Higher prices had the obvious effect of motivating farmers to produce more grain by increasing their acreage. America's westward expansion made more land available, but farmers had to contend with the high cost of hired labor, for many workers were being recruited for railroad and building construction in the expanding Middle West. The purchase of a mechanical reaper was an expensive proposition for most farmers, but what really mattered was its cost relative to labor; as the latter rose, it made good economic sense to purchase a reaper—a classic case of substituting capital for labor under the influence of changed factor prices.

At the same time, this substitution only made sense if a farmer had enough land to keep his reaper fully employed; it would have been foolish to invest a large amount of money in an implement that would spend a large amount of time in the barn because there was no work for it to do. What counted here was what David describes as a farm's "threshold size"—the number of acres that would justify the purchase of the reapers. The ownership of a mechanical reaper would confer the greatest benefits when a farm was large enough to keep the reaper in continuous operation during the harvest season. Average farm size in the American Midwest was indeed growing during the 1850s, the result of new land being open to cultivation. This tendency was reinforced by the rising grain prices during this period, for an expansion of cultivated acreage resulted in higher incomes for farmers. The farming of larger acreages meant that farmers were moving up to and surpassing the threshold size, thereby justifying the purchase of mechanical reapers at a time when their relative cost in comparison to the hiring of farm workers was steadily declining. According to David, mechanical reapers were purchased in large numbers only when average farm size had reached the threshold, more than twenty years after the invention of the reaper.

David's analysis of the diffusion of the reaper is neat and elegant, but it has not gone unchallenged. Some of the most telling criticisms appear in an article by

An early McCormick reaper in operation. (State Historical Society of Wisconsin.)

Alan L. Olmstead.[29] In the first place, Olmstead found that David's estimates of interest rates and the life span of the reaper were in error, rendering his calculations of threshold acreage highly suspect. Second, David's threshold model assumes that all farmers worked in complete independence and did not engage in cooperative ventures with fellow farmers. Yet there is substantial evidence that they did just that. Several farmers could economically harvest their individual fields by jointly buying a reaper, even though their farms were individually too small to justify purchasing this expensive piece of equipment. Finally, Olmstead does not agree that the mechanical reaper was a truly practical device from the start. Although the mechanical reaper was invented in the 1830s, twenty years elapsed before it evolved into a workable implement. Farmers were quite justified in foregoing the purchase of reapers until they did what they were supposed to do.

This last point is particularly significant, for it agrees with a point made in the previous chapter: a new technology often requires extensive detail improvements before it actually works. Even if David's analysis of the factors that shaped the demand for reapers was correct, it would not tell the whole story. A technology diffuses not just because there is an abstract demand for it; it also has to work. What may appear to be a hidebound resistance to a useful technology may in fact be motivated by a realistic appraisal that the technology is not yet ripe.

A RISKY BUSINESS

Although there are many instances of technological diffusion being shaped by economic considerations such as the cost of using a new technology, the decision to use a new technology can rarely be governed by them exclusively. Along with the opportunities presented by a new technology come uncertainties that cannot be assessed through conventional economic analysis. No new technology is a "sure thing." The immediate advantages of a new technology may be slight, and the benefits may be a long time in coming. It is often the case that the *potential* is what matters, and this may be the hardest thing of all to assess.[30] Many years may elapse before a technology passes from laboratory feasibility to commercial value, or from *invention* to *innovation*. Although some inventions made the transition rapidly—only one year in the case of Freon refrigerants—others have taken much more time. In the case of the mechanical cotton picker it was fifty-three years, and for the zipper twenty-seven years. The fluorescent lamp took no less than seventy-nine years to make the passage from laboratory demonstration to commercial viability.[31]

The decision to adopt a technology represents a firm's assessment of the likely benefits accruing from the use of the new technology, weighed against the uncertainties that attend its use and the speed at which these uncertainties can be dissipated.[32] Given the nature of technological innovation, uncertainties will rarely vanish completely, but as information is gathered and experience gained, it is often possible to convert uncertainties into calculable risks.[33] Success or failure cannot be assured, but at least business managers can come to an informed judgment of their likelihood. At the same time, not all of the risks are technological in nature. Although any new way of doing things can be presumed to have its share of "bugs" and other unanticipated problems, these may be minor when compared to the commercial uncertainties of the new venture. These can be particularly problematic, for business managers usually find it easier to assess the probability of technical success than the probability of marketing success.[34] "Will it fly?" is easier to determine than "will anybody pay money to fly in it?"

This has been the literal problem for the Anglo-French supersonic airliner, the Concorde. A stunning technological achievement that required the solution of countless complex problems, the Concorde has nonetheless been a commercial failure. An extraordinarily expensive aircraft to fly and maintain, it has never attracted a sufficient number of customers willing to pay the premium prices necessary to operate the airplane at a profit. No private business could have survived this sort of failure. Only the financial support of two governments, motivated by considerations of national pride as much as anything else, has kept the Concorde aloft.

THE NIH SYNDROME

The inherently risky nature of many technological innovations may explain the presence of one persistent block to the diffusion of new technologies, the Not Invented Here syndrome, or NIH for short. People and organizations exhibiting this syndrome are reluctant to make use of technologies that were invented elsewhere. The corrosive effects of the NIH mentality were noted by Henry Ford II, who bemoaned the fact that "There's too much NIH—not invented here [in the automobile industry]. . . . Lots of times a guy brings something in, and unless the improvement is rather dramatic, nothing happens. The status quo is a hell of a lot easier than making changes."[35] Ford's chief domestic rival, General Motors, seems no less prone to this affliction. Despite having a New Devices Section for the acquisition of technologically innovative components from outside inventors, GM bought the rights to only eleven patents during the 1970s.[36]

Taking on the risks that go along with technological innovation often makes people uncomfortable, and their discomfort is heightened when the new technology is not of their own making. It is also the case that internally generated technologies are likely to be more compatible with established organizational structures and activities.[37] Moreover, an internally generated technology will probably have a "champion" within the organization who seeks its utilization, and management may be more inclined to pay attention to the ideas of their own personnel.

These are explanations for the presence of the NIH syndrome, but not excuses. A dynamic firm can ill-afford to ignore useful new technologies just because they were developed elsewhere. We have already seen that a receptivity to "foreign" technologies has been crucial to the development of Europe, the United States, and Japan, and that a resistance to them was a major reason for economic retardation of imperial China. An individual firm that chooses to resist or ignore new technologies because of their source may ultimately find that security and pride have been purchased at a very great price. Its comfortable routines and organizational structures are retained, right up to the day that it goes out of business.

EFFORTS TO RESTRICT THE DIFFUSION OF TECHNOLOGY

The exclusive possession of a particular technology can confer great advantages on those that have it. Individual firms or whole countries may go to great lengths to prevent the diffusion of these technologies. During the sixteenth century the glassmakers of Venice sought to prevent others from learning the secrets that went into the production of Venetian glass, and they even sent assassins to poison Venetian expatriates who had set up glassworks abroad.[38] England enacted a law in 1719 that forbade the emigration of skilled artisans, such as the aforementioned Samuel Slater, whose know-how was a crucial element in the early development of the American textile industry. The export of machinery also was forbidden from the 1780s until the middle of the nineteenth century. Today, technologically advanced nations attempt to prevent the export of advanced technologies through export licensing requirements. Licenses may be denied if national security could be compromised; in one case the export of home computers to the Soviet Union was forbidden because they supposedly had military applications. There has even been an effort by the U.S. Central Intelligence Agency to prevent the free circulation of scientific papers that, in their judgment, may adversely affect national security.

Past efforts to suppress the outflow of technology were rarely successful, and there is little likelihood that contemporary efforts will fare any better. Clandestine transfer of plans, equipment, and products are extremely difficult to stop, even with rigorous attempts at policing. Also, as exemplified by the case of Samuel Slater, the movement of people has historically been the main vehicle for the transfer of technologies.[39] This sort of movement is very hard to stop, especially in a free society.

Attempts to stem the outward flow of technology may even be counterproductive. A country that attempts to bar the export of technology may end up damaging itself more than anyone else. Technological development often requires the interchange of information from numerous sources. The attempt to seal off a technology will restrict the flow of information from both domestic and foreign sources, and thereby result in a slower pace of technological advance.

Finally, it is often the case that the most important "secret" is that a technology *exists*. During the late 1940s there was a great deal of anguish in the United States when the Soviet Union exploded its first nuclear bomb. The general belief was that the Soviets had obtained our "atomic secrets," and two Americans were executed for their alleged role in passing on these "secrets." But the Soviets had long known the most important thing: that a massive explosion could be produced through the fissioning of an atomic nucleus. Once this fact was known, it was only a matter of time before the Soviets learned how to do it themselves, with or without the clandestine transfer of American technology. Knowing that a problem *can* be solved is often the most important step in its solution.

PATENTS AND THE DIFFUSION OF TECHNOLOGY

An individual or a business naturally wants technological diffusion to take place on its own terms. If it has developed a novel technology it wants the benefits of that technology to accrue only to itself, and not to others who could easily copy an invention that has required a great deal of effort and expense. This is a legitimate desire, and it has been recognized as such by the establishment of the patent system. The possession of a patent confers exclusive use of an invention; it is a legal monopoly. Without the prospect of such a monopoly, it is widely believed, there would be a diminished motivation to invent, and in the long run society as a whole would suffer from a slower rate of technological advance.

At the same time, however, patent protection may retard technological development. A well-entrenched firm may suppress inventions that could seriously shake up existing routines or even threaten the firm's existence. It seems as though everybody has a brother-in-law who knows a man who invented a way to make his car run on water straight from the garden hose, but "of course the big oil companies bought up all the patents." One shouldn't put much credence in these stories, but it is true that dominant companies have used their control of patents to restrict innovation as happened with the design of light bulbs when General Electric was in this position.[40].

Even so, there may be considerable scope for "inventing around" another firm's patents. This has been a regular feature of the microcomputer industry, to name only one. There are no industries where the basic technology is covered by a single "master patent" that serves as an effective barrier to potential competitors. During the early twentieth century American automobile manufacturers paid a licensing fee to the holders of the Selden patent, which covered the automobile as a whole, but Henry Ford mounted a successful legal challenge to the Selden patent, and won a great deal of public acclaim in the process.[41]

There are even some ways in which the existence of patents stimulates technological diffusion. The filing of a successful patent application makes the invention public, for the basic design and specifications are open to inspection by anyone willing to pay a small fee. In fact, the word "patent" is derived from the Latin verb *pateo*, which means "to open." One of the chief justifications for awarding a patent is that the inventor has not attempted to keep his or her invention secret, but has revealed its workings to the public. As the U.S. Supreme Court ruled in 1933, an inventor ". . . may keep his invention secret and reap its fruits indefinitely. In consideration of its disclosure and the consequent benefit to the community, the patent is granted."[42]

Although direct copying is of course forbidden, access to the public record may give a clearer sense of how a technological problem may be addressed and may stimulate alternative approaches to its solution.[43] Then, too, patent holders often pass their inventions to other firms in return for a licensing fee. This arrangement may allow for the more rapid diffusion of a technology because the licensing agreement often facilitates the transfer of supplementary material necessary for the best use of the device or process.[44] In some cases, a licensing arrangement may result in a better product, as happened when Dupont developed waterproof cellophane after being licensed to produce the original version.[45]

Perhaps not too much should be made of the patent system's supposed value in stimulating the creation and diffusion of new technologies. A patent never confers iron-clad protection for inventors; it is, as cynics have pointed out, little more than "a license to sue." A vast amount of time and money can go into litigation, in many cases leaving the inventor poorer than he or she was at the start. As a result, many individuals and firms are content to exploit their invention until the copiers are able to seize a large share of the market. In any event, many patents are quickly outmoded by the advance of technology; as a result the monopoly position is quickly lost.

Whether the patent system serves as a stimulus or a hindrance to technological advance is a matter of considerable debate, and one not likely ever to be settled. Like many social institutions, it owes its existence to inertia as much as anything else. It is retained simply because it is virtually impossible to determine if its advantages outweigh its disadvantages, or vice versa. Under these circumstances, the status quo usually prevails.

Questions for Discussion

1. What sort of skills are required for the successful copying of a technology developed elsewhere? Why do you think that the Japanese have been so successful at making effective use of technologies that first appeared elsewhere? How might other countries duplicate their success?
2. Most economically advanced countries have technical assistance programs that are used to upgrade the technological levels of poorer countries. Under what circumstances might these programs be harmful? How might these programs be shaped to better meet the needs of poor countries?

3. If you were the leader of a poor nation, what general policies would you formulate in regard to the importation of technology? Where might you expect to find opposition to these policies?
4. Suppose that you are the president of a firm that has a choice between two new technologies: one that promises a modest profit with very little risk, and another that may yield a very high profit but at considerable risk. What would your choice be? Who in your company might support the first technology, and who might support the second?
5. Does the patent system encourage or discourage technological innovation? What would happen if it were abolished? Can you think of an alternative that would better serve the needs of both inventors and the general public?

Notes

1. See Robert Temple, *The Genius of China: 3,000 Years of Science, Discovery, and Invention* (New York: Simon and Schuster, 1989).
2. Lynn White, Jr., *Medieval Technology and Social Change* (New York: Oxford University Press, 1966), pp. 14ff.
3. Jean Gimpel, *The Medieval Machine: The Industrial Revolution of the Middle Ages* (New York: Penguin, 1977), p. 33.
4. A. Rupert Hall, "The Rise of the West," in Charles Singer, et al. (eds.), *A History of Technology*, vol. III (Oxford: Clarenden Press, 1957), pp. 716–717.
5. Robert C. North, *Chinese Communism* (London: Weidenfeld and Nicolson, 1966). p. 10.
6. Xin Xiangrong, "Open Policy, Import of Culture," *Beijing Review* 29, 46 (17 November 1986): 4.
7. Elting E. Morison, *From Know-How to Nowhere: The Development of American Technology* (New York: New American Library, 1977), pp. 21–36.
8. Angus Maddison, *Economic Growth in Japan and the USSR* (New York: W.W. Norton, 1969), pp. 16–17.
9. William Wirt Lockwood, *The Economic Development of Japan: Growth and Structural Change* (Princeton, N.J.: Princeton University Press, 1968), pp. 331–332.
10. E. E. Bauer, *China Takes Off: Technology Transfer and Modernization* (Seattle: University of Washington Press, 1986), pp. 82–86.
11. Nathan Rosenberg, *Inside the Black Box: Technology and Economics* (Cambridge: Cambridge University Press, 1982), p. 271.
12. Tuvia Blumenthal, "A Note on the Relationship between Domestic Research and Development and Imports of Technology," *Economic Development and Cultural Change* 27, 2 (January 1979): 303–306.
13. Christopher Freeman, *The Economics of Industrial Innovation*, 2d ed. (Cambridge, Mass.: The MIT Press, 1982), p. 61.
14. Zhou Jiahua, "Gunpowder and Firearms," in *Ancient China's Technology and Science* (Beijing: Foreign Languages Press, 1983), pp. 184–191.
15. Nathan Rosenberg, *Perspectives on Technology* (Cambridge: Cambridge University Press, 1976), pp. 192–193.
16. Yujiro Hayami and Vernon W. Rutan, *Agricultural Development* (Baltimore: The Johns Hopkins University Press, 1971), pp. 174–182.

17. The example is based on Nakaoka Tetsuo, "Imitation or Self-Reliance: A Lesson from the Early History of Modern Japanese Iron Manufacturing," in *The Japanese Foundation Newsletter* 8, 4 (October–November 1979): 1–6.
18. Colin Norman, "Soft Technologies, Hard Choices," in Albert H. Teich, *Technology and Man's Future*, 3d ed. (New York: St. Martin's Press, 1981), p. 389.
19. Robert C. Toth, "Fitting Technology to Need Held Critical in Third World," *Los Angeles Times* (18 June 1978), part 1: 1, 32.
20. The most influential book on this subject is E. F. Schumacher, *Small Is Beautiful: Economics as If People Mattered* (New York: Harper & Row, 1973).
21. George McRobie, *Small Is Possible* (New York: Harper & Row, 1981), p. 200.
22. Louis T. Wells, Jr., "Economic Man and Engineering Man: Choice and Technology in a Low-Wage Country," *Public Policy* 21, 3 (Summer 1973): 319–342.
23. Sarah Jackson, "Economically Appropriate Technologies for Developing Countries: A Survey," in Pradib K. Ghosh, *Appropriate Technology in Third World Development* (Westport, Conn.: Greenwood Press, 1984), p. 87.
24. Christian P. Tanon and Everett M. Rogers, "Diffusion Research Methodology: Focus on Health Care," in Gerald Gordon and G. Leonard Fisher (eds.), *The Diffusion of Medical Technology* (Cambridge, Mass.: Ballinger, 1975).
25. Mira Wilkins, "The Role of Private Business in the International Diffusion of Technology," *Journal of Economic History* 34, 1 (March 1974): 176.
26. Hugh G. J. Aitken, *Syntony and Spark: The Origins of Radio* (Princeton, N.J.: Princeton University Press, 1985), pp. 329–335.
27. Ibid., p. 335.
28. Paul David, "The Mechanization of Reaping in the Ante-Bellum Midwest," in Henry Rosovsky, *Industrialization in Two Systems* (New York: John Wiley & Sons, 1966), pp. 3–39.
29. Alan L. Olmstead, "The Mechanization of Mowing and Reaping in American Agriculture, 1833–1870," *Journal of Economic History* 35, 2 (June 1975): 327–353.
30. Carlo M. Cipolla, *Guns, Sails, and Empires: Technological Innovation and the Early Phase of European Expansion, 1400–1700* (New York: Pantheon, 1966), p. 131.
31. See John Enos, "Invention and Innovation in the Petroleum Refining Industry," in Richard Nelson (ed.), *The Rate and Direction of Inventive Activity* (Princeton, N.J.: Princeton University Press, 1962).
32. Edwin Mansfield, *Technological Change* (New York: W.W. Norton, 1971), p. 88.
33. Donald A. Schon, *Technology and Change: The New Heraclitus* (New York: Dell, 1967), pp. 24–25.
34. Freeman, op. cit., p. 154.
35. James C. Jones, "Dr. Schweitzer's Dilemma: Not Invented Here," *Ward's Auto World* (November 1974): 41.
36. Patrick Bedard, "New Devices," *Car and Driver* (May 1982): 144.
37. J. E. S. Parker, *The Economics of Innovation: The National and Multinational Enterprise in Technological Change*, 2d ed. (London: Longman, 1978), p. 111.
38. John U. Nef, *The Conquest of the Material World* (Chicago: University of Chicago Press, 1964), p. 153.
39. Carlo M. Cipolla, "The Diffusion of Innovations in Early Modern Europe," *Comparative Studies in Society and History* 14 (1972): 48.
40. Leonard S. Reich, "Lighting the Path to Profit: GE's Control of the Electric Lamp Industry, 1892–1941," *Business History Review* 66 (Summer 1992), pp. 310, 312.
41. See James Flink, *America Adopts the Automobile, 1895–1910* (Cambridge, Mass.: The MIT Press, 1970), pp. 318–328.

42. *United States* vs. *Dubilier* (289 U.S. 178), quoted in C. Lyle Cummins, *Internal Fire* (Lake Oswego, Ore.: Carnot Press, 1976), p. 47.
43. Parker, op. cit., p. 310.
44. Ibid., p. 327.
45. Freeman, op. cit., p. 61.

Chapter 6 Technology, Energy, and the Environment

Although technological advance has been blamed for a variety of ills, its most obvious and long-lasting consequence has been the destruction of the natural environment. For most of human existence people left the environment pretty much as they found it. But beginning with sedentary agriculture and accelerating with industrialization, the use of new technologies has at times left a ruined environment as its legacy. And there is a strong possibility that environmental problems will worsen as more nations make greater use of industrial technologies. The use of coal, oil, and gas by the developing countries is likely to triple by the year 2020, resulting in an intensified assault on the environment.[1] To take just one example, a complex of electric power plants being built in Thailand is expected to produce more carbon dioxide than all of western Germany.[2]

The deleterious consequences of technological advance can be grouped into two broad categories: pollution and depletion. In the former, the environment is damaged by the addition of harmful substances, resulting in polluted skies, acid rain, contaminated water, and possible climate change. In the case of the latter, the unchecked application of technology leads to the permanent loss of resources, deforestation, and the extinction of plant and animal species. On many occasions, of course, the two go together; for example, the use of coal to generate electricity can devastate an area through strip mining while fouling the air and water with airborne pollutants.

But technology is not just a source of environmental problems; it can also be part of the solution. In this chapter we will look at some of the environmental consequences of technological advance, review some technologies that can alleviate the degradation of the environment, and briefly consider how the appropriate policies can help make this happen.

FOSSIL FUELS, AIR POLLUTION, AND CLIMATE CHANGE

The use of fossil fuels (petroleum, coal, and natural gas) has vastly extended the amount of available energy. At the same time, the emissions of cars, power plants, factories, and other sources have polluted the environment with immense quantities of carbon compounds, sulfur, and oxides of nitrogen. In 1988, 5.66 billion tons of carbon fuel exhaust were dumped into the earth's atmosphere—more than a ton for every human being on earth. Moreover, the amount of fossil-fuel exhaust is steadily growing; if nothing is done to address the problem, carbon emissions will double by the year 2010 and triple by 2025.[3]

The blanket of smog now choking many of our cities makes the consequences of fossil fuel use all too evident. Potentially more dangerous in the long run is the accumulation of the main product of carbon combustion: carbon dioxide (CO_2). Carbon dioxide is produced when the carbon in gasoline, diesel fuel, coal, or natural gas unites with oxygen during the process of combustion. This creates quantities of CO_2 significantly greater than the original carbon; burning one pound of gasoline generates more than three pounds of carbon dioxide.[4] To look at it in a different way, a car travelling 100 miles produces about 100 pounds of CO_2.

Carbon dioxide is a clear, odorless gas that by itself presents no threat to health. It is the normal product of any animal's respiration, and without it drinks like beer and soda pop would be unpalatably flat. The danger presented by CO_2 emerges only when large quantities of it accumulate in the atmosphere along with other products of an industrial society (ozone, methane, chlorofluorocarbons, and oxides of nitrogen). Light rays from the sun can easily pass through a layer of these gases. The radiant energy then warms the surface of the earth, but the heat cannot pass back through this layer. This is known as the "greenhouse effect". Just as a greenhouse is warmed by the light of the sun on a cold winter's day, the earth's temperature begins to rise as greenhouse gases accumulate in the atmosphere. CO_2 makes up about half of these gases, and along with the others its presence in the atmosphere is steadily increasing. The earth's atmosphere now contains CO_2 in a ratio of 350 parts per million, an increase of 25 percent over immediate pre-industrial levels.[5] At current rates of increase, CO_2 will occur in a ratio of 550–600 parts per million by the next century.[6]

The precise effects of global warming remain unclear, for the forces driving the earth's climate are exceedingly complex. Attempting to understand their consequences requires sophisticated models and a great deal of computer time, along with the introduction of a number of unproven assumptions, to simulate the interaction of all the governing factors. Yet for all the uncertainties, it still seems likely that fundamental changes to the earth's climate will take place if present trends continue. Natural fluctuations have caused the earth's temperature to warm in the past, but not at the speed taking place today. Many scientists are convinced that global warming has already begun and will accelerate in the future. Already, average world temperatures are 0.6 degrees centigrade warmer than they were a hundred years ago, and the buildup of man-made greenhouse gases seems to be primarily responsible. The worst may be yet to come, for several computer models indicate that temperatures could rise 2.5 to 5.5 degrees centigrade by the end of the next century.[7]

Should something like this happen, different regions of the earth will be affected in different ways. Some places would become significantly hotter. To take one example, Washington, D.C., might go from having one summer's day with temperatures over one-hundred degrees to twelve such days, and from thirty-six days when the temperature is above ninety degrees to eighty-seven of them.[8] Global warming would also affect rainfall patterns, leaving some areas wetter and others drier. Major crop growing areas could suffer severe droughts, although northernmost croplands might benefit from warmer temperatures. These areas could grow more food crops, but the gain would be more than offset by declines elsewhere. Maintaining agricultural production in many parts of the world would require capital

Deforestation in Brazil. (© Luiz C. Marigo, Peter Arnold Inc.)

investment of as much as $200 billion just to make the necessary changes to irrigation systems.[9] Most disturbingly, significant global warming would lead to a partial melting of glaciers and polar ice caps, as well as an expansion in the volume of water due to its greater warmth. This would bring potentially catastrophic changes to many coastal areas; one study by the U.S. Environmental Protection Agency projected a rise in sea level by the year 2100 of 1.4 to 2.2 meters. Many coastal cities would face inundation, while food supplies would decline as low-lying parts of Asia were lost to cultivation.[10]

The prospect of global warming presents a number of disturbing scenarios. Still, it must be emphasized that much remains unknown about the accumulation of greenhouse gasses and their long-term consequences. Many scientists are reasonably certain that there is a historical correlation between the buildup of CO_2 and warmer temperatures, but it is possible that the latter is the cause of the former, as increased ocean temperatures have led to the release of greater amounts of carbon dioxide. More generally, much is still unknown about the ocean's role in regulating the volume of atmospheric CO_2. Uncertainties here are particularly vexing, for the world's oceans contain fifty times as much CO_2 as the atmosphere. Accordingly, any comprehensive understanding of the greenhouse effect will require a greatly expanded knowledge of the dynamic relationship between the oceans and the atmosphere.

The accumulation of that knowledge may take many decades. In the meantime, the buildup of greenhouse gasses may reach a critical point or, conversely, may be of minor significance. This creates a major dilemma of policy-making. Addressing the CO_2 problem—if in fact it is a problem—will be very expensive; it has been estimated that the cost of removing only carbon dioxide from the exhaust of power plants and industrial facilities in the United States would approach $1 trillion.[11] Heading off global warming and all that accompanies it would require substantial changes in the way energy is produced and used. At the same time, coping with the consequences of global warming might entail even more sweeping adjustments. In either case, decisions (or nondecisions) will have to be made amidst a great deal of uncertainty, as scientific understanding struggles to catch up with technological application.

A PLANET UNDER STRESS

Global warming and the hazards and discomforts of air pollution are not the only consequences of burning fossil fuels. The nitrogen and sulfur byproducts of combustion readily combine with water to produce acid rain, which kills fish and other marine life in lakes and ponds. It also poses a serious threat to trees, leaving them weakened and vulnerable to diseases and insect infestations. In some parts of the world this already is a serious problem; acid rain has damaged half of the forested areas of the Netherlands, Germany, and Switzerland.[12]

While the burning of fossil fuels is indirectly leading to a widespread loss of trees, many of the world's forests are suffering from a more direct attack, as trees fall victim to chainsaws and bulldozers. Tropical forests currently are being cut down at the rate of 42,000 square miles a year as expanding rural populations attempt to increase cultivated acreage.[13] This comes to nearly 74,000 acres a day, or more than 50 acres each minute. And trees are not the only victims of intentional deforestation, for many plants and animals are killed at the same time. If present trends continue to the end of this century, the area covered by Latin American forests will contract by nearly half, and 15 percent of plant species—13,600 separate species—and 12 percent of the bird species will be forever lost.[14] The environmental consequences go beyond the destruction of plant and animal species. The process of clearing and burning forests produces large amounts of CO_2—23 percent of the total emissions according to one estimate.[15] Making matters worse, this loss of forest cover exacerbates the problem over the longer term, for trees and plants absorb substantial amounts of CO_2.

At the same time that a variety of technologies have been assaulting the land and air, the ground below is also being changed for the worse. Many industrial processes generate hazardous wastes, sometimes in lethal quantities. In the United States, 255 to 275 million metric tons of hazardous wastes are produced every year—about one ton per person.[16] Many of these wastes have accumulated over time in inadequate disposal facilities where leakage could eventually threaten surrounding communities. Already, some areas have been so badly contaminated that they had to be abandoned for a number of years while the mess was being cleaned up. Many other places contain dump sites that are only slightly less hazardous. Restoring them to safety will be an expensive process; cleaning up the 10,000 worst dump sites in the United States is expected to cost $100 billion.[17]

The toxic-waste problem is not confined to areas situated near long-established industrial sites. Seemingly clean, high-tech industries generate substantial quantities of hazardous wastes. The production of semiconductors requires the use of many dangerous substances that can pose severe health hazards if not disposed of properly. In California's famed Silicon Valley, drinking water has been contaminated with suspected carcinogens. Nineteen production sites in the area were so contaminated that they appeared on the list of hazardous sites targeted by the federal "Superfund" for intensive cleanup.[18] Those counting on the semiconductor industry to create a pollution-free, "post-industrial" society are doomed to disappointment.

Pollution, global warming, and the other negative consequences of industrial production cause problems on one front; depletion of the resources that provide energy and raw materials pose another set of challenges. As recent energy crises have demonstrated, disruptions of oil supplies can create severe dislocations in industrial societies. Fossil fuels have been essential to the process of industrialization since at least the middle of the nineteenth century, but high and growing rates of consumption will eventually result in their exhaustion, forcing massive changes in transportation systems, production methods, and general lifestyles.[19] In the United States, oil and gas consumption has doubled every fifteen to twenty years for the last hundred years; at this rate these crucial sources of energy will be 80 percent depleted within three or four decades.[20] More pessimistic projections indicate that the United States will have virtually no reserves of oil by then.[21]

The proximate causes of the energy crises of the 1970s were political events, and not the sudden depletion of oil. Still, the temporary loss of substantial amounts of imported oil heightened the American public's realization that stocks of domestic oil and gas are dwindling. There is less recognition that other, no less vital, resources are also diminishing. To cite only one example, a study conducted in the early 1980s found that each year the world was losing 26 billion tons of topsoil. Losses in the United States alone amounted to 2 billion tons.[22]

The situation is even worse in many other parts of the world. Climate change, deforestation, and poor agricultural practices have turned many regions into dust bowls inhabited by undernourished (and in some cases actually starving) people. Given the inescapable fact of steady population growth in much of the world, the loss of topsoil, coupled with climatic change induced by global warming, could produce a famine of ghastly proportions.

IS TECHNOLOGY THE PROBLEM OR THE SOLUTION?

The previous pages have presented a depressing catalogue of environmental ills. Pollution, climate change, species extinction, and resource depletion pose multiple threats to our standard of living and, perhaps, even to our continued existence. The reliance on fossil fuels has left our atmosphere contaminated by various pollutants, and the earth threatened by global warming. Irresponsible agricultural practices have poisoned groundwater with pesticides and other chemicals while depleting the topsoil. Industrial processes produce thousands of tons of toxic wastes, while nuclear power plants leave behind radioactive wastes that will pose potential health hazards for thousands of years. In the opinion of some critics, the technological advances of the past few centuries seem to have produced only temporary benefits that will ultimately be overwhelmed by the consequences of environmental stress. In the long run, it may be argued, technology generates more harm than good.

Is a modern economy, supported by advanced technologies, doomed to destruction, leaving us with no choice but to retreat to the simpler technologies of the past, perhaps even back to the era of hunting and gathering? Or can technology itself provide us with the solutions to technologically induced problems? In addressing these questions it is useful to first gain some historical perspective. If nothing else, some acquaintance with the past should convince us that damage to the environment is not solely a phenomenon of modern times and modern technologies. Entire civilizations have collapsed due to excessive land clearance, overgrazing, withdrawal of ground water, and eventual desertification. Epidemic diseases of catastrophic proportions have been spread by the careless disposal of household wastes. There are some scholars who believe that a major factor in the collapse of the Roman Empire was sterility and premature death brought on by the widespread use of lead in pipes and utensils. Air pollution existed long before the invention of fossil-fuel consuming engines; those suffering from twentieth century smog can find a seventeenth-century counterpart in John Evelyn, who decried the growing use of coal for energy and warmth:

> [I]n London we see people walk and converse pursued and haunted by that infernal smoake. The inhabitants breathe nothing but an impure and thick mist, accompanied by a fuliginous and filthy vapour, which renders them obnoxious to a thousand inconveniences, corrupting the lungs and disordering the entire habit of their bodies, so that catarrs, phtisicks, coughs and consumption rage more in that one city than in the whole earth besides.[23]

SOME TECHNOLOGICAL FIXES OF THE PAST

These examples of past environmental ills provide little comfort, for they can be seen as a preview of what might be in store for us. But there are other examples of environmental threats that have been successfully countered. One example is the deforestation that had become a serious problem in England during the sixteenth century. By that time, vast quantities of wood had been consumed by the demands of an expanding population and the growth of shipbuilding, construction, and iron manufacture (which required large quantities of charcoal). Within a century the depletion of timber was perceived as a serious problem, as seen in the complaint of one contemporary writer that ". . . at this present, through the great consuming of wood as aforesaid, and the neglect of planting of woods, there is so great a scarcitie of wood through the whole kingdom."[24]

England's forests were never fully restored, but fuel shortages were alleviated by burning coal in the place of wood. Although there were misgivings about the noxious vapors given off by burning coal, it came to be widely used for domestic heating, and as a source of process heat for the production of beer, sugar, bricks, soap, glass, and iron. More than simply a substitute for wood, by the end of the

The development of coal resources provided a new source of energy. At the same time, new energy technologies, such as this mine's steam engine, increased the need for coal. (Bettmann/Hulton.)

nineteenth century coal had become the basis of industrial civilization, as the rich coal deposits of Britain significantly contributed to that country's unique position as "the Workshop of the World." Much of the industrial age was the era of coal, as coal-fired steam engines powered factories, hauled railroad trains, generated electricity, and propelled ships to distant destinations.

Yet, just when coal had established its primacy as the most important energy source for industrial society, hard questions were being asked about the continued viability of coal-based technologies. By the end of the nineteenth century it was becoming evident that stocks of coal, while still large, were being depleted at ever-increasing rates. The projection of established trends seemed to offer indisputable proof that the day of reckoning was not far off: Britain was running out of coal. In the words of the contemporary English economist, W. Stanley Jevons, "there is no reasonable prospect of any relief from a future want of the main agent of industry. We must lose that which constitutes our particular energy."[25]

Coal was king, and in Jevons' estimation, as well as those of other informed students of the British economy, there was no hope that anything could take its place. In Jevons' gloomy appraisal, "All things considered, it is not reasonable to suppose or expect that the power of coal will ever be superseded by anything better."[26] His pessimistic assessment of potential substitutes is exemplified by his quick dismissal of petroleum as a fuel: "Its natural supply is far more limited and uncertain than that of coal, and an artificial supply can only be made by the distillation of some kind of coal at considerable cost. To extend the use of petroleum, then, is only a new way of pushing the consumption of coal. It is more likely to be an aggravation of the drain than a remedy."[27] Natural gas, another possible substitute, was an equally forlorn hope. Jevons approvingly quoted the assessment of an

American steel executive: "Of late years the supply of gas has been decreasing . . . and it would seem that before many years this fuel would cease to be a factor in the large operations of a steel works."[28]

One can smile at the remarkable wrongheadedness of these assessments, but it is easy to be wise after the fact. After all, the true extent of oil and gas reserves were only dimly perceived at the time Jevons was coming to his gloomy conclusions. The third edition of Jevons' book that contains the passages quoted above was published in 1905, four years after the Spindletop field demonstrated the vast oil reserves of East Texas. There, a single well produced twice as much oil as the state of Pennsylvania, until then the center of the American oil industry.[29] And it was not until three decades later that the immense oil deposits of the Middle East began to be explored.

The essential point here is that a problem such as resource depletion can often be solved by the use of substitutes, just as coal substituted for wood, and oil replaced a great amount of coal. This does not happen easily or automatically, of course; it requires the invention, development, and application of many new ways of doing things. The large-scale employment of petroleum fuels required a host of new technologies: seismic exploration devices, casings and bits for rotary drilling, new compounds for the cementation of bore holes, and so on. Equally important, the use of a new source of energy must be complemented by the emergence of new energy-using technologies. In the early 1900s, the coal-fired reciprocating steam engine was a proven technology, while the gasoline-fueled internal combustion engine was cranky and unreliable, and the diesel engine had scarcely emerged from the laboratory. The rapid strides made by these new engine types in the ensuing years was both a product of the availability of new fuels, as well as a stimulus to their accelerated extraction.

ALTERNATIVES TO FOSSIL FUELS

The example of the widespread substitution of petroleum-based energy for coal-based energy applies to many other diminishing resources. It demonstrates that at one level of analysis, the cure for depletion and environmental damage can be found in new technologies that successfully address the problems generated by old technologies. Still, this is not the end of the matter. Historical analogies have their place, but like all analogies they must be treated with caution. The fact that technological solutions have been found in the past is no guarantee that they will be found in the future.

The limitations of technological solutions to energy shortages are all too apparent when nuclear energy is examined. For the last thirty years or so, nuclear energy has been heralded as the next stage in the evolution of energy sources, the logical solution for fossil-fuel depletion. This may come to pass, but it is far from certain. Many problems still attend the widespread use of nuclear energy.

First, the generation of nuclear energy itself requires considerable expenditure of energy for uranium mining and transportation, equipment manufacture, plant construction, maintenance and administration, and waste disposal. When all of these energy inputs are taken into account, nuclear plants make a much smaller contribution to total energy supplies than their output indicates. The computation of the net energy supplied by any power system is difficult, and it requires certain assumptions that cannot always be empirically grounded. But there is good reason to believe that, whatever nuclear energy's contributions, they take a long time to be realized. According to one calculation made in the mid-1970s, an average of twelve years passed before the cumulative contributions of a nuclear plant exceeded the energy that went into its construction and operation.[30]

Second, the product of nuclear plants is electricity, and electricity has its limitations. Electricity is not a source of primary energy; it is a means of transmitting energy, and about two-thirds of the energy used to generate electricity is lost in transmission. This means that, if possible, it is always better to produce energy close to where it will be used. If you want to take a hot shower, it makes more sense to heat the water at the source than it does to run a resistance heater with electricity that has been generated hundreds of miles away. Also, electrical energy cannot be used for all purposes. Although fossil fuels are an important source of electrical power, the substitution of conventional generating facilities by nuclear plants would not come close to eliminating fossil fuel use, for only about 25 percent of the coal, oil, and gas consumed today is used for the generation of electricity.[31]

Third, conventional nuclear energy plants require uranium for their fuel, and supplies of uranium are not inexhaustible. If the nuclear reactors currently under construction were added to those already in operation, they could generate close to 400 gigawatts (a gigawatt is 1,000 megawatts, the output of a large power plant). This rate of output would exhaust global uranium supplies in one hundred years. If the long-term goal of nuclear enthusiasts was met and 2,500 reactors were in operation, they could produce two terawatts (a terawatt equals 1,000 gigawatts). This would result in the depletion of uranium within twenty years.[32] The long-term solution to the depletion of uranium is the breeder reactor, which actually creates more fuel than it consumes. But a major product of the process is plutonium, one of the most toxic elements on earth, and one that could be readily used to make nuclear weapons.

Finally, nuclear wastes present serious disposal problems. The radioactive wastes produced by nuclear reactors in use and currently under construction would total 8,000 tons yearly.[33] Most of this waste is uranium, a difficult enough substance, but the most severe challenge is to properly dispose of plutonium. While plutonium comprises only one percent of total nuclear wastes, it has a half-life of 24,000 years. It is difficult to be certain that any waste storage facility will remain safe for so long a time.

Lest it be thought that nuclear energy is being singled out as an especially problematic source of energy, it should also be noted that other alternative energy sources have their own shortcomings. The energy of the sun offers a fantastically large source of potential energy; the solar energy that strikes the earth every year amounts to 178,000 terawatt-years—15,000 times the energy used in the world

Southern California's 10-megawatt Solar One generating facility operating during the 1980s. Southern California Edison plans to construct Solar Two on the former site of Solar One using a new molten salt heat transfer technology. Solar Two will use 1,800 heliostats, or mirrors, to focus the sun's rays on a central tower, where the heat will be captured in molten salt. (Southern California Edison Company.)

today.[34] The trick, of course, lies in effectively using even a modest fraction of that energy. Significant strides have been made in that direction. In 1970, electricity produced by photovoltaic cells cost $60 per kilowatt-hour. By 1980 the cost had fallen to $1 per kilowatt-hour. Today, it costs 20 to 30 cents per kilowatt-hour. Still, this amounts to five times the present cost of electricity from conventional sources. Thin-film solar cells hold out the promise of much cheaper photovoltaic power (one-tenth current costs) because they lend themselves to mass production. But this technology has not made it out of the laboratory, and many questions remain concerning its reliability and efficiency.

Wind power, a very old energy technology, may also play a larger role in the future. Wind-generated electricity is still expensive, although at 7 cents per kilowatt-hour it is only two cents above the cost of electricity provided by a new coal-fired plant. One U.S. Department of Energy study claimed that within a couple of decades the cost could fall to 3.5 cents per kilowatt-hour.[35] Whether or not this occurs remains to be seen, and even if it does, the irregularity of wind supplies will require some means of storing the energy produced. There is no sure way on the horizon of doing this. Batteries, for example, have a long history of resisting efforts to improve their efficiency and longevity as energy storing devices.

At present, modern wind machines seem to work best when electrical demands are modest and the system can remain relatively small. Some experts think that inherent technological and economic limitations will constrain wind-generated power to installations of about one megawatt—three orders of magnitude less than some nuclear and fossil-fuel fired generating plants.[36] If this is true, wind power may be best suited to the generation of electricity in places that are difficult to tie into a large power grid and where demands are low. This has been the case in China, where there has been a substantial increase in wind-powered electricity to serve households where electrical needs are confined to running recently acquired television sets.[37]

Many other sources of energy might become significant as reserves of fossil fuels are depleted and their environmental consequences become intolerable: biomass (plants and other organic sources), geothermal, methanol (methyl alcohol) produced from coal and gas, ethanol (ethyl alcohol) produced from plants, and nuclear fusion (although probably not before the middle of the next century).[38] Still, for the immediate future, none of them can rival petroleum as a relatively cheap and convenient source of energy. Petroleum-based energy sources will be of central importance for many decades. The key issue will be using them efficiently.

DOING MORE WITH LESS

The generation and use of massive quantities of energy is the cornerstone of industrial society. Much of the expansion of production (and of course consumption) that has taken place since the Industrial Revolution has come through the development of technologies dependent on external sources of energy. Countering the depletion of fossil fuels as well as their adverse consequences will require the use of new sources of energy. Still, it must be remembered that their use will generate new problems, as will the employment of effective pollution-control strategies; in both cases, there are few, if any, cost-free technological fixes. In the early twentieth century, the replacement of the horse by the automobile was widely applauded, for it promised a far cleaner urban environment—in those days horses in New York City deposited 2.5 million tons of manure annually.[39] One might speculate on which source of pollution is preferable, the emissions of cars or of horses.

Instead of substituting one source of pollution for another, a better course of action is to try to reduce pollution by cutting back on energy use. This would not be a new effort, for the long-term historical trend has been to use energy more efficiently. In the United States, energy intensity (the ratio of energy used to productive output) peaked around 1915, and has been declining ever since.[40] The total amount of energy used has increased, but the use of more fuel-efficient sources of energy (primarily the shift from wood and coal to oil and natural gas), changes in the structure of the economy, and the development of more energy-efficient technologies allowed the rate of economic growth to outstrip by a comfortable margin the rate of energy use. A heightened interest in promoting energy efficiency emerged in 1973, when the first oil embargo put a serious crimp in energy supplies.

The world's industrial economies went into a temporary tailspin as a result of skyrocketing oil prices, but major efforts were made to use energy more efficiently. From 1972 to 1985 the U.S. economy grew by 40 percent while energy consumption remained level.[41] As a whole, the economy produces more output with each unit of energy; energy intensity has fallen by 28 percent since 1973.[42]

A portion of the improved output/energy ratio can be attributed to structural changes in the economy, especially to the growth of services (which use relatively little energy), and to the relative decline of industrial production (which uses a lot of energy). One-third of the gain in energy intensity can be attributed to this shift.[43] The rest of the improvement in energy intensity has been the result of increased efficiency. Taken together, the market-oriented industrial societies have improved their energy efficiency 20 to 30 percent since 1973.[44] In the United States, the savings in oil and gas over 1973 rates of consumption equals 13 million barrels of oil per day—half of OPEC's daily production capacity.[45]

These improvements have not been confined to the industrial sector. Conservation measures applied by American homes and offices have saved 1.2 million barrels of heating oil per day relative to 1973 consumption levels. In monetary terms, $45 billion has been trimmed from energy expenditures through the use of new technologies along with better management of heating, lighting, and ventilation systems.[46] Relatively modest changes in everyday items such as refrigerators and freezers have produced large benefits. Running these appliances currently requires the energy equivalent of thirty 1,000 megawatt power plants, but if they were as inefficient as the ones in use in 1973, the United States would need twenty additional power plants.[47] These improvements are cumulative, and by the end of the century more efficient household appliances will have saved $28 billion in electricity and 342 million tons of carbon that would otherwise have been spewed into the atmosphere.[48]

A multitude of energy conservation measures can also be applied to commercial and residential buildings. Many of them require nothing in the way of advanced technology. Intelligent site selection and proper orientation of buildings can result in buildings that stay warmer in the winter and cooler in the summer. Adequate sealing and insulation prevents the loss of hot air in the winter and cool air in the summer, while the provision of thermal storage in buildings allows the use of electricity for air conditioning to be shifted to off-peak hours, thereby obviating the need to build new generating facilities.[49] At the same time, there are many new technologies that can produce significant energy savings. Microprocessor-based integrated controls for heating, ventilation, and cooling can keep temperatures within narrow limits that are optimal for different rooms in a building, so during the summer a computer room can be kept cooler than a storage room.[50] Many industrial processes can also benefit from the adoption of energy-saving technologies. According to the Congressional Office of Technology Assessment, "Implementation of state of the art technologies in all petroleum refining, pulp and paper, steel, aluminum, cement and glass plants would save 16 to 37 percent of the energy used in these industries."[51] These technologies will not be installed overnight, of course, but they do indicate that stable or even increased industrial production can be compatible with diminished energy consumption.

MORE MILES TO THE GALLON

Even greater savings are possible when the energy requirements of cars and trucks are scaled down. Cars and trucks currently use nearly half the world's petroleum supplies, and in the developed world the fraction is considerably higher. At the same time, they are major sources of environmental damage. The world's 400 million automobiles annually discharge about 550 million tons of carbon into the atmosphere, ten percent of the total discharge from fossil fuels.[52] In developed industrial countries, motor vehicles contribute an even larger share of pollution, emitting nearly half the oxides of nitrogen, two-thirds the carbon monoxide, and nearly half the unburned hydrocarbons.[53] Improving automotive technology to produce greater efficiency would thus be a major step in arresting both resource depletion and pollution.

There are essentially three ways of enhancing the fuel economy of cars and trucks: reducing weight, improving aerodynamics, and making engines and accessories perform more efficiently. All of these have contributed to the impressive gains in fuel economy scored by American automobiles since the first energy crisis in 1973. In that year, American cars averaged about twelve miles per gallon; fifteen years later, fuel mileage had more than doubled.[54] The largest share of this improvement was effected by reducing weight. Cars became smaller and lighter, but the shift to front-wheel drive designs for many models maintained interior space. Better aerodynamics also helped, while under the hood, fuel injection, turbochargers, computerized spark advance and fuel metering, multivalve engines, and detail improvements allowed engines to squeeze more miles per gallon with few if any performance losses.

It is not certain that the improvements to fuel mileage that occurred after the first round of energy shocks can continue at the same rate. Still, there are many promising technologies that could result in considerably more fuel-efficient vehicles. Greater use of aluminum and plastics can make cars lighter, as can the use of more space-efficient designs (aided perhaps by the use of compact two-stroke engines). More aerodynamic improvements will be made, for they promise substantial rewards; a ten percent reduction of a car's air resistance can improve fuel economy by 3.5 percent, while a ten-percent weight reduction yields only a 2.5 percent gain.[55] Today's automobiles are aerodynamically "cleaner" than those produced ten years ago, but there is still considerable room for improvement; experimental cars have already been made that present less air resistance than an F-15 jet fighter. Normal passenger cars are not likely to match this in the near future, but it does show that major aerodynamic gains are still possible.[56]

Although the internal combustion engine is not likely to be replaced any time soon, other parts of the automobile's powertrain could undergo great changes, such as the widespread use of continuously variable transmissions that allow engines to run at the most efficient rotational speeds. Even more radical concepts have already been tested, such as flywheel mechanisms that use the energy otherwise lost by deceleration or braking to run accessories or even to power the car. Experimental cars using a variety of innovative technologies are capable of achieving 60 to 70 miles per gallon in city driving, and even more on the open highway. Cars produced in the immediate future will lag behind these figures, but there seems ample room for improvement without pushing beyond existing technological frontiers.

ECONOMIC SYSTEMS, GOVERNMENT POLICIES, AND THE ENVIRONMENT

Just as technologies impact energy supplies and energy-conservation measures, so too are there a substantial number of innovative technologies that can help to control toxic wastes, pollutants, and other undesirable by-products of modern technology. Many new technologies can be used to substitute for hazardous materials and processes, make more efficient use of potentially toxic materials, and counteract existing hazards. But as important as these technological advances are, it would be a mistake to think that new and improved technologies are the solution to all of our environmental ills. As Chapter 1 indicated, technological fixes have their place, but when major problems are involved, they must be complemented by social, political, and cultural changes. Some of these changes may occur spontaneously, but many of them must be encouraged by laws, regulations, and other government interventions. And even when a technological fix can be found, it is often necessary to stimulate its use through the formation and implementation of effective policies.

Chapter 17 will consider how governmental policies shape the course of technological change in general. Here, we will confine our remarks to some general policy issues directly affecting energy use and the environment.

During the energy crisis brought on by the OPEC oil embargo in 1973, some scholarly observers came close to panic. As they saw things, the massive reduction in oil supplies was the latest symptom of a massive crisis. Combined with population growth and environmental destruction, the contraction of energy supplies was forcing major changes in industrial societies. Democratic procedures would have to be replaced by powerful, even authoritarian, government mandates. One influential economist looked toward a dangerous future and reluctantly concluded that "the passage through the gantlet ahead may be possible only under governments capable of rallying obedience far more effectively than would be possible in a democratic setting."[57] His assumption, one shared by many others, was that in the face of a massive crisis, the only salvation lay in a centralized government's ability to force people to sacrifice their individual interests in favor of the collective good. A highly centralized government was inescapable, for only such an authority could bring sufficient expertise to bear on the shaping of the necessary policies, while at the same time stifling individual actions contrary to these policies.

There may be a surface plausibility in these ideas, but a little reflection should serve to demonstrate their falsity. The centralization of governance does not necessarily improve the ability to forge and administer effective policies, especially those relating to something as complex as the natural environment. Even with the best intentions, a centralized decision-making body would be overwhelmed by the amount of information necessary for the management of a fantastically complex ecosystem. The second assumption, that governmental actions will be largely directed toward the public good, is even less tenable. The self-serving tendencies of governments when they are insulated from any oversight by the citizenry seem evident enough.

The shortcomings of authoritarian governments have been made amply evident by the performance of centrally planned economies ruled by Communist regimes. To take one example, all of them have used energy far less efficiently than market-oriented industrial economies. China, one of the worst in this regard, uses three times as much energy per unit of gross national product as does Japan.[58] This can be partially excused on the grounds that a poor nation like China cannot afford to rapidly deploy new, energy-saving technologies, but the wastage of energy has been universal in the Communist world. Even more disturbing, these countries also exhibited extraordinarily high levels of environmental degradation:

> In the Soviet Union . . . energy production had, by 1989, increased concentrations of toxic air pollutants to a level ten times the maximum permissible in 88 Soviet cities with a combined population of 42 million people . . . Twenty large hydro-electric power stations erected on lowland rivers have transformed flowing rivers into stagnant reservoirs and reduced the ability of these rivers to assimilate wastes. The Chernobyl nuclear reactor accident killed several dozen people and contaminated 10,000 square kilometers with levels of radioactivity that exceeded 15 curies per square kilometer, affecting more than 250,000 people.[59]

The countries of eastern Europe that endured more than four decades of Soviet domination did no better. In Poland and Czechoslovakia, sulfur dioxide emissions from the uncontrolled burning of coal are four to eight times higher than in most Western European countries. Toxic elements have accumulated in the soil and contaminated food to such a degree that the bone growth of one-third of the children in the worst-affected parts of Czechoslovakia is retarded by ten months or more.[60] In Poland, only one percent of the water is safe to drink, the life expectancy for men aged forty to sixty has been dropping, and one-third of the population is expected to suffer from respiratory disease, cancer, skin disease, or afflictions of the central nervous system as a result of environmental pollution.[61]

While environmental catastrophes have become endemic in centrally planned economies, one shouldn't jump to the conclusion that an unfettered market economy would be environmentally benign. The history of capitalist economic development is also marred by numerous examples of pollution, contamination, and heedless exploitation of natural resources. This is to be expected, for a pure market system, engine of economic growth though it may be, is by its nature a threat to the environment. Markets generally do a good job of coordinating production and consumption, but they are not effective in dealing with effects that lie outside the transactions of individual buyers and sellers. The price paid for a commodity sold by one party and bought by another does not take into account the costs borne by other parties, such as the damages suffered by downstream residents when a paper mill dumps its wastes into a river.

In similar fashion, a transaction may not take into account the loss of an irreplaceable resource to future generations. The price of a commodity may reflect some depletion costs, but time lags may prevent adequate corrective actions. In many cases, the price of a dwindling resource will increase at too slow a rate to signal its rapid depletion. The result is an overshoot-and-crash situation. This situation resembles a car driven by someone with slow reflexes; by the time the driver perceives a problem ahead and decides to apply the brakes, it is too late to avoid a collision.[62]

As will be noted in greater detail in Chapter 17, the imperfections of a market system can be alleviated by intelligent regulations and tax policies, government-sponsored research programs, and educational efforts. It must be noted, however, that government interference with market mechanisms has at times made matters worse. Many conservation efforts have been blunted by government programs aimed at keeping the price of energy artificially low. To cite one example, it has been estimated that from 1918 to the mid-1970s, federal subsidies for energy production amounted to $217 billion (in 1977 dollars), 60 percent of which went to the petroleum industry.[63] The result has been lower prices than would exist in a perfectly competitive market, leading to excessive consumption and to stunted development of alternative energy sources.

On occasion, the market system can combine with shortsighted government policies to thwart the use of potentially valuable energy technologies, as happened in the mid-1980s when short-term price movements and changes in government policy were nearly fatal to solar energy. From 1984 to 1986, cheaper oil and a change in the tax code led to a 70 percent decline in sales of solar collectors and to the loss of 28,000 of the industry's 30,000 employees.[64] Many promising developments were curtailed and solar technology languished, despite the obvious fact that oil would eventually have to be supplanted by alternative sources of energy.

Putting things more positively, there have been times when government programs have brought major benefits. One striking example is the development of new lighting technologies. Currently, 20 percent of all the electricity generated in the United States is used for lighting. Some of this energy is now being used more efficiently through the use of fluorescent lights with high-frequency ballasts. This new technology is expected to save $25 billion dollars over a five-year period. Yet the cost of the federally sponsored research project that supported it was less than $3 million, of which half was supplied by industry. This example of Federal sponsorship of energy-related research is expected to yield savings in energy costs that will pay off at an 8,000 to one ratio.[65]

While fluorescent lighting research has been an unqualified success, many other promising energy-related technologies have languished as a result of governmental inattention. Federally sponsored research on renewable sources of energy dropped from $900 million (in 1986 dollars) in 1980 to $177.2 million in 1986. The latter figure amounts to only 7.8 percent of the total energy R&D budget.[66]

The relative neglect of research into alternative energy sources is not confined to the United States, although the U.S. does trail most industrial nations in terms of alternative energy research expenditures relative to population and gross national product (GNP). In 1986, the governments of Japan, the United States, Italy, West Germany, the United Kingdom, Canada, Sweden, Greece, and Denmark collectively spent a bit over $1.1 billion for research on conservation and renewable sources of energy. This is an impressively large sum of money, but it is still only 15.5

percent of the total governmental expenditures for energy research. Research on nuclear energy by itself accounted for about $4.5 billion.[67]

There is no guarantee, of course, that higher levels of research support will produce an array of technological breakthroughs, but it seems likely that we would be in better shape today if research on conservation and renewable sources of energy had been supported at the level nuclear energy enjoyed for many years. Still, it must be recalled that the availability of a better technology does not insure that it will be used. The development of technologies that help us use energy more efficiently, control pollution, and safely dispose of harmful wastes is only part of the solution. Individual choices, corporate and governmental policies, the general distribution of income and power, and the willingness to forsake short-term advantages for longer-term benefits are more important than any combination of technological fixes. These will be the ultimate determinants of our ability to preserve and even improve our material standard of living while at the same time preserving the environment that has sustained us through centuries of carelessness, neglect, and exploitation.

Questions for Discussion

1. What do you consider to be the greatest environmental threat facing the world today? What sort of measures need to be taken to counter it? What will be the major sources of resistance to these measures?
2. The long-term environmental consequences of CO_2 emissions cannot be known for sure; computer models and a few empirical studies are the source of current projections. Are we justified in enacting laws to enforce significant reductions in the use of fossil fuels on the basis of imperfect scientific information? At what point can we decide that the assessment of risk is exact enough to warrant taking firm actions?
3. A distinction has to be drawn between conservation and curtailment. The former implies doing the same sorts of things with fewer inputs of materials and energy, while the latter implies an actual loss of output and consumption. Are the energy-saving measures taken in the last fifteen years primarily examples of conservation or curtailment? Will future energy-saving strategies be based on the former or the latter?
4. The continual extraction and use of natural resources does not result in their being completely "used up," but eventually the costs of extracting a diminishing resource exceed the value of the resource; barring improvements in extraction technologies, the resource is as good as gone. This may not happen for a long time, perhaps not until we as individuals are long departed from this earth. Is there anything immoral about using large quantities of the earth's resources for our own benefit? Do we owe anything to future generations? If so, by how much should we restrict our use of resources? How should these restrictions be mandated?
5. A considerable improvement in the fuel economy of automobiles has been the result of "downsizing." Yet all other things being equal, smaller cars are not as

safe as larger ones. Can a substantial savings in fuel justify the likelihood of more traffic-related injuries and fatalities? At the same time, more fuel-efficient automobiles also produce fewer pollutants, leading to fewer pollution-induced deaths and illnesses. Is it possible to construct a balance sheet that takes into account all of these factors in order to determine if smaller cars improve or threaten our physical health?

Notes

1. United States Congress, Office of Technology Assessment, *Energy in Developing Countries* (Washington, D.C.: U.S. Government Printing Office, 1991).
2. Charles P. Wallace, "Pollution Is Price of Asia Boom," *Los Angeles Times* (15 February 1991): 23.
3. Christopher Flavin, "Slowing Global Warming," in Linda Starke (ed.), *State of the World 90* (New York: W.W. Norton, 1990), p. 18.
4. Newton H. Copp and Andrew W. Zanella, *Discovery, Innovation, and Risk: Case Studies in Science and Technology* (Cambridge, Mass. and London: MIT Press, 1993), p. 299.
5. Ibid., p. 302.
6. Paul E. Gray, "The Paradox of Technological Development," in Jesse H. Ausubel and Hedy E. Sladovich (eds.), *Technology and Environment* (Washington, D.C.: National Academy Press, 1989), p. 196.
7. Flavin, op. cit., p. 17.
8. Sandra Hackman and Marc S. Miller, "The State of the World: An Interview with Lester Brown," *Technology Review* 91, 5 (July 1988): 52.
9. Lester R. Brown and Christopher Flavin, "The Earth's Vital Signs," in Linda Starke (ed.), *State of the World 88* (New York and London: W.W. Norton, 1988), p. 17.
10. Ibid.
11. Copp and Zanella, p. 315.
12. Ibid., p. 14.
13. John Maxwell Hamilton, *Entangling Alliances: How the Third World Shapes Our Lives* (Cabin John, Md.: Seven Locks Press, 1990), p. 78.
14. Edward C. Wolf, "Avoiding a Mass Extinction of Species," in *State of the World 88*, op. cit., p. 103.
15. Amulya K. N. Reddy and Jose Goldemberg, "Energy for the Developing World," *Scientific American* 263, 3 (September 1990): 112.
16. Gordon F. Bloom, "The Hidden Liability of Hazardous-Waste Cleanup," *Technology Review* 89, 2 (February–March 1986): 60.
17. Ibid.
18. Lenny Siegel and John Markoff, *The High Cost of High Tech: The Dark Side of the Chip* (New York: Harper & Row, 1985), pp. 164–165.
19. It is of course possible that human knowledge, manifested in new technologies, can overcome scarcities of any sort. See Julian L. Simon, "Resources, Population, Environment: An Oversupply of False Bad News," *Science* 208, 27 (June 1980). At the same time, however, the author's assertion that ". . . copper can be made from other metals . . ." (p. 1435) demonstrates the limits to human knowledge.
20. John P. Holdren, "Energy in Transition," *Scientific American* 263, 3 (September 1990): 158.

21. John Gever, Robert Kaufmann, David Skole, and Charles Vörösmarty, *Beyond Oil: The Threat to Food and Fuel in the Coming Decades* (Cambridge, Mass.: Ballinger, 1986). pp. 54–65.
22. Brown and Flavin, op. cit., p. 5. Some estimates of future topsoil loss go even higher. See Wes Jackson, *New Roots for Agriculture* (San Francisco: Friends of the Earth, 1980), pp. 17–20.
23. Carlo M. Cipolla, *Before the Industrial Revolution: European Society and Economy, 1000–1700*, 2d ed. (New York: W.W. Norton, 1980), p. 138.
24. Quoted in Ibid., p. 288.
25. W. Stanley Jevons, *The Coal Question: An Inquiry Concerning the Progress of the Nation, and the Probable Exhaustion of our Coal-mines*, 3d ed. (London: Macmillan and Co., 1906), p. 9.
26. Ibid., p. 187.
27. Ibid., pp. 184–185.
28. Ibid., p. 185.
29. James R. Chiles, "Spindletop," *American Heritage of Invention and Technology* 3, 1 (Summer 1987): 34.
30. Kenneth E. F. Watt, *Understanding the Environment* (Boston: Allyn and Bacon, 1982), pp. 94–96.
31. Calculated from the figures in Ged R. Davis, "Energy for Planet Earth," *Scientific American* 263, 3 (September 1990): 59.
32. Wolf Haefele, "Energy from Nuclear Power," *Scientific American* 263, 3 (September 1990): 141–142.
33. Ibid., p. 138.
34. Davis, op. cit., p. 55.
35. Carl J. Weinberg and Robert H. Williams, "Energy from the Sun," *Scientific American* 263, 3 (September 1990): 148.
36. Robert D. Kahn, "Harvesting the Wind," *Technology Review* 87, 8 (November–December 1984): 61.
37. Cynthia Pollock Shea, "Shifting to Renewable Energy," in *State of the World* 88, op. cit., p. 76.
38. United States Congress, Office of Technology Assessment, *Starpower: The U.S. and the International Quest for Fusion Energy* (Washington, D.C.: U.S. Government Printing Office, 1987).
39. James J. Flink, *The Car Culture* (Cambridge, Mass.: The MIT Press, 1975), p. 34.
40. As indicated graphically in Reddy and Goldemberg, op. cit., p. 112.
41. United States Congress, Office of Technology Assessment, *Energy Use and the U.S. Economy* (Washington, D.C.: U.S. Government Printing Office, 1990).
42. Arnold P. Fickett, Clark W. Gellings, and Amory Lovins, "Efficient Use of Energy," *Scientific American* 263, 3 (September 1990): 65.
43. *Energy Use and the U.S. Economy*, op. cit.
44. Christopher Flavin and Alan Durning, "Raising Energy Efficiency," in *State of the World* 88, op. cit., p. 41.
45. Arthur H. Rosenfeld and David Hafemeister, "Energy-efficient Buildings," *Scientific American* 258, 4, (April 1988): 78.
46. Ibid.
47. Ibid.
48. Christopher Flavin, "Creating a Sustainable Energy Future," in *State of the World* 88, op. cit., p. 30.

49. Rosenfeld and Hafemeister, op. cit., p. 81.
50. Rick Bevington and Arthur H. Rosenfeld, "Energy for Buildings and Homes," *Scientific American* 263, 3 (September 1990): 78.
51. "Industrial Energy Efficiency," *OTA Report Brief* (April 1993).
52. Christopher Flavin, "Slowing Global Warming," in *State of the World* 90, op. cit., p. 23.
53. Deborah L. Bleviss and Peter Walzer, "Energy for Motor Vehicles," *Scientific American* 263, 3 (September 1990): 103.
54. Ibid.
55. James J. Flink, "The Path of Least Resistance," *American Heritage of Invention and Technology* 5, 2 (Fall 1989): 42.
56. Deborah L. Bleviss, "Saving Fuel: Time to Get Back on Track," *Technology Review* 91, 8 (November–December 1988): 50.
57. Robert L. Heilbroner, *An Inquiry into the Human Prospect* (New York: W.W. Norton, 1974), p. 110.
58. William U. Chandler, Alexei A. Makarov, and Zhou Dadi, "Energy for the Soviet Union, Eastern Europe, and China," *Scientific American* 263, 3 (September 1990): 125.
59. Ibid., p. 122.
60. Ibid.
61. Bevington and Rosenfeld, op. cit., p. 7.
62. Kimon Valaskakis, et al., *The Conserver Society: A Workable Alternative for the Future* (New York: Harper & Row, 1979), p. 166.
63. Michael D. Yokell, "The Role of the Government in the Development of Solar Energy," in Lewis J. Perelman, et al. (eds.), *Energy Transitions: Long-Term Perspectives* (Boulder, Colo.: Westview Press, 1981), p. 130.
64. Shea, op. cit., p. 72.
65. Rosenfeld and Hafemeister, op. cit., p. 85.
66. Ibid., pp. 80–81.
67. Computed from figures in Flavin, "Creating a Sustainable Energy Future," op. cit., p. 37.

Chapter 7 Medical and Biological Technologies

Some people may doubt the overall benefits of technological advance, but most would admit that improvements in medical technologies have made our lives better. Who would want to go back to the not-so-good old days, when vast numbers of children died in infancy, epidemic plagues wiped out millions, and pain and suffering were an inevitable part of everyday life? Not too long ago, medicine's ability to cope with sickness was severely limited at best. In many—perhaps most—cases, medical intervention only made bad situations even worse. The knowledge base of medical technology was pathetically underdeveloped, rendering the majority of therapies ineffective or worse. As Lewis Thomas describes this unhappy situation:

> Bleeding, purging, cupping, the administration of infusions of every known plant, solutions of every known metal, every conceivable diet including total fasting, most of these based on the weirdest imaginings about the cause of disease, concocted out of nothing but thin air—this was the heritage of medicine up until a little over a century ago.[1]

More recently, technological advance supported by an expanding knowledge base has extended the lives of many, while at the same time eliminating or at least reducing a great deal of agony. How could anyone quibble with such positive results? And yet, it should be apparent by now that, in aggregate, technological advance is never an unmixed blessing. Problems are solved or at least alleviated, but at the cost of a new set of difficulties, many of them unforeseen. Medical technology is no exception. As we shall see, for all its unquestioned benefits, technological advance has generated some vexing problems. Understanding the nature of these problems is important in its own right, and at the same time it may help us to come to a deeper understanding of how technological advance can take with one hand as it gives with another.

In assessing the significance of modern medicine, it is important to bear in mind that not all historical improvements in mortality and morbidity are the result of advancing medical technologies. During the last three hundred years, dramatic advances have taken place in medical care: vaccination for the prevention of infectious diseases, antiseptic surgery, antibiotics, and even the transplantation of organs. However, the positive impact of all these advances has been comparatively small. The real gains in life span and improvements in the physical quality of life have been due to better nutrition, sanitation, and personal hygiene, along with higher standards of living in general.[2]

This is not to say that technological advances in medicine have been of no value. Many people are alive today because of kidney dialysis, computerized tomography, and antibiotics. Modern medical technologies have generated new possibilities and new hopes. Instead of accepting sickness and death as the inevitable product of fate or God's will, we increasingly expect that cures will be found, and that they will be made available when we need them. But along with rising expectations come rising costs. We do not have to suffer and die quietly (and inexpensively); new medical technologies allow something to be done. As a result, medical advances and their associated costs have the potential to expand without limit, for as long as people are subject to sickness and death, there will be no lack of demand for new medical interventions. Moreover, unlike other goods and services, expenditures on medical care are not usually weighed against other possible expenditures. You may decide to defer the purchase of a new car so that you can make a down payment on a house, but indefinitely deferring a quadruple bypass operation is not an option when it is all that stands between you and a life of progressive debilitation.

Paradoxically, while the costs of medical care have been escalating, they have become largely irrelevant to most individuals because government and private insurance programs have paid for the bulk of medical expenditures. In 1950, private insurance paid for less than half of the nation's hospital expenses. Today, private insurance, Medicare, and Medicaid cover 90 percent of them.[3] Under these circumstances, there is a strong tendency for increasing amounts of medical care to be provided, and for a country's medical costs to increase with no apparent endpoint. As one analyst put it, ". . . when benefits do not have to be weighted against costs—when the only criterion is that there be some benefit—the number of good things that can be done in medical care is, for all practical purposes, unlimited."[4]

Although it is difficult to do so, especially when life and well-being are at stake, choices eventually have to be made. Increasingly sophisticated medical technologies offer many benefits, but they can also generate costs that threaten to spiral out of control. Items of high-technology medicine should not be applied indiscriminately; like any other claim on our financial and other resources, they need to be evaluated in terms of the benefits they provide and the costs they incur. A useful starting point for evaluating the appropriateness of particular medical technologies has been provided by Bryan Jennett, a British neurologist. According to his analysis, a medical technology can be deemed inappropriate for a number of different reasons:[5]

Unnecessary. The patient has a condition too advanced to respond to treatment, as would be the case with attempting an extended treatment of metastatic cancer.

Unsafe. Complications outweigh the probable benefit, as with some invasive investigations and dangerous therapies.

Unkind. The quality of life after treatment is not good enough or its duration long enough to have justified the intervention.

Unwise. It diverts resources from activities that would yield greater benefits to other patients.

The Dilemmas of New Technologies

It is not always easy to scrutinize a particular medical technology in terms of costs, benefits, and general appropriateness—especially when it applies to us or someone close to us. Still, it is worth making the effort, if only to better comprehend the dilemmas often posed by the advance of medical technologies. The brief studies that follow should help to make the key issues a bit more concrete.

THE CASE OF KIDNEY DIALYSIS

Sophisticated technologies may be of limited importance to the population as a whole, but for many individuals they are literally the difference between life and death. The benefits of recent technological advances are nowhere more evident than they are to people suffering from kidney disease. In a healthy individual, the kidneys regulate the body's acid-base concentration, maintain proper water balance, and concentrate metabolic wastes that are subsequently excreted as urine. Complete or substantial kidney failure (known in medical jargon as "end-stage renal disease") usually results in incapacitation and premature death. Many are afflicted: urinary diseases are the nation's fourth largest killer, right behind cardiovascular diseases, cancer, and pneumonia.

In the early 1940s a Dutch physician used a bathtub and parts salvaged from a foundry to construct the first device to successfully take the place of the kidneys. The inventor, Dr. Willem Kolff, later emigrated to the United States, and his device inspired the development of more refined versions at a number of American hospitals. At that time, cleansing the blood by means of an artificial kidney (a process known as dialysis) was confined to short periods of time. Use of an artificial kidney as a permanent replacement for a real kidney became possible in the early 1960s through equipment improvements and the invention of a connecting tube that obviated the need to use a new artery and vein every time that the machine was hooked up. This made long-term dialysis a practical proposition, but at the same time it generated a host of nonmedical problems that have yet to be resolved.

When dialysis became an accepted medical practice, the number of patients that could potentially benefit from it far exceeded the number of available machines. It was therefore necessary to select some patients for dialysis and to reject others. In order to make these choices, the nation's pioneering dialysis institution, the Seattle Artificial Kidney Center, established an Admissions and Policy Committee to screen applications and determine who would get dialyzed and who would not. The committee was intended to reflect the community as a whole, being initially comprised of a lawyer, a minister, a housewife, a labor leader, a government official, a banker, and a surgeon, as well as two physician-advisors.[6] The committee made their first selective cut by only accepting patients from the state of Washington, and by eliminating children along with adults over the age of forty-five. Having done this, the committee then applied a set of criteria that took into account the personal characteristics of prospective patients. According to one report, these included the "sex of patient, marital status and number of dependents; income; net worth; emotional stability, with regard to patient's ability to accept the treatment; educational background; nature of occupation; past performance and future potential, and names of people who could serve as references."[7] As might be expected, making life-or-death decisions on the basis of the presumed worth of the patient generated a fair amount of indignation in some quarters. According to two critics, a psychiatrist and a lawyer, the published accounts of the selection criteria ". . . paint a disturbing picture of the bourgeoisie sparing the bourgeoisie, of the Seattle committee measuring persons in accordance with its own middle-class suburban value system: scouts, Sunday school, Red Cross. This rules out creative nonconformists, who rub the bourgeoisie the wrong way but who historically have contributed so much to the making of America. The Pacific Northwest is no place for a Henry David Thoreau with bad kidneys."[8]

The problem of controlling access to a potentially life-saving technology was mitigated as the number of machines increased and they were made easier to operate. It even became possible for a patient to routinely dialyze himself or herself at home. But formidable cost problems remained. In 1970 a twice-weekly dialysis cost $3,000 to $5,000, but required start-up expenditures of $9,000 to $13,000. For most patients these expenses produced grave financial pressures, and for some they were impossible to meet. The problem was apparently solved in 1972 when the U.S. Congress authorized payment for dialysis treatments through the government's Medicare insurance program.

The process whereby this policy was enacted illustrates the capricious way in which technologies are sometimes supported. The federal government's assumption of payments for dialysis treatments was in large measure the result of intense lobbying, an effort that was made all the more effective by connecting an abstract ailment with actual people. In one instance, the vice-president of the National Association of Patients on Hemodialysis and Transplantation not only testified, but dialyzed himself in the presence of members of the House Ways and Means Committee.[9] The provision of Medicare payments for dialysis was only a small portion of a larger piece of legislation; only thirty minutes of debate on the floor of the Senate took place, culminating in a lopsided vote in support of the measure at a time when close to half of the Senators were absent. The conference committee of the House and the Senate took only ten minutes to discuss the measure, which was only slightly modified as a result. Both houses subsequently accepted the conference committee's report, and the bill including the dialysis provision was signed into law by President Nixon two weeks later.[10]

Payment for dialysis through the Medicare program addressed a real need, but at considerable cost. Initial estimates of annual expenditures of $135 to $250 million were far too low; by 1987 dialysis was costing the federal government $2.4 billion a year. Of course, it is not the federal government that pays for this; the costs are ultimately borne by individual taxpayers, the vast majority of whom are not afflicted by end-stage renal disease. There is nothing inherently unreasonable or unfair about this. A catastrophic organ failure could happen to any of us, and we shouldn't begrudge helping people who have been less fortunate than ourselves.

Still, questions remain about the equity of paying for dialysis and not for other ailments. Dialysis patients comprise only one-quarter of one percent of Medicare beneficiaries, yet their treatment accounts for 4 percent of total Medicare payouts.[11]

Currently, about a third of the patients receiving dialysis are over the age of sixty-five.[12] As the population ages it can be expected that the demand for dialysis will increase apace, putting further financial strains on the health-care system. In some countries this possibility has already been foreclosed. In Great Britain, the government-run National Health Service always has operated under tight financial constraints. Consequently, certain treatments are restricted. Dialysis is one of them: a person over the age of fifty-five is rarely given the opportunity to receive dialysis. The ostensible reason for this is that older patients do not have the kind of constitution that would allow them to survive and flourish under a regimen of lifetime dialysis; one British physician candidly stated that people over the age of fifty-five are not suitable candidates, for they are all "a bit crumbly."[13] This may be true in a statistical sense, but in fact there is greater variation in overall levels of health among the elderly than there is in any other age group. Using age as a means of limiting access to an expensive medical technology is administratively convenient, but it dodges the issue of making choices on the basis of more relevant criteria.

Is the American policy of giving universal access to dialysis the correct one, or is the British policy more defensible? There are no easy answers. At the same time, the problems of distributive justice posed by the use of artificial kidneys are only a preview of what will have to be faced as new and even more expensive medical technologies become available and medicine gains the ability to successfully address hitherto untreatable conditions.

REPLACING BROKEN HEARTS

The dilemmas engendered by the advance of medical technologies are further illustrated by the prospect of using new technologies to counteract the number one killer in the United States, heart disease. For the last twenty-five years, many victims of clogged arteries and weakened hearts have gained a new lease on life through bypass surgery, a process through which a segment of a vein is removed from the leg or chest and spliced into one or more of the five coronary arteries that transport blood from the heart. In extreme cases, all five arteries may receive this treatment. This procedure is now routine in many hospitals, but once again, questions of cost, benefits, and distributive justice have to be confronted. In the first place, as with many other medical procedures, not all bypass operations can be justified; one informal survey of cardiologists indicated that 25 percent of these operations are performed with no clear indication of their necessity.[14] A more structured survey found that 94 percent of patients who underwent bypass surgery were alive five years later, but for heart patients who had been treated by other means, the survival rate was only slightly lower, 84 percent.[15] Second, as with kidney dialysis, a disproportionately small share of the patient population is receiving a financial benefit; bypass surgeries account for 1 percent of national medical expenditures, yet serve only .04 percent of the population.[16]

For many heart patients, a bypass operation is not sufficient, for the heart has been so weakened as to be barely functional and is in danger of failing completely. The transplantation of another person's heart offers one solution, but demand far exceeds the supply; as many as 75,000 people have conditions that require a heart transplant, but only 2,000 donor hearts are available annually.[17] Not so long ago the implantation of artificial hearts seemed to offer a way out of this impasse. In the mid-1960s a consulting group reporting to the federal government assumed that the major technical impediments were well on the way toward solution, so that before too long many people would have a diseased heart replaced by a mechanical one. According to this optimistic scenario, recipients of these hearts would be able to return to the work force, adding $19 billion to the gross national product over a ten-year period. It was even asserted that the taxes paid by artificial heart recipients would more than offset the federal government's expenses in supporting the program.[18]

In the late 1970s these hopes seemed close to realization as a result of the development of the Jarvik-7 artificial heart at the University of Utah. As it turned out, optimistic expectations were dashed, for the technology was far from being ripe. The first recipient of the device, a Seattle dentist named Barney Clark, underwent great suffering as a result of the immaturity of this technology. Seizures, severe nosebleeds, pneumonia, kidney disease, gout, epididymitis, and an intestinal ulcer followed the implantation.[19] Clark's postoperative existence for 112 days could hardly justify the expense and pain of the operation. Subsequent implants met with hardly any greater success. One patient lived for 620 days, but during that span he suffered four strokes and a series of infections that severely eroded his physical and mental capacities.[20]

No mechanical substitute for a terminally diseased heart is imminent. A compact, reliable, and safe implantable power supply has yet to be developed, nor have nonclotting surfaces for the heart and its associated plumbing.[21] At best, the artificial heart has some value as a "bridge" for patients needing the transplant of a human heart. But here too its availability has created problems by pushing recipients of artificial hearts to the head of the queue, even though they may be less suited for a transplant than other patients.[22]

As a result of the all-too-evident difficulties with artificial hearts, early optimism gave way to a more realistic assessment. But even if the technology had been ready, financial and ethical issues would have quickly intruded. At a per-patient cost of at least $160,000 for each implant, the extensive use of artificial hearts would have placed substantial financial burdens on society, perhaps as much as $5 billion per year.[23] Under these circumstances, a viable artificial heart program would engender some very difficult choices. The primary beneficiaries of the technology would be those who could afford it, either through insurance coverage or personal resources. Alternatively, some sort of rationing system could be set up, using criteria such as the likelihood of the patient's long-term survival. This is clearly a distasteful prospect—only 21 percent of Americans indicate a willingness to ration costly new medical technologies.[24] And if this were to occur despite public resistance, what sort of criteria might be used? Ominously, criteria could include those invoked by the Seattle Admissions and Policy Committee three decades ago:

the "worth" of a prospective recipient, that is, how much they had contributed or were likely to contribute to society. Whatever the criteria invoked, the decision to use an expensive medical technology in a world of finite resources necessitates making decisions. In many cases this can be akin to "playing God," a role that few humans are eager to assume.

The selection or deselection of recipients is not the only equity issue that the use of expensive medical technologies pushes to the forefront. In a world of necessarily limited resources there is no escaping the fact that an expenditure in one area means the foreclosing of an expenditure in another. The $50,000 a year cost of dialysis or the $160,000 that would have been required for the implantation of an artificial heart could be used to produce significantly higher levels of overall well-being. To cite one example, according to Dr. David Eddy, the director of Duke University's Center for Health Policy Research and Education, the funds that might have been absorbed by an artificial heart program could instead be used for an antitobacco education campaign. Smoking is a major contributor to heart disease; consequently, even a 1 percent reduction in tobacco use would produce benefits considerably in excess of those offered by artificial hearts.[25] Other examples of effective tradeoffs could be cited. In 1983, the cost of one liver transplant and a year of postoperative care exceeded by $34,000 the total budget of an inner-city health clinic that had provided 29,000 office visits by poor people in the area.[26] Heroic measures to sustain severely premature infants can easily cost more than $100,000, yet prenatal programs go severely underfunded, even though they offer the prospect of millions of dollars of savings by heading off subsequent medical problems. At the other end of the life span, increasingly sophisticated technologies are used to keep people going even though there is no meaningful life to be preserved. In 1985, total expenditures of the Medicare program were $70 billion, of which 28 percent went to maintaining patients during the last year of their life and 30 percent of this sum was expended during the last month of the patients' lives.[27] In the mid-1970s $4 billion was spent for new technologies used for Medicare patients. Some of these technologies were surely of great value, but it can be argued the money would have been better spent on raising all aged people above the poverty line or providing funds to allow 2 million of them to upgrade or move out of their substandard housing units.[28]

HALFWAY TECHNOLOGIES

These dilemmas demonstrate the decidedly mixed blessings bestowed by medical technologies that are only palliative. A kidney dialysis machine does not cure the disease that led to kidney failure in the first place, just as a heart bypass operation comes into play only after the damage has been done. Many of the apparent breakthroughs of recent years are nowhere near as efficacious and cost-effective as the dramatic advances of the past. For example, relatively inexpensive vaccination campaigns completely eradicated smallpox as a human disease, significantly contributing to longevity and a better quality of life. In contrast, despite two decades of intensive efforts, we have made very little progress in overcoming cancer, the second largest cause of death in the industrial world.[29] Many of today's technological advances in medicine have extended life spans and eased pain but have done little to address the causes of disorders themselves.

For a patient suffering from organ failure, the availability of dialysis or a heart transplant can be the difference between life and death. But for the society as a whole, the escalating costs of medical care driven by these "halfway technologies" constitute a formidable problem. No one is comfortable with putting a dollar value on a human life, but the years to come may see the emergence of more explicit efforts to balance the costs of new technologies with their benefit to society as a whole. Far from offering across-the-board advantages, the development of new medical technologies will force us to grapple with increasingly difficult choices concerning their application.

DIAGNOSTIC TECHNOLOGIES

The escalation of medical costs is one of the most problematic consequences of the advance of medical technologies, but hardly the only one. A variety of diagnostic technologies have substantially advanced physicians' ability to determine the source of medical problems. At the same time, however, they have altered the nature of medical practice, sometimes with unfortunate results.

For centuries, medical diagnosis had been based on the observation of external symptoms, patients' narratives of their illness, and the application of dubious theories that explained illness in terms of such things as imbalances in four bodily "humors." A major reorientation began in the 1820s when doctors began to use a simple tube to hear heartbeat and other sounds within the patient's chest. Dubbed the "stethoscope," the new instrument allowed more accurate diagnosis by amplifying and focusing the sounds produced by the circulatory system. Within a few years, other instruments for apprehending hitherto hidden parts of the body were invented, such as the ophthalmoscope for observing the interior of the eye and the laryngoscope for the throat. A series of sophisticated diagnostic technologies such as X-ray machines and electrocardiographs followed. By the beginning of the twentieth century, physicians were using instruments for measuring everything from blood pressure to the electronic currents generated by the nervous system. The accuracy of diagnosis was further advanced by the development of laboratory tests for everything from cholesterol levels to the presence of the organism that causes syphilis.

The use of diagnostic tests exemplifies the tendency, mentioned in Chapter 1, of technologies to create their own needs. Many of the tests routinely performed today seem to be unnecessary; according to one study of a teaching hospital, 47 percent of tests performed could have been eliminated with no evident loss in the quality of patient care.[30] Fetal heart monitors have had no discernable positive benefit; their main consequence seems to have been an increase in the number of deliveries using a cesarean section.[31] On the other hand, the increase in lab tests is not solely

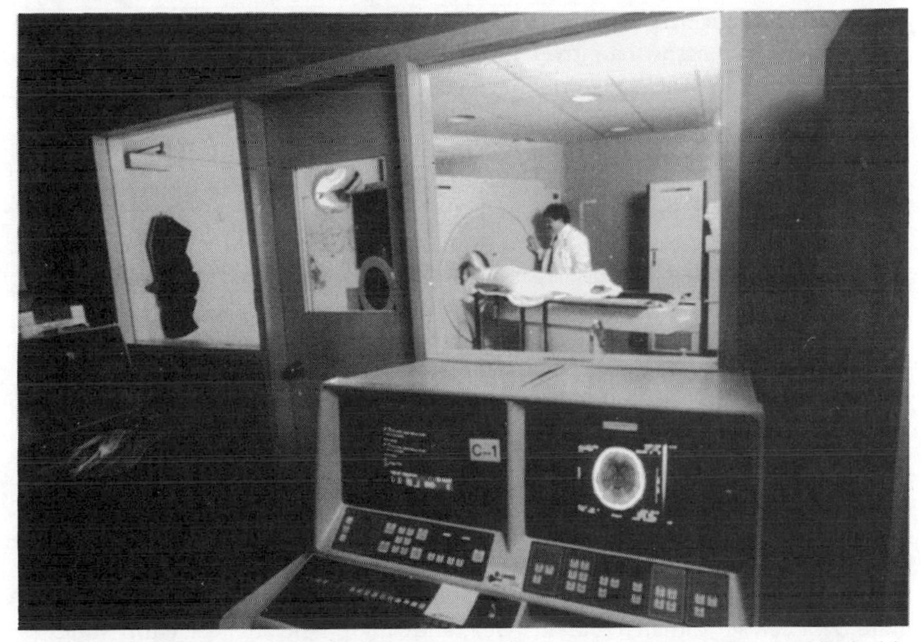

The CAT scanner makes possible more accurate diagnoses, but it also has contributed to spiraling medical costs. (Photo Researchers.)

the result of the availability of these tests. Some of the increase is directly attributable to the practice of "defensive medicine." Doctors are justifiably nervous about the threat of malpractice suits, so the safest course seems to be the overprescription of tests in order to avoid accusations of not having used every available diagnostic tool. According to one estimate provided by the American Medical Association, an expenditure of $15 billion per year results from this practice.[32]

The overuse of tests can produce more than financial pain. According to a past president of the Association of American Physicians, "As our interventions have become more searching, they have also become more costly and more hazardous. Thus, today it is not unusual to find a fragile elder who walked into the hospital, [and becomes] slightly confused, dehydrated, and somewhat the worse for wear on the third hospital day because his first forty-eight hours in the hospital were spent undergoing a staggering series of exhausting diagnostic studies in various laboratories or in the radiology suite."[33]

It can be reasonably argued that the physical and financial discomfort produced by diagnostic tests is outweighed by the greatly expanded ability to understand and treat illnesses. But another, more subtle consequence of sophisticated diagnostic technologies is that they have encouraged the treatment of specific disorders rather than human patients. In many cases, a physician by examining an X-ray plate or a microscope slide can produce an accurate diagnosis even when the patient is not physically present; indeed, the physician does not even have to be personally acquainted with the patient.

In going about their diagnostic work, few physicians today trouble themselves with obtaining lengthy narratives from their patients regarding their patients' ailments. The physician's personal diagnostic skills are often slighted in favor of more "objective" scientifically based knowledge attained through the use of sophisticated tests and instruments. This is a process that began with the stethoscope; although a very simple device, it changed the nature of the interaction between doctors and their patients. As Stanley Joel Reiser explains, the use of the stethoscope "helped to create an objective physician, who could move away from involvement with the patient's experiences and sensations, to a more detached relation, less with the patient but more with the sounds from within the body."[34]

A similar story could be told of the spygmanomometer, the ubiquitous instrument used for the measurement of blood pressure. After its introduction in the late nineteenth century, physiologists praised the instrument's capacity to produce objective, quantitative, and precise data. Many practicing physicians saw things differently, however, fearing that the spygmanomometer would ". . . intervene between patients and doctors, dehumanize the practice of medicine, and replace the delicate art of medicine with mere technical proficiency."[35]

The use of increasingly sophisticated diagnostic technologies illustrates the mixed blessings of technological advances in medicine. On the positive side, far more data can be gathered and analyzed, allowing physicians to identify illnesses that might otherwise have been misdiagnosed or to catch a disease in its early stage when it can be more easily treated. Standard tests also help to promote a standardized medical practice, removing the biases and blind spots of individual practitioners. Consequently, medicine is more "scientific" and less tied to the variable skills of individual practitioners. At the same time, however, other kinds of medical skills are in danger of being lost. Not every aspect of a functioning organism can be reduced to "objective" data. Treating illnesses often requires more than the ability to process information of this sort and use it as the basis of treatment. Judgment, experience, and even "intuition" can be of vast importance for a successful medical practice. Ideally, the use of sophisticated diagnostic technologies frees doctors from tedious work, leaving them with more time for personalized interactions with patients and more accurate diagnoses. Unfortunately, this is not what always happens. A fixation on the use of modern diagnostic technologies can easily result in the denigration of diagnostic skills based on more personal relationships with patients. Consequently, a focus on "objective" data to the exclusion of all else can lead a physician to miss some crucial clues regarding the source of a patient's illness.

This has a significance for more than diagnosis. Sick people usually have something physically wrong with them, but organic disorders may be only part of the problem. Illnesses often have psychological components as well, and the successful treatment of these requires more than the application of medical techniques; communication, the ability to build a bond of trust with the patient, and some measure of empathy are also necessary. An overreliance on "objective" data can obscure and even drive off these qualities, to the detriment of medical care as a whole. As Reiser notes, a physician must resist the tendency to place an excessive reliance on diagnostic technologies and the information they provide: ". . . accuracy, efficiency, and security, are purchased at a high price when that price is impersonal medical care and undermining the physician's belief in his own medical powers."[36]

The Genetic Fix

Many of the dilemmas that surround the use of procedures such as dialysis and heart transplantation are a consequence of their status as "halfway technologies." They offer the prospect of survival and some relief from a potentially fatal disease, but they do not cure it. The practice of medicine is badly in need of technologies that are more than palliatives. This is not a vain hope, for we may be on the threshold of medical advances equal to vaccination or antiseptic surgery. In recent years, much has been learned about the structure and mechanisms of cellular reproduction, giving rise to the prospect of eventual cures for a host of ailments. Yet when we consider the possible impact of technological breakthroughs in this area, we are confronted with a familiar theme: every advance is both a solution and a problem.

The fundamental scientific insight underlying many emerging medical technologies is that the growth and reproductive mechanisms of each cell are governed by the basic unit of heredity, the gene. Humans have 50,000 to 100,000 genes arrayed along the forty-six chromosomes found in every cell. Some of them are defective, and can give rise to a variety of disorders. Genetic screening or actually intervening in an organism's genetic code offers revolutionary new ways of diagnosing, curing, and even preventing the emergence of many disorders.

Addressing the medical problems induced by faulty genes became a possibility as more came to be known about their composition. In 1871, before the existence of genes was known, scientists discovered that the nuclei of cells contained a complex chemical substance called deoxyribonucleic acid, or DNA for short. Considerably later, in 1944, came the discovery that DNA was the substance from which genes were made. Less than ten years later James Watson and Francis Crick made one of the most important scientific discoveries of the twentieth century, that the molecular structure of DNA is shaped in the form of a double helix. The discovery of this structure provided a key clue to how cells reproduce. Scientific advances in genetics came in rapid order. A major milestone was reached in 1972 when Stanley Cohen of Stanford University and Herbert Boyer of the University of California produced the first organisms containing genetic material transferred from another organism. To do this they used special enzymes to cut apart the DNA extracted from the cells of a toad. The resulting strands of DNA were spliced together to make new DNA molecules that were subsequently introduced into the host organism, the common bacterium *Escherichia coli*. Bacteria with the new genetic endowment were selected from ordinary bacteria by linking the DNA fragments of the former to genes that conferred a resistance to antibiotics. Accordingly, when all the bacteria were exposed to an antibiotic, only those containing the new genetic material survived. The process, which has been refined in the ensuing years, is the foundation of a powerful technology, for it allows such accomplishments as the cataloging of the genetic endowments of specific organisms, including humans. It also is used to determine the ordered sequence of base pairs found in each gene, the first step in cracking the unique code embedded in each gene.

Far from being solely a pure scientific enterprise, gene splicing is the basis for many emerging technologies: modification of food crops for faster growth and greater production, new vaccines, improved diagnosis of diseases, new techniques for manufacturing antibiotics and hormones, and eventually the possibility of directly addressing hereditary diseases. Yet as with all significant technological advances, the application of genetic engineering will pose major challenges. The greatest potential problem centers on the modification of those genes that control inheritance patterns, for modifying them could change the biological nature of the human race. In altering the mechanisms governing our inheritance we would be affecting genetic information extending from generation to generation, with no evident endpoint. As Suzuki and Knudtson warn, "Genetic manipulation of human reproduction has the potential to multiply medical errors exponentially—sending ripples that radiate far beyond the finite lifetimes of gene therapist and consenting patient."[37] Many of our existing technologies have left unpleasant residues: toxic dump sites, acidified lakes, and radioactive wastes, to name but a few; genetically altered humans could be a technological legacy far surpassing any of these.

Screening for genetically based disorders does not pose potential hazards of this sort, but it does give rise to some troublesome issues. As genetic research advances, it is likely that an increasing number of diseases will be tied to specific genetic defects. Through an examination of a pregnant woman's amniotic fluid it is already possible to discover the presence of genes responsible for such things as Down's Syndrome and sickle-cell anemia. The indication of these disorders may prompt the parents to seek an abortion, a difficult moral choice for many. As genetic screening develops the choices may become even more difficult. It is possible that in a few years an expectant mother will be told that her unborn child will have a 75 percent chance of developing colon cancer before he or she turns fifty. For a person with no moral abhorrence of abortion, would this be an adequate reason for terminating the pregnancy? The results of genetic screening could also be used to bar people from certain jobs, increase their insurance premiums, or limit potential marriage partners. The development and application of genetic technologies will bring many benefits, but at the same time they will generate cultural, social, and moral problems, the solution of which may require radically different ways of thinking about life, death, and health.

THE COMMERCIALIZATION OF DNA TECHNOLOGY

While tinkering with the genetic endowment of human beings may constitute the greatest potential hazard posed by the development of DNA technology, there are a number of more immediate problems that also have to be confronted. In recent years, biotechnologies based on genetic discoveries have emerged as commercially significant enterprises; fifteen biotechnology based drugs are already on the market, and over one hundred others are in various stages of human clinical tests.[38] The commercial value of genetically engineered products was given a major boost in 1980 when by a 5–4 margin the United States Supreme Court ruled in the case of *Diamond* v. *Chakrabarty* that human-made organisms were entitled to full patent protection. In the realm of patent law at least, a genetically engineered entity is no different from one constructed out of aluminum or plastic.

Some firms have already reaped significant gains from biotechnology. The most famous was the case of Genentech, which in 1980 had its initial offering of stock rise from $35 to $89 during the first twenty minutes of trading. The two founding partners, who had each put $500 into the fledgling company in early 1976, were suddenly worth $82 million each.[39] Some educational institutions have also profited handsomely. By 1985 the licensing fees paid to Stanford University and the University of California for Cohen and Boyer's recombinant technology came to $3.5 million.[40]

One shouldn't begrudge the amassing of profit from the development of a new technology; the hope of eventually earning a profit is a prime motivator for taking the risks inherent in developing new technologies and bringing them to market. Still, it must be noted that biotechnology is not solely the creation of free enterprise; in fiscal year 1987 the federal government spent $2.7 billion for biotech research and development, a figure significantly greater than the $1.5 to $2 billion spent by private industry.[41] To a certain extent, the commercial application of genetic research should be regarded as a public trust, and not simply a source of profits. Second, an excessive emphasis on the commercial exploitation of emerging technologies may hinder the course of future research. Firms pursuing the development of potentially valuable biological technologies might be unwilling to publish or otherwise share their research findings. They may also be more reluctant to engage in basic research, preferring instead to pursue development projects with a more certain payoff. As we shall see in Chapter 17, a market economy is a powerful generator of new technologies, but it does not provide everything that is necessary for technological advance in general.

CONTROLLING BIOTECHNOLOGY

Few technologies have generated the furor that genetic engineering did as it emerged in the 1970s. Unlike all other technologies, the application of recombinant DNA research promised—or threatened—to change not our external environment, but our own biological attributes. One critic denounced the application of DNA research in no uncertain terms: "Our time is cursed with the necessity for feeble men, masquerading as experts, to make enormously far-reaching decisions. Is there anything more far-reaching than the creation of new forms of life? . . . My generation, or perhaps the preceding one, has been the first to engage, under the leadership of the exact sciences, in a destructive colonial warfare against nature. The future will curse us for it."[42]

Efforts to prevent dangerous applications of genetic engineering have been made, but their adequacy is open to question. In early 1976 the National Institutes of Health (NIH) took a first step by establishing guidelines for recombinant DNA research that categorized different levels of danger and stipulated the measures that had to be taken to counteract the inherent hazards of genetic research. But in fact these guidelines are not directly operative, for government regulators cannot supervise every laboratory engaged in genetic research. Institutions have their own safety committees, but these too are not monitored by NIH.[43] Questions of safety aside, biological technologies have emerged through a process similar to the development of most modern technologies: through the efforts of individuals, university laboratories, and commercial enterprises. The process by itself does not guarantee the development of technologies attuned to our most significant needs. Genetic engineering has an immensely powerful capacity for both good and ill, yet it still lacks adequate mechanisms for concentrating on and responding to public concerns, an inadequacy that has been exacerbated by the growing influence of commercial enterprises over its development.[44]

The need to guide and control genetic engineering can only grow in importance in the years to come, as more medical, psychological, and social problems are linked to genetic disorders. It will be a major task to simply sort out the genetic component of such varied human attributes as propensity to certain ailments, aggressive behavior, and physical stature. As more discoveries are made there will be a growing tendency to attribute individual and societal problems to faulty genetic programming. It is even possible that the years to come will show a resurgent enthusiasm for eugenics, the attempt to perfect humanity through the reinforcing of "desirable" traits and the suppression of "undesirable" ones. This is a dangerous course of action; the horrific example of Nazi Germany shows what can happen when eugenic principles motivate and justify the extreme oppression of people who are defined as "inferior."

Even if these excesses are avoided, many pitfalls remain. Although it has already brought undeniable benefits to some, the use of recombinant DNA technologies does not offer a panacea for everything that ails humanity. Even in the struggle against the myriad diseases that afflict the human race, genetic engineering will not have all the answers. Genetic defects are not always clear-cut; a gene that malfunctions under one set of environmental, nutritional, or other external conditions may be perfectly adequate in different circumstances.[45] It cannot even be said with complete confidence that the presence of a seemingly harmful gene is an unmixed hazard; genetic "imperfections" provide the raw material of genetic variability upon which natural selection operates.[46] Also, a search-and-destroy campaign waged against "defective" genes is likely to be a quixotic affair, for it may not always be possible to identify them. Each of us carries between five and ten defective genes that have no consequence for us individually, but could at some future time become a problem for our progeny.[47] Finally, an exclusive focus on genetic endowment can obscure efforts to deal with nongenetic sources of disease that may be of equal or greater importance. As Thomas H. Murray cautions, "All disease depends on the interaction of organism and environment. Which of the two we focus on is a social and political choice with important ethical consequences."[48]

Misguided efforts to address a large number of illnesses solely through genetic manipulation are only one troubling aspect of an overly enthusiastic application of genetic science. In a time of widespread frustration over social problems such as crime and welfare dependency, it is very tempting to look to "defective" genetic endowments as the root cause of these problems. It may yet be proven that there is a genetic component to a wide variety of individual behaviors, but it is not likely that this is the whole story. Fastening upon genetic endowments to the exclusion of everything else exemplifies the naive belief that science has all the answers and that technology offers the best solutions for what troubles us as individuals and as members of society.

Questions for Discussion

1. In 1990 Helga Wanglie, a 86-year-old woman, suffered cardiopulmonary failure. She was put on a respirator, fed intravenously, and injected with antibiotics. Although she remained in a vegetative state, her family insisted that she be kept alive. $800,000 was expended on her maintenance for the next 17 months, at which time the hospital petitioned the county district court to have a conservator appointed to consider the termination of treatment. What arguments might have been made in favor of termination, and what arguments might have been made against it?
2. When the demand for a new medical technology exceeds the supply, what should be used to determine who gets it? A lottery? The ability to pay? The "merit" of the recipient? Might it be better to limit the development of new technologies in order to forestall the need to make these choices?
3. The Jarvik-7 artificial heart cannot be viewed as a success, but many technologies require a long development period before they are successfully employed. Should R&D for the artificial heart resume? If so, how should it be financed? Are there some ongoing medical programs that are less deserving of financial support than the development of an artificial heart?
4. In the years to come we will likely see genetic engineering used for the treatment of many disorders. Few would argue that this technology should not be used to treat or prevent disabling diseases, but would it be proper for it to be used for such things as increasing height or changing eye color? If not, should such applications be prohibited? If so, how would such prohibitions be enforced?

Notes

1. Lewis Thomas, *The Medusa and the Snail: More Notes of a Biology Watcher* (New York: Bantam Books, 1980), p. 133.
2. Thomas McKeown, *The Role of Medicine: Dream, Mirage, or Nemesis?* (Princeton, N.J.: Princeton University Press, 1979).
3. Louise B. Russell, *Technology in Hospitals: Medical Advances and Their Diffusion* (Washington, D.C.: The Brookings Institution, 1979), p. 156.
4. Ibid., p. 3.
5. Adapted from Bryan Jennett, *High-Technology Medicine: Burdens and Benefits* (Oxford: Oxford University Press, 1986), p. 174.

6. Renee C. Fox and Judith P. Swazey, *The Courage to Fail: A Social View of Organ Transplants and Dialysis* (Chicago: University of Chicago Press, 1978), p. 230.
7. Shana Alexander, "They Decide Who Lives, Who Dies" *Life* 53 (9 November 1962), quoted in Ibid., p. 231.
8. Fox and Swazey, op. cit., p. 233.
9. Ibid., p. 348.
10. Ibid., p. 349.
11. John C. Moskop, "The Moral Limits to Federal Funding for Kidney Disease" *Hastings Center Report*, 17, 2 (April 1987): 12.
12. United States Congress, Office of Technology Assessment, "Life Sustaining Technologies and the Elderly" (Washington, D.C., 1987).
13. Henry J. Aaron and William B. Schwartz, *The Painful Prescription: Rationing Health Care* (Washington, D.C.: The Brookings Institution, 1984), p. 35.
14. Marcia Angell, "Cost Containment and the Physician" *Journal of the American Medical Association* 254, 9 (6 September 1985): 1204.
15. Jennett, op. cit., p. 162.
16. Judith Randal, "Coronary Bypass Surgery," *The Hastings Center Report* 12, 1 (February 1982): 14.
17. Nancy G. Kutner, "Issues in the Application of High-Cost Medical Technology: The Case of Organ Transplantation," in Howard D. Schwartz (ed.), *Dominant Issues in Medical Sociology*, ed. 3 (New York: McGraw-Hill, 1994), p. 406.
18. Diana B. Dutton, *Worse than the Disease: Pitfalls of Medical Progress* (Cambridge: Cambridge University Press, 1988), p. 99.
19. Ibid., p. 118.
20. Gideon Gil, "The Artificial Heart Juggernaut," *Hastings Center Report* 19, 2 (March/April 1989): 24.
21. Dutton, op. cit., p. 125.
22. Gil, op. cit., pp. 29–31.
23. Dutton, op. cit., p. 295.
24. Howard D. Schwartz, "Rationing and the Ideology of Exclusion," in Schwartz, op. cit., p. 417.
25. Dutton, op. cit., p. 300.
26. Ibid., p. 300.
27. Ibid., p. 304.
28. C. Gaus and B. Cooper, "Technology and Medicare: Alternatives for Change," paper presented at the Conference on Health Care Technology and Quality of Care (Boston University, 1976), cited in Kenneth E. Warner, "The Cost of Capital-Embodied Medical Technology," in National Research Council, *Medical Technology and the Health Care System* (Washington, D.C.: National Academy of Sciences, 1979), p. 273.
29. Tim Beardsley, "A War Not Won," *Scientific American* 270, 1 (January 1994).
30. United States Congress, Office of Technology Assessment, *The Changing Health Care System*, in Schwartz, op. cit., p. 286.
31. Warner, op. cit., p. 288.
32. Schwartz, op. cit., p. 421.
33. David E. Rogers, "On Technologic Restraint," *Archives of Internal Medicine*, 135 (1975), p. 1395, cited in Stanley Joel Reiser, *Medicine and the Reign of Technology* (Cambridge: Cambridge University Press, 1978), p. 161.
34. Ibid., p. 38.
35. Hughes Evans, "Losing Touch: The Controversy over the Introduction of Blood Pressure Instruments into Medicine," *Technology and Culture* 34, 4 (October 1993): 803.

36. Ibid., p. 231.
37. David Suzuki and Peter Knudtson, *Genethics: The Clash between the New Genetics and Human Values* (Cambridge, Mass.: Harvard University Press, 1989), p. 207.
38. United States Congress, Office of Technology Assessment, *Biotechnology in a Global Economy* (Washington, D.C., 1991).
39. Nicholas Wade, "Gene Splicing Company Wows Wall Street," *Science* 210 (31 October 1980): 506–7.
40. Dutton, op. cit., p. 209.
41. United States Congress, Office of Technology Assessment, *U.S. Investment in Biotechnology* (Washington, D.C., July 1988).
42. Erwin Chargaff, "On the Dangers of Genetic Meddling," *Science* 192 (4 June 1976), pp. 938, 940.
43. Diana B. Dutton and Nancy E. Pfund, "Genetic Engineering: Science and Social Responsibility," in Dutton, op. cit., p. 199.
44. Ibid., p. 225.
45. Suzuki and Knudtson, op. cit., p. 205.
46. Ibid.
47. Ibid.
48. Thomas H. Murray, "Warning: Screening Workers for Genetic Risk," *The Hastings Center Report*, 13, 1 (February 1983): 6.

PART THREE

TECHNOLOGY AND THE TRANSFORMATION OF WORK

Chapter 8 Work in Nonindustrial Societies

For most people the greatest part of their waking hours is spent doing some sort of work. Be it as an assembly line worker, a sales clerk, or a psychiatrist, we gain our livelihood and a good measure of our individual identity from the work that we do. Technological change has been a major influence on the way we go about our work. Today, few of us are directly engaged in the actual production of things. Nor do we find it necessary to work from sunup to sundown in order to secure our livelihood. All in all, technological change appears to have lightened our workload substantially. Yet by now it should be apparent that the consequences of technological change are not always simple and straightforward. The development of technology over time has not always resulted in diminished workloads; in fact, the opposite has sometimes been the case. In this chapter we will examine the relationship between work and technology before the onset of industrialization. The facts and concepts developed here will provide perspectives that should be helpful when we consider subsequent changes in work and technology.

WORKING WITH THE EARLIEST TOOLS

The history of work is virtually synonymous with the history of the human species. To trace the influence of technology on work we must therefore begin with the first humans. Paleontologists generally agree that primate evolution resulted in the appearance of the first human, *Homo habilis*, about two million years ago. Less than half a million years later, a more advanced human, *Homo erectus*, appeared. The use of technologies that allowed some control over the environment was a distinguishing characteristic of these early people; archeological evidence clearly shows that *Homo erectus* made use of fire and produced simple tools.[1] These tools were improved over the ages, but the techniques used by early humans for artistic and religious purposes were far more impressive than the ones they used for production. The first *Homo sapiens*, who emerged perhaps 40,000 years ago, produced cave paintings of great beauty, reflecting a level of skill not seen in the technologies used to produce the necessities of life.

Although early tools have a crude appearance, this does not mean that the productive technologies early humans employed were primitive. It is important here to draw a distinction between the apparent crudity of the product and the skill that went into making it. Technologies in "primitive" societies are simple because only a few tasks are involved in the productive process, and a single person usually performs all of them.[2] In these societies the division of labor is limited; a worker does everything required to produce something, and the work process is not fragmented into separate tasks.[3] There may be some degree of specialization; a few jobs may be the exclusive work of women or men, or a worker may specialize in the production of a particular good, although again the task is his or hers from start to finish.[4] But most workers in "primitive" societies do not specialize in a single task, so they are not likely to be virtuosos in a particular area. They are, as the old adage has it, jacks of all trades and masters of none.

This lack of specialized skills does not mean, however, that workers in "primitive" societies are lacking in technical ability. Although the overall work process is simple, the individual workers may be highly dexterous and ingenious as they go about their work. Take, for example, the production of stone tools. What could be more primitive than "Stone Age" technology? But in point of fact, the production of stone tools is a considerable achievement. In the first place, the idea of making an implement by chipping a stone is hardly obvious. Although sea otters commonly use a stone for breaking clam shells, modifying a stone so that it can serve as a better tool is unknown in the animal world. Second, the fabrication of stone tools is a task that calls for considerable skill and practice. The effective chipping of a stone tool requires the preparation of a properly sited striking platform on the stone to be chipped, and the delivery of the striking blows at the proper angle. In some cases the stone is chipped by the use of a pressure chipping tool, such as a piece of antler, which forces a flake off as pressure is applied to it. Whatever the method employed, the production of stone tools is not a simple process; most of us would experience considerable difficulty in duplicating the work of early humans and protohumans. At the same time, new generations of toolmakers did not have to develop these techniques by themselves. The regularity of the tools' shape and design suggests that the techniques of toolmaking were transmitted from person to person. An educational process was an integral part of Stone Age technologies.

The use of stone tools by prehumans and early humans was a crucial factor in the steady development of a species of animals that would forever be characterized by its ability to make and use tools. This ability was not simply the result of biological evolution; it was at the same time a cause of it. As Sherwood Washburn describes the process, "Man-apes—creatures able to run but not yet walk on two legs, and with brains no larger than those of apes now living—had already learned to make and use tools. It follows that the structure of modern man must be the result of the change in the terms of natural selection that came with the tool-using way of life."[5]

WORK AND LEISURE IN TECHNOLOGICALLY PRIMITIVE SOCIETIES

Significant as the first tools were for human development, the control over the environment that they offered was modest indeed. It is easy to visualize a life that, in the classic formulation of the seventeenth-century philosopher Thomas Hobbes, was "nasty, brutish, solitary, and short." A "primitive" technology seems to be associated with a precarious and insecure existence, with the daily struggle for mere subsistence occupying all of a person's time and energy. But once again we fail to give technologically unsophisticated people the respect they deserve. Recent inquiries into the actual lives of these people indicate that Hobbes may have been wrong, and that a life of unremitting toil is not a necessary consequence of a low level of technological development.

Societies that sustain themselves through hunting and gathering are rare today. Still, there are parts of the world where agriculture and industry have not taken over, and the ways of earning a living strongly resemble the ones employed by our prehistoric ancestors. One such place is the Kalahari Desert of southern Africa, the home of the !Kung Bushmen. It is a harsh, difficult environment, one that would seem the natural locale for the kind of life described by Hobbes. Since they possess only the simplest technologies, it might be expected that the life of the !Kung is one of unceasing hardship, and that the bare necessities of life are obtained only through continuous toil. Yet this is not the case. As the field research of Richard Lee has demonstrated, the !Kung do not have to work especially hard.[6] By gathering protein-rich mongongo nuts and over eighty other species of food plants, and supplementing these with the meat of game animals, the !Kung provide themselves with a nutritionally adequate diet. The work of gathering and hunting is not particularly arduous. On average, adults put in a six-hour workday about two and a half days a week. Earning a living thus requires only a total of twelve to nineteen hours of labor each week. The remainder of the !Kung's working hours are taken up with such activities as visiting and entertaining friends and relatives, and engaging in dances that put the participants into a trance.

The modest work required to meet the needs of the !Kung has led at least one commentator to characterize the !Kung community as "the original affluent society."[7] Here, affluence is not the same thing as abundance, for the !Kung lack the material possessions common to more technologically advanced societies. A nomadic people, they have no permanent dwellings. Their clothing, as befits life in a warm climate, is minimal. They lack all but the most basic domestic utensils, and they have no way of conveying these except on their own backs.

Besides not producing many material goods, they have not produced many of their own kind. They have controlled their rate of reproduction and have avoided putting excessive pressure on their environment. The population density in their territory is only 41 persons per square mile. In contrast, agricultural societies have densities ranging from a few hundred to several thousand, while urban centers in industrial societies may pack tens of thousands of people into a square mile of space.

What the !Kung do have in great supply is leisure time. This leads to the seemingly paradoxical conclusion that technological progress may result not in the saving of labor but in its increase, or, as Marshall Sahlins puts it, "the amount of work per capita increases with the evolution of culture, and the amount of leisure per capita decreases."[8] Why is this so? The answer lies in the equivocal connection between technological change and the requirements of work. A technologically dynamic economy generates labor-saving devices, but at the same time it produces a steady stream of new goods that are eagerly sought after. This means that labor-saving technologies are generally used to increase income, not to reduce the hours of work. As workers in a technologically advanced society, we often find ourselves on a treadmill, working long and hard to obtain the material goods that we scarcely have the time to enjoy.

The !Kung live lives of marked deprivation when measured against the material standards of most other societies. Their static economy and society have not produced the continually expanding and changing outflow of goods characteristic of technologically dynamic ones. Instead, they have maintained an equilibrium between their wants and their capacity to achieve them. Modest requirements and simple technologies have produced a way of living that lacks physical amenities, but leaves much more leisure time than can be found in our more "advanced" way of life.

WORK AND THE DEVELOPMENT OF AGRICULTURE

The !Kung way of life is a rarity today. Still, it must be remembered that hunting and gathering has been the dominant mode of existence over most of the history of humankind. This age-old pattern began to change less than ten thousand years ago, when people in the Tigris-Euphrates Valley of the Middle East first began to cultivate their food instead of foraging for it. Within a few thousand years farming independently appeared in other parts of the world, most notably in Egypt's Nile Valley, the Yellow River Valley in China, the Indus Valley in India, and parts of Central and South America. These were epochal developments, for in many ways the development of agriculture marked the most fundamental change in human life and the way it is sustained through work.

Why, after tens of thousands of years of successful foraging, did humans take up farming? Much debate still surrounds this question. Some hold to the view that changes in climate, associated with the retreat of glaciers, decimated the wild animal and plant populations. Hunting and foraging became far more precarious, and forced humans to find a new basis of subsistence.[9] Critics of this explanation point out that similar climatic changes had occurred at other times, but had produced no comparable response among early humans. Equally important, they argue, the climate did not radically change during the period when agricultural production was initiated in the hills alongside the Tigris-Euphrates Valley.[10]

Whatever the cause, the ability of people to produce their own food grew rapidly, if not always intentionally. Indeed, the domestication of food grains provides a very early example of technological development proceding partially by fortunate accident. The wild grasses that were the progenitors of early wheat and barley reproduced through the dispersal of their seeds by the wind. But some of these wild plants had seed-holding spikes that did not easily detach themselves, and because they had not already blown away, a disproportionate number of these were harvested by human gatherers. When humans began to sow wheat and barley, the resulting crops were increasingly made up of plants with tough spikes and intact heads, which made them far more suitable for regular harvesting. Early farmers found it advantageous to sow their crops on level ground, rather than on the higher-elevation hillsides where the grasses normally grew. This gave a further stimulus to the survival of the mutated forms of the wild grains that could prosper in the new environment.[11]

The result was a new symbiosis between plants and people. Domesticated grains could not reproduce themselves as they had done before humans intervened.

People had to assiduously cultivate these descendents of wild grasses, and their lives increasingly depended on regular harvests of grain. Technological progress thus resulted in a greater regularity of work patterns. Also, people had to work harder and more consistently because there were more of them. Agriculture allowed considerably higher population densities than those found in hunting-and-gathering societies, but at the cost of a greater workload for each individual.

FARMING TECHNIQUES AND PATTERNS OF WORK

The interconnections between the development of agricultural technologies, population expansion, and increases in the amount of work performed can be seen through an examination of successive stages of agricultural technology. The earliest form of farming, known as slash-and-burn (or swidden) cultivation, is based on cutting down indigenous vegetation and then burning it on the spot. This not only clears the land, but it also puts nutrients into the soil. A variety of crops are then planted on the cleared land. After harvesting, the plot is abandoned for a period of up to twenty years, which allows the land to replenish itself before the cultivators return to begin the process once again. Slash-and-burn cultivation stands in an intermediary position between foraging and more intensive forms of farming, and it likely antedated the more settled forms of agriculture that were practiced in the aforementioned river valleys of India, China, the Middle East, and Central and South America.[12]

Although slash-and-burn cultivation alters the environment more than foraging does, on the whole it is less destructive than other forms of farming. The soil is given ample time to recover, and every effort is made to nurture the tree seedlings that will regenerate the forest. The natural variety of plant life is preserved, thus avoiding the radical simplification of the environment characteristic of settled forms of farming.[13] At the same time, slash-and-burn agriculture supports population densities greater than those found in hunting-and-gathering societies—up to 150 people per square mile—but with little increase in work effort. Slash-and-burn agriculture requires an annual average of 500 to 1,000 person-hours of labor be expended on all phases of crop production.[14] This comes to a little more than nine and a half to nineteen hours of work a week, figures that compare quite favorably with the work schedules of the !Kung. In some societies, such as the Bemba of Zimbabwe, even less work is required. Only three to four hours of work is performed each day even during the busiest agricultural seasons, and averaged over a year, only one to two hours per day are required.[15]

Slash-and-burn agriculture requires less effort than more "advanced" forms of farming, but its drawback is that it cannot support large populations. One study of three Mexican farm communities that employed different mixes of slash-and-burn and sedentary farming technologies showed that almost fourteen times as many families could be supported on the same acreage when irrigated farming was used instead of slash-and-burn methods.[16] This, in fact, may be the primary reason for the development of settled, intensive agriculture. People began to practice this form of cultivation (perhaps while continuing to employ slash-and-burn techniques on other plots of land) in order to supply food to a growing population when no new land was available for cultivation. Settled forms of agriculture allowed more people to be fed because farm work was not dependent on the availability of vast tracts of infrequently cultivated land. The same farm plot could feed much larger numbers of people whose labor was in turn required by the much greater demands of sedentary farming. In extreme cases, such as that of twentieth-century China, with labor-intensive agricultural technologies the land could support enormous numbers of people—6,000 per square mile in some places.[17]

The paradox underlying the development of agricultural technologies until recently is that as the land becomes more productive, people have to work much harder, for the increased productivity of the land is the result of an increase in the number of workers and the amount of work that they do. Settled forms of agriculture require labor not just for the direct tasks of planting, weeding, and harvesting. If the land cannot be left fallow for long periods of time, it is imperative that the farmer take an active role in maintaining the fertility of the soil; this requires gathering, storing, and applying organic wastes that range from animal droppings to the mud scraped from the bottom of ponds. Stubborn weeds and grasses must be removed at regular intervals. Unless rainfall is totally adequate, irrigation networks have to be built and maintained. Animals that provide power and manure have to be cared for. Storage buildings have to be constructed. Farm implements have to be built and kept in repair. Plants have to be protected against insects, even if this means manually picking them off the growing plants.

The adoption of settled agriculture thus signifies a steep increase in the amount of work that has to be performed. Irrigated agriculture in particular requires sharp increases in the amount of work performed. Surveys of farms in India show that with the introduction of irrigation much more labor is required while the crops are growing; work is no longer largely confined to plowing, planting, and harvesting.[18] Twice as much labor may be required per unit of irrigated cropland than in the case of dry farming.[19] All in all, as sedentary agriculture develops, the amount of work required goes up sharply. In Southeast Asia, for example, an agricultural economy based on the transplantation of rice seedlings into irrigated fields requires an annual average of nearly 300 person-days of work. This situation changes only when industrial products such as farm machines and chemical fertilizers are introduced into the farm sector. This began to happen in Europe and North America a century-and-a-half ago, and in most parts of the world it is yet to occur. In these places, the Biblical injunction that "in the sweat of thy face shalt thou eat bread"[20] is a harsh reality.

THE IRONIES OF PROGRESS

At this point, one may question the benefits of an "advanced" technology. The three productive systems just examined represent successive stages in economic development, but it is by no means certain that they represent across-the-board progress, especially when the amount of time that has to be devoted to labor is considered. One cannot even say with certainty that they represent progressive increases in one critical component of technology, the amount of skill exercised by individuals, for the skills employed by foragers seem no less sophisticated than those employed by sedentary farmers. To be sure, the total skills found within the society as a whole may be greater in the latter case, for the permanent communities that accompany sedentary agriculture allow specialization and a more extensive division of labor. Permanency of habitation also stimulates the development of roads and the elements of infrastructure, giving further stimulus to the development of artisan and merchant activities.[21] Merchants and artisans can be supported by the increase in the total productive output of the community, although as we have just seen, much of the productive gain generated by sedentary agriculture is absorbed by population increases.

These comparisons of foraging, slash-and-burn, and sedentary agriculture give a strong indication that technological advance does not necessarily result in the saving of human labor. We are accustomed to thinking that new technologies result in the diminution of human drudgery, but the example of the epochal shift from hunting and gathering and slash-and-burn agriculture to settled crop cultivation shows that this need not be the case. We are left with the sobering realization that the development of agriculture, one of the greatest technological advances in human history, resulted in dramatic increases in the duration, pace, and extent of human labor.

ARTISAN AND CRAFT WORK

Not all of the consequences of sedentary agriculture were so grim. The development of towns and cities was made possible by settled farming, and this in turn allowed some segments of the population to pursue specialized occupations. No longer was it necessary for individual households to do everything necessary for their subsistence. Denser permanent settlements also gave rise to specialized religious practitioners, as well as a more complex hierarchy of political offices. In the economic realm, occupational specialization allowed the production of goods and services that had not even existed before. New and more complex buildings, weapons, foodstuffs, and articles of clothing could now be made.

In some cases, division of labor extended beyond workers specializing in a particular product; the work process was itself broken up. The Greek historian Xenophon noted how on occasion shoe manufacture was subdivided: "one man earns a living by only stitching shoes, another by cutting them out, another by sewing the uppers together, while there is another who performs none of these operations but only assembles the parts."[22] A similar pattern can be observed in Rome during the days of the Republic, when the metalworking industry contained such occupational specialities as pattern makers, smelters, turners, metal-chasers, and gilders.[23]

The change in economic relationships necessitated by specialization was even more important than the products themselves. Specialized producers need to exchange their goods and services with others. This is commonly done through some kind of market. Production for a market instead of for the direct use of family and community members implies a different kind of relationship with others. For most of human existence work was performed as part of an extensive set of mutual obligations shared by members of the same group. But when work is done in order to make goods for sale in a market, it ceases to be embedded in particular social relationships. Instead, it is done as a freestanding activity, governed by self-interest rather than a sense of obligation to members of one's group. The spirit of this kind of work was captured by a famous passage in Adam Smith's eighteenth-century classic, *The Wealth of Nations*: "It is not from the benevolence of the butcher, the brewer, or the baker, that we expect our dinner, but from their regard to their own interest. We address ourselves, not to their humanity, but to their self-love, and never talk to them of our necessities, but of their advantages."[24]

A market-based exchange is one in which both parties participate in a transaction through which each expects to gain. When the transaction is completed, that is the end of the relationship. A car salesman may engage in some friendly conversation about your spouse and kids, but what he is really interested in is your money, just as your interest is getting a good price on the car. The relationship ends when he pockets your money and you drive off with the car.

An exchange system such as this, cold-blooded as it may be, has vast consequences for the way work is done. When, as often was the case in the past, work is firmly rooted in a larger set of social relationships, the way that work is organized will not be governed solely by the technical requirements of the job.[25] More will be at stake then efficiency, innovation, and the general development of production. It may be far more important to see to it that all members of the family or community are employed, or that jobs are parceled out not according to the abilities of the workers, but in accordance with their social position. For example, in India only members of the lowest castes engage in the tanning of leather, for this is considered an "unclean" occupation, and therefore improper for a high-caste individual.

Although urbanization results in more occupational specialization and some disentangling of economic and social spheres, this separation may not be fully realized. Particular occupations may be the special province of specific social groups. In the African city of Timbuctoo, until recently, butchers, barbers, slipper-makers, masons, tailors, and smiths were all hereditary tradesmen, and in the case of the smiths, almost all marriages were between members of families involved in the same occupation.[26] These alignments may be firmly buttressed by religious beliefs and practices that prevent the free movement of individuals into specific occupations. In Timbuctoo it is widely believed that if someone other than a member of the Arna tribe tries to sew up a leather slipper, supernatural forces will cause the needle to jab his hand. In similar fashion, it is believed that one who is not a member of a mason family will suffer a fatal fall if he is so unwise as to attempt to build his own house.[27] In societies such as this, the division of labor and the market system have not completely taken over. Although the economy is fairly complex, economic relationships are still kept within the context of the established social order.

GUILD ORGANIZATION AND TECHNOLOGICAL CHANGE

Families and tribes are not the only social groupings that regulate economic activity in traditional societies. Of considerable importance are organizations known as guilds. Guilds are groupings of people engaged in the same kind of occupation, be it manufacturing, trade, the provision of a service, or even begging and thievery.[28] One of the primary purposes of guilds is to restrict the practice of a particular craft to members of the guild, who are often recruited exclusively from within guild families. Solidarity among guild members is further reinforced by regular festivals and ceremonial functions, such as observance of the feast day of the guild's patron saint. These are not the only sources of solidarity; guilds also have a hierarchy of political offices that set rules, adjudicate disputes among members, and generally regulate the work activities of the members.

Guild organization and its restrictions had significant consequences for the pattern of technological change and for the work activities of guild members. Since the practice of a trade required membership in the appropriate guild, recruitment regulations and stipulated training programs set limits on who could practice a trade. These regulations often resulted in the exclusion of individuals who were not already members of a guildsman's family.[29] Although many guild-regulated occupations did not require an especially high degree of skill, the exclusion of outsiders meant that the potential skills that they could have brought to the craft were never realized. Moreover, individuals who had not been steeped in traditional craft lore throughout their lives were probably more likely to develop innovative ideas and practices, and their exclusion put a brake on innovation.

Many other guild regulations and procedures worked against technological change. In order to maintain their monopoly positions and keep prices high, guilds set exacting standards for workmanship and attempted to restrict output. Limits were placed on the number of apprentices and other workers that could be employed in a guild-member's shop, and deviations from normal work practices, such as working under artificial light, were prohibited. Technological innovations were discouraged, for the prime concern of the guild was the maintenance of the existing way of life.

The inhibitions created by guild organization meant that economic and technological progress often required the establishment of enterprises in areas remote from guild control. This occurred in the Roman Empire, where craft workshops were established on great estates in the outlying provinces, far from the cities and guild regulation. Here could be found such technological and economic novelties as an establishment near Arles that boasted eight water mills, as well as glass and textile factories in eastern Gaul where work was done by wage laborers rather than independent craftsmen.[30] It probably is no coincidence that the rise of modern industry first occurred in England, where guild restrictions were weaker than they were in the rest of Europe.[31]

SLAVERY AND THE INHIBITION OF TECHNOLOGICAL DEVELOPMENT

One of the most extreme means of tying occupation to social status is through the operation of an institution that is unfortunately all too common in human societies: slavery. The institution of slavery has had a number of unfortunate consequences for the work and the technologies used. When slaves are readily available, there are few incentives to invent and use labor-saving machinery. The economies of Greece and Rome rested upon the work of large numbers of slaves, perhaps a third of the total population, and few machines were invented to ease their labors. In contrast, during the Middle Ages substantial technological development took place, especially in the use of water and wind power to drive machinery that milled grain, sawed wood, forged iron, fulled cloth, and tanned leather. Not coincidentally, this was an era in which slavery had all but disappeared, for the medieval church forbade the enslavement of fellow Christians.[32]

Perhaps an even more important consequence of slavery in both the ancient and the modern world was the attitude toward work that it engendered. The Greeks and Romans of the classical age often manifested a strong respect for work when it was part of a self-sufficient life style. But work was despised when it was done for another—be he a master or simply a customer—because it signified that the worker was locked into a relationship of servitude that prevented him from being a free citizen, whose proper activities were directed toward political life and the pursuit of pure knowledge.[33] This attitude is exemplified by the assertion of the Greek philosopher Aristotle that "No man can practice virtue when he is living the life of a mechanic."[34] When slavery became widespread, as in the days of the Roman Empire, this attitude toward work hardened. Since more and more of it was being done by slaves, manual work was seen as an inherently degrading activity not fit for a free man.

THE MEASUREMENT OF TIME AND CHANGED WORKING PATTERNS

There are two ways to consider the relationship between work and time: in terms of the amount of time that is expended on work, and in the way that it is scheduled. Technological change altered both of these. As we have seen, in technologically simple societies where subsistence comes through foraging or slash-and-burn agriculture, relatively little time is expended on work. With the introduction of settled farming and the beginning of urbanization, the hours of work lengthened. Still, the sharp separation between work and other activities that we take for granted today was not nearly as pronounced. Work was intertwined with social events, religious observances, and community social activities.

Moreover, the scheduling of work activities was much more lax. The workweek was punctuated with religious feast days or other excuses for the avoidance of work. For the farmer, certain times of the year, such as sowing and harvesting, required long hours of work, but during other times of the year some leisure time could be enjoyed. Although a craftman might spend long hours at his place of business, not all of his time was taken up with productive activities. Many hours were spent socializing with neighbors and customers. A good deal of time was spent on haggling over prices with prospective customers, an activity often done more for the sheer pleasure of bargaining than for any real economic purpose. Concern about time did not pervade workers' consciousness. There was little of the frenetic desire to "get ahead" that is so typical of many inhabitants of the modern world. Jacques Le Goff's description of the Middle Ages can be taken as typical of the general attitude: "On the whole, labor time was still the time of an economy dominated by agrarian rhythms, free of haste, careless of exactitude, unconcerned by productivity—and of a society created in the image of the economy, *sober and modest*, without enormous appetites, undemanding, and incapable of quantitative efforts."[35]

Above all, the idea of keeping to a fixed work schedule rarely took hold. As Gideon Sjoberg describes work patterns in a preindustrial city, "Merchants and handicraft workers generally do not adhere to any fixed schedule. Shopkeepers open and close their shops as they see fit. They may open one morning at nine, the next at ten, and so on. The lunch hour is likely to be longer on some days than others. Ambulatory merchants, likewise, are apt to keep rather irregular schedules."[36]

These attitudes toward time and work seem to indicate that the producer was not concerned with the maximization of income. Work was not pursued with the kind of single-mindedness we take for granted in the modern world. Nor was the time devoted to work considered a scarce commodity to be carefully utilized in order to bring the maximum economic return. These conceptions of work developed slowly and only in conjunction with major changes in the way people viewed themselves and their surroundings. In the Western world, a major impetus to these changes may have been religious. The rise of Protestantism, especially in its Calvinist form, gave a new centrality to work, for it made work into a quasi-religious "calling." According to Calvinist doctrine only those predestined to salvation could hope to enter heaven, and material success was taken as a sign of being one of the chosen few. Equally important, this "calling" was manifested by a willingness not just to work hard in order to gain material success, but also by a systematic and methodical approach to work.[37] This spirit was clearly congenial to the regulation of work activities according to the dictates of a precise schedule.

THE CLOCK

Important as these changed attitudes might have been, "modern" attitudes toward work, especially the way it was organized and scheduled, also were influenced by a key invention that allowed the precise scheduling of work activities: the clock. To be sure, methods of telling time had existed throughout the ancient world. Sundials, candles with marked segments, and vessels that discharged water at a regular rate were all employed in antiquity. But each had drawbacks: sundials were useless at night and on cloudy days, candles could blow out, and water clocks froze in the wintertime. Equally important, most work activities in the ancient world required little in the way of precise timing and scheduling. These requirements arose only in conjunction with a new kind of social organization: the medieval monastery.[38] By the standards of their day, these monasteries were very large enterprises, many of them containing hundreds of monks and other workers. Their cooperative efforts required precise scheduling, as did their patterns of religious observance, which required regular times for prayers, masses, and other religious observances. This was most pronounced in the monasteries following the Rule of St. Benedict, which divided days and nights into intervals of twelve hours each and required prayers to be said at sunrise, the third hour of daylight, at noon, at the ninth hour of daylight, and at sunset.[39]

The first clocks were built in the thirteenth century, their construction motivated by a growing awareness of the value of accurate timekeeping.[40] The scheduling of activities through the use of special timepieces had brought economic benefits to the monasteries and secular enterprises, and over time the idea of orderly routines and schedules had become an integral part of city life in late medieval Europe. There is the example of Philip VI of France giving to the city of Amiens in the fourteenth century a charter that allowed the issuance of an ordinance "concerning the time when the workers of the said city and its suburbs should go each morning to work, when they should eat and when to return to work after eating; and also, in the evening, when they should quit work for the day; and that by the issuance of said ordinance, they might ring a bell which has been installed in the Belfry of said city, which differs from the other bells."[41] It was in these urban centers during the early fourteenth century that the mechanical clock began to exert its influence. The ringing of the town bell could be specifically tied to the needs of emerging occupations, as when the governor of the county of Artois granted to the government of the city of the same name the right to construct a special belfry because of the "cloth trade and other trades which require several workers each day to go and come to work at certain hours."[42]

In addition to laying the foundation for a more regularized pattern of work, the clock embodied all of the key characteristics of a machine. It used an external source of energy (a spring or a falling weight), unlike traditional tools and devices that required human or animal muscle power. The even distribution of this energy required some sort of regulator, as typified by the pendulum discovered by Galileo, and first brought to practical application in Holland. The clock's operation was automatic, requiring little human intervention. Finally, the clock put out a standard "product"—hours, minutes, and seconds.[43]

In producing this standard product, the clock had made time into a kind of substance. Without getting into deep philosophical waters, we can assert that time as we tend to think of it does not exist; when we think or speak about time what we are actually dealing with are *measures* of time: seconds, hours, decades, or centuries. These units of time are then taken to have real substance, and are treated as valuable commodities. Hence, we use phrases like "saving time," "wasting time," "spending time," and "buying time." Note too the commonly heard assertion that "time is money."

The conversion of time into a commodity reinforces the rational spirit that was briefly described in Chapter 1. It stimulates the conscious choice of specific means for the attainment of given ends, and provides a basis for judging the worth of a particular activity relative to others. Individuals in a society obsessed with time find it difficult to go about their work in an unplanned and irregular fashion.

The precise scheduling of work received further emphasis as new mechanical technologies began to make their appearance. Spinning jennies, power looms, flour mills, and refractory furnaces, as well as the water wheels and steam engines that ran them, called for a mechanically paced way of working that was profoundly different from the old ways of doing things. The process that began with medieval monasteries and thirteenth-century clocks moved inexorably toward the conver-

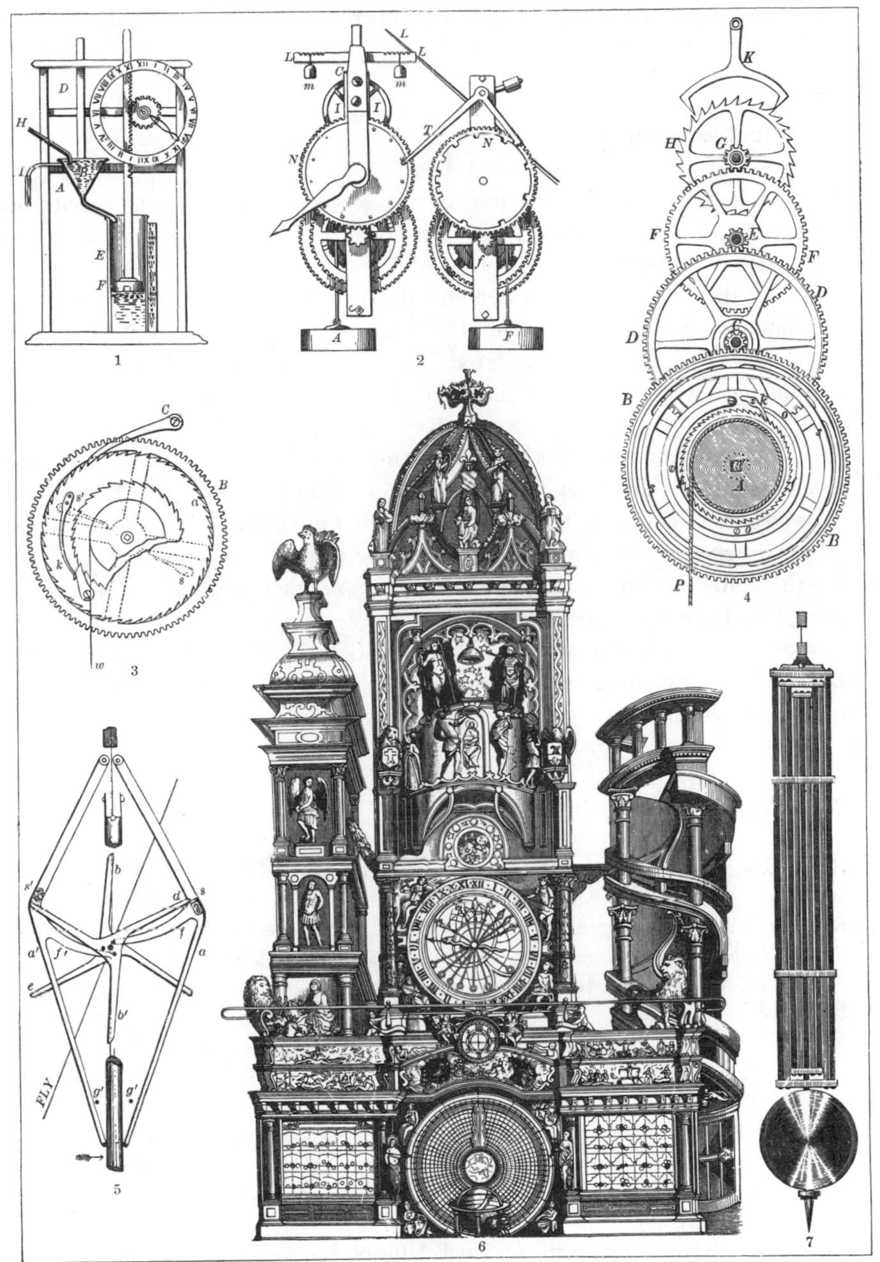

MEASUREMENT OF TIME:—1. Egyptian clepsydra (200 B. C.). 2. Turret-clock of Heinrich von Wick, or De Vick (A. D. 1370). 3. Harrison's maintaining power as attached to weight-clocks. 4. Train of a pendulum-clock. 5. Denison's double three-legged gravity escapement. 6. Great clock of the Strasburg Cathedral. 7. Compensated clock-pendulum.

Early time-keeping devices. (Culver Pictures, Inc.)

sion of most work into routinized procedures governed by artificial time schedules. In Chapter 10 we will look at these historical developments and attempt to bring them up to date. But before analyzing the effects of technological development on the quality of work today, we will examine an issue even more fundamental: the prospect that technological advance might leave many people with no work to do.

Questions for Discussion

1. After reading about the work habits of the !Kung, does trading places with one of them begin to seem attractive? How would a member of !Kung society feel about such a swap? All in all, who would be better off?
2. Which of the following two statements do you find more satisfactory: (1) "the development of agricultural technologies was beneficial to mankind because it allowed more people to be supported," or (2) "the development of agricultural technologies was harmful to mankind because it required people to work harder"?
3. The development of a market economy stimulated technological and economic development, but at the expense of a rather cold-blooded approach to human relationships. Is this a fair trade-off? Under what circumstances can marketplace exchanges coexist with emotionally satisfying interpersonal relationships?
4. Are there any modern occupations that retain characteristics of guild organization? Why have these persisted?
5. How often do you look at a clock or your watch during the day? Is there anything unnatural about being governed by clocks and watches? What would happen to modern society if all the timepieces stopped, never to run again?

Notes

1. Roger Lewin, *Human Evolution: An Illustrated Introduction* (New York: W.H. Freeman, 1984), pp. 55–56.
2. Manning Nash, "The Organization of Economic Life," in George Dalton (ed.), *Tribal and Peasant Economies: Readings in Economic Anthropology* (Garden City, N.Y.: The Natural History Press, 1967), pp. 4–6.
3. Ibid, p. 4.
4. Melville J. Herskovits, *Economic Anthropology: The Economic Life of Primitive People* (New York: W.W. Norton, 1965), pp. 124–152.
5. Sherwood L. Washburn, "Tools and Human Evolution," *Scientific American* 203, 3 (September 1960): 63.
6. Richard B. Lee, "What Hunters Do for a Living, or How to Make Out on Scarce Resources," in Richard B. Lee and Irven DeVore (eds.), *Man the Hunter* (Chicago: Aldine-Atherton, 1968).
7. Marshall Sahlins, "Notes on the Original Affluent Society," in Ibid., p. 85.
8. Ibid., p. 86.
9. Marvin Harris, *Cannibals and Kings: The Origins of Culture* (New York: Random House, 1977), pp. 21–30.
10. Robert J. Braidwood, "The Agricultural Revolution," *Scientific American* 203, 3 (September 1960): 134.
11. Hans Helbaek, "Domestication of Food Plants in the Old World," *Science* 130, 3372 (14 August 1959): 365–372.

12. Ester Boserup, *Population and Technological Change: A Study of Long-Term Trends* (Chicago: University of Chicago Press, 1981), p. 17.
13. Roy A. Rappaport, "The Flow of Energy in an Agricultural Society," in *Energy and Power* (San Francisco: W. H. Freeman, 1971).
14. Robert McN. Netting, *Cultural Ecology* (Menlo Park, Calif.: Cummings, 1977), p. 62.
15. Boserup, op. cit., p. 46.
16. Netting, op. cit., p. 68.
17. R. H. Tawney, *Land and Labor in China* (London: George Allen & Unwin, 1932), p. 24.
18. Boserup, op. cit., p. 52.
19. Ibid., p. 39.
20. *Genesis* 3:19.
21. Boserup, op. cit., pp. 71–72.
22. Quoted in Claude Mosse, *The Ancient World at Work* (New York: W.W. Norton, 1969), p. 79.
23. Ibid., p. 104.
24. Adam Smith, *An Inquiry into the Nature and Causes of the Wealth of Nations* (New York: Random House, 1937), p. 14.
25. Stanley Udy, Jr., *Work in Traditional and Modern Society* (Englewood Cliffs, N.J.: Prentice-Hall, 1970), p. 66.
26. Horace Miner, *The Primitive City of Timbuctoo* (Garden City, N.Y.: Doubleday, 1965), pp. 53–55.
27. Ibid., p. 57.
28. Gideon Sjoberg, *The Preindustrial City: Past and Present* (New York: The Free Press, 1960), p. 187.
29. Mack Walker, "Hometowns and Guilds in Early Modern Germany," in Edward Shorter (ed.), *Work and Community in the West* (New York: Harper & Row, 1973), pp. 40–41.
30. Mosse, op. cit., pp. 106–107.
31. Ibid., p. 45.
32. Marc Bloch, *Land and Work in Medieval Europe* (New York: Harper & Row, 1969), pp. 181–182.
33. Mosse, op. cit., p. 45.
34. *Politics* 1278. Quoted in Alison Burford, *Craftsmen in Greek and Roman Society* (Ithaca, N.Y.: Cornell University Press, 1972), p. 34.
35. Jacques Le Goff, *Time, Work, and Culture in the Middle Ages* (Chicago: University of Chicago Press, 1980), p. 44 (author's emphasis).
36. Sjoberg, op. cit., p. 209.
37. Max Weber, *The Protestant Ethic and the Spirit of Capitalism* (New York: Charles Scribner's Sons, 1958), p. 161.
38. Lewis Mumford, *Technics and Civilization* (New York: Harcourt, Brace and World, 1934), pp. 12–18.
39. Arnold Pacey, *The Maze of Ingenuity: Ideas and Idealism in the Development of Technology* (Cambridge, Mass.: The MIT Press, 1976), p. 63.
40. David S. Landes, *Revolution in Time: Clocks and the Making of the Modern World* (Cambridge, Mass.: Harvard University Press, 1983), pp. 53–82.
41. Le Goff, op. cit., pp. 45–46.
42. Ibid., p. 46.
43. Mumford, op. cit., pp. 14–16.

Chapter 9 Technology and Jobs: More of One and Less of the Other?

Mention technological change to some workers, and a cold chill is likely to go down their backs. Lurking behind many an innovation is the threat of a job lost. This is not an irrational fear. Throughout history, but especially in the modern era, many production technologies have been explicitly motivated by the desire to increase productivity. Since productivity is usually measured in terms of output per worker, the consequences of productivity increases seem clear: the reduction or elimination of labor. After all, if production levels remain constant, productivity improvements must reduce the number of workers used.

There can be no question that technological development has led to spectacular improvements in worker productivity. In 1900, an hour's labor resulted in the equivalent of a dollar's worth of goods and services; seventy years later, that hour produced the equivalent of more than six dollars, even after adjusting for inflation.[1] But while productivity gains have made our material lives much richer, at the same time they have raised the specter of mass unemployment. If a relatively few workers can perform all the necessary labor, large numbers of people will end up with no jobs and no work to do. At least that is the fear. Fortunately, the situation is more complicated. Technology's effects on employment are substantial, but they are also contradictory. In this chapter we will trace the complex connections between technological change and unemployment by looking at the historical record, and by considering where the productivity improvements wrought by technological change seem to be leading us.

THE THREAT OF MACHINE-MADE UNEMPLOYMENT

A cursory examination of technological change reveals many examples of jobs lost because of technological advance. Between 1966 and 1978, computerized typesetting equipment resulted in the reduction of typesetters' jobs from 8,000 to 3,800 in New York alone.[2] Through the use of computers, one bank in St. Louis was able to increase transactions by 35,000 each day while employing 10 percent fewer tellers.[3] During the early 1980s studio musicians saw recording jobs erode by more than a third as a result of the growing use of musical synthesizers.[4] The Ford Motor Co. made about as many vehicles in 1988 as it did in 1978, but employed only half as many production workers.[5]

The effects of technological advance have not been confined to specific firms and occupations; entire industries have undergone a contraction of their work force while their production has increased. Between 1947 and 1957, 10,000 fewer oil refinery workers were employed, even though operating capacity rose from 5.3 to

8.4 million barrels per day.[6] From 1955 to 1960, production in the U.S. chemical industry rose 27 percent but the number of workers declined by 3 percent. During this same period, steel production increased 20 percent, yet the number of workers declined by 17,000, while in the meat packing industry 28,000 jobs were eliminated despite increasing levels of production.[7] 180,000 jobs were lost in the lumber industry between 1950 and 1960, again despite a substantial increase in production.[8] From 1972 to 1977, 21 percent more telephone calls were made throughout the Bell system, yet the number of operators fell by 32 percent.[9]

THE TECHNOLOGICAL THREAT IN HISTORICAL PERSPECTIVE

Striking as these developments have been, they are not unique to our times. Beginning in the late eighteenth century, the power loom forced many handloom weavers out of their traditional craft, while those that stayed on did so only because they were willing to accept miserably low wages.[10] Many similar examples could be extracted from the history of the Industrial Revolution. The introduction of machinery threatened the livelihood of many workers and produced a considerable amount of social unrest. Attempts to introduce the spinning jenny into the English woolen trade during the late eighteenth century resulted in numerous riots. By the early nineteenth century worker resistance to new machinery was widespread in several regions of England, culminating, as we saw in Chapter 2, in the machine-smashing Luddite outbreaks.

Opposition to new technologies because of their consequences for employment has not been confined to the workers whose jobs were threatened. The fear of the effects of new technology on employment has on occasion moved public officials to take drastic actions. In 1638 the British government banned the use of "engines for working of tape, lace, ribbon, and such, wherein one man doth more amongst them than seven English men can doe."[11] A more extreme (although perhaps apocryphal) example comes from the Polish city of Danzig, where in 1661 the municipal authorities destroyed a mechanical ribbon loom and drowned its inventor, for fear that the new device would put hand weavers out of work.[12]

While officials acted in response to the threat of technologically induced unemployment, scholars agonized over the presumably negative consequences of improved productivity. To take one example, in 1878 W. Godwin Moody, in a pamphlet sponsored by the American Social Science Association, expressed the fear that consumption patterns were not keeping up with improved methods of production, thereby raising the specter of high levels of unemployment. Moody's proposed solution to this problem was the enactment of legislation that would expand employment by limiting the hours of work.[13]

A CASE FOR OPTIMISM

Examples, however numerous, do not conclusively prove that technological advance is incompatible with high levels of employment. Although there can be no dispute that particular technological changes have resulted in job losses and attendant personal tragedies, this does not tell the whole story. The effects of technological change are complex, and it is easy to fix upon one particular result and ignore others. Rather than dwell exclusively on the fate of particular industries and occupations, it is essential to consider the larger picture.

When considering the effects of technological change on the overall level of employment, one thing must always be kept in mind: even if all available technologies were used to their utmost, there would still be plenty of work to do. In many parts of the world today, including many "developed" countries, there are still large numbers of people who lack the basic necessities and amenities of life. More and better food needs to be produced and distributed, more houses built, more clothing manufactured. People need to be educated, healed, and entertained. Even with high levels of affluence needs and wants do not decrease. When it comes to the consumption of goods and services, it is likely that high levels of consumption do not diminish the desire for still higher levels of consumption; appetite may grow with eating. Although it may not be a particularly noble trait, the desire to acquire more and more is present in most human beings. As a result, progressively higher levels of economic development may generate the need for continued high (or perhaps higher) levels of production.

This tendency can be seen clearly in the field of medicine. Paradoxically, despite stupendous achievements in medical technology, we find that the monetary and human resources being devoted to health care have increased rather than diminished. Indeed, the rising cost of health care, which now takes more than 11 percent of our gross national product, has become a major social and political issue. New medical technologies have generated new possibilities and new hopes. Instead of accepting sickness and death as the inevitable working of fate or God's will, we expect that cures will be found and that medical personnel will be available to administer them. It should therefore come as no surprise to find that employment projections indicate that the demand for health care workers will steadily rise for the remainder of the century and beyond.

As this example indicates, technological advances will not lead to job losses if the demand for products or services increases at the same pace as increases in productivity. It is also important to bear in mind that the level of demand is affected by the price of a particular product or service. If productivity increases bring down the cost, more people can afford it, thereby increasing the level of demand. This in turn may motivate an employer to take on more workers so that the increased demand is met. Much of the history of mass production conforms to this pattern, as exemplified by the automobile industry and the interrelated increases in productivity, employment, production, and purchases that began when Henry Ford began to produce large numbers of Model T cars during the second decade of the twentieth century.

In other cases, lower prices may not result in significantly higher levels of demand because people will buy only so much of a particular product irrespective of its price (in the jargon of economists such a product is "price inelastic"). Even so, since consumers pay a lower price for this product, they have more money left over to spend on other things. This increases the effective demand for other products and enhances the employment prospects in the industries that make them.

As a final point, it is also important to remember that technological change often generates new problems that require more work to be done. Hazardous wastes have to be cleaned up, accident victims need to be restored to health, and a degraded environment returned to its original state. Just to clean up the radioactive mess left by the accident at the Three Mile Island nuclear plant will require the labors of an estimated 10,000 workers for ten years.[14] Reclaiming the many nuclear and other toxic sites situated throughout the country will require the efforts of even more workers.

HOW TECHNOLOGY CREATES JOBS

Fixing upon particular examples of jobs lost to technological advance can lead to a kind of tunnel vision. While some technologies destroy existing human jobs, others produce jobs that had not existed earlier. No one worked as a locomotive engineer before the coming of the railroad, and there was no need for X-ray technicians before the twentieth century. In fact, most of the occupations held today did not even exist a hundred years ago. This trend can be expected to continue. Occupations such as genetic engineers and technicians, holographic inspection specialists, and laser process technicians could hardly be imagined a generation ago, but will become increasingly significant as the technologies that gave rise to them are developed and deployed.[15]

While the historical record shows that new technologies can give rise to whole new industries, it is often difficult to determine in advance the occupational consequences of a new invention. When the transistor was invented, it was at first thought that its primary application would be simple substitution: the replacement of vacuum tubes in amplifiers.[16] Because of its small size and low power requirements, the transistor's first practical application was to make hearing aids more compact and efficient. Today, transistors and other solid-state devices are the foundation of industries unknown a few decades ago—personal computers, cellular telephones, and precision controls, to name but a few.

General technological advance has created a host of new products and the jobs necessary to make them. And in this regard too, actual events have run well ahead of many predictions. In 1963 one economist warned that American households were saturated with domestic appliances, resulting in a stagnant market for manufactured goods; the only significant new product was the electric can opener.[17] Today, the man who made this prediction is likely to live in a house equipped with a food processor, microwave oven, videocassette recorder, compact disc player, and telephone answering machine. These products, all of them virtually nonexistent as consumer items three decades ago, have generated many new jobs in manufacturing, marketing, and servicing.

THE INDIRECT EFFECTS OF NEW TECHNOLOGIES ON EMPLOYMENT

In the past, a single new technology could be responsible for a large proportion of the new jobs that came into existence. Today's economy is much more complex, so no particular new technology is likely to have the same impact. It is highly doubtful that any recent technological innovation will equal the greatest job generator of the twentieth century: the automobile. The automobile is directly or indirectly responsible for at least one out of every seven jobs in the American economy. Hundreds of thousands of people are directly employed in the manufacture of automobiles, and to their ranks can be added legions of mechanics, sales personnel, and insurance agents, as well as those whose jobs are indirectly tied to an automotive culture, such as workers in fast-food restaurants and oil refineries.

None of the new industries that have emerged as a consequence of recent technological change is likely to match this record. The jet airliner, as emblematic of our present transportation system as the automobile, has not led to the direct creation of many jobs; relatively few people are employed by America's airlines and passenger aircraft manufacturers.[18] The computer, often hailed as the most significant invention of our age, has not added a great number of manufacturing and service jobs, and many of the former have been relocated to the low-wage countries of the Third World. The same can be said of the television industry, and just about any of the industries that owe their origin to recent technological changes.

At first glance it looks as though the pessimists have got it right: technological development in established industries destroys jobs, and the occupations created by new technologies do little to offset the losses. But these conclusions follow only if one looks at the specific industries without considering indirect effects on employment. Consider the airline industry again. It is true that despite the industry's dramatic growth in recent decades, employment has gone up only slightly. But at the same time, air travel has stimulated a number of other sectors. Tourism has grown dramatically during the jet age, bringing with it a great many job opportunities in hotels, restaurants, travel agencies, tourist attractions, and the like. Even in many advanced industrial countries tourism has become a very significant industry in terms of money spent and people employed. Television has had a considerable, if unmeasurable, impact on economic growth by stimulating a demand for new products and services that are presented on programs and commercials. In similar fashion, the spread of computers into our economy has been essential to the development of new industries, videogames being the most obvious. Some jobs are lost as computers increase productivity in clerical and other occupations, but at the same time new jobs are created as computers stimulate the growth of new or existing industries.

Since the effects of jet aircraft, television, computers, and other advanced technologies are often indirect, their role in job creation is often obscured. It is much easier to take note of those jobs that have been directly eliminated by a developing technology, for they are dramatic and often well publicized. The indirect effects of technological change on job creation are less easily noted. For example, it is occasionally claimed that the spread of e-mail and other computer-based communication systems will result in diminished use of the postal system and the loss of jobs for hundreds of thousands of postal workers. But at the same time, computerized mailing lists have resulted in a great expansion of junk mail, complete with "personal" messages: "Just think, George Piarkowitz, you and your family at 1128 Majestic Drive may have already won a trip to the Bahamas, a Boeing 767, and a scholarship to dental school!" Not everyone enjoys receiving these letters, but they certainly have kept letter carriers busy, while creating a fair number of new jobs for other people.

THE MACHINES AREN'T READY TO TAKE OVER

When people visualize technological change in the workplace, they often conjure up an image of legions of robots doing what people used to do. There is no denying that robots are beginning to make an impact on factory production. 15,000 robots are at work in U.S. industry—one for every 900 manufacturing production workers.[19] Moreover, robots are only the final step in a more general process of automating production. It is now possible to dispense with a great deal of routine engineering and drafting work through the use of computer-aided design (CAD) systems. These systems make it possible to "draw" a design on a computer screen, change its size and shape, call up stored shapes, and rotate the design to see how it looks from different perspectives. The design can even be tested through the use of computerized procedures. When the design work is completed, the new product can be built by a computer-aided manufacturing (CAM) system, in which computerized instructions run the robots and other machinery that fabricate and assemble the parts that have been designed. Computerized systems can also be used to organize the productive process by scheduling production, directing the movement of parts, and generally coordinating the production process.

Industrial computers are also the basis of flexible manufacturing systems that alter the basic rules of mass production. Instead of using special-purpose machines that do only one thing, it is possible to use general-purpose machines that can be quickly reprogrammed to do a variety of things. For example, whereas a conventional automated system might machine parts for a V-8 engine only, through the use of programmable machine tools and ancillary equipment it is possible to quickly change over to the production of four-cylinder engines, or possibly even engines of an entirely different type. This is especially important for the manufacture of products that are made in small quantities or undergo rapid changes in specifications.

At present, these technologies have appeared piecemeal in most industrial settings; they are "islands of automation" on the shop floor. Over time, it is hoped, they will become part of an integrated system that governs the total manufacturing process from initial design to the testing of the completed product. When this occurs, a great deal of production will take place with very little direct human intervention.

As exciting as these developments are, they are not likely to affect manufacturing substantially for many years to come. Flexible manufacturing systems are still very expensive, with each machining workstation costing between $600,000 and $800,000. The minimum cost of a flexible manufacturing system is between $3 and $4 million. At the same time, many technical difficulties remain unsolved. There are still many problems surrounding the transport of materials from one station to another without human intervention. A good deal of development work still needs to be done on sensors to monitor the wear in cutting tools so that necessary adjustments can be made automatically. There are still no machines that can reliably remove metal chips during the cutting process.[20] Perhaps far in the future manufacturing processes will be completely automated, but for now heed must be taken of warning given by two experts in this field: "Manufacturing operations may seem simple—until you try to reduce them to computer programs. Then cleaning up metal shavings, aligning odd-shaped parts, and routing work around broken equipment do not turn out to be so easy."[21]

Even more problematic, manufacturing systems that rely on automated production often operate on the thin edge of disaster. If a robot or any other computer-controlled process fails, production comes to a complete stop as everything stacks up behind it. Automated manufacturing systems promise lower labor costs, but there are limits to how far they can go in eliminating human labor. Flesh-and-blood workers are of critical importance for the operation, repair, and unjamming of these sophisticated production technologies.[22] The dream (or nightmare) of factories operating without human workers is not likely to be realized for many years to come.

TECHNOLOGY, JOBS, AND THE CHANGING STRUCTURE OF THE ECONOMY

Although completely automated industrial processes are a long way from realization, it still must be admitted that they have already reduced employment levels in the manufacturing sector. According to one estimate by the Congressional Budget Office, 3 million manufacturing jobs, representing 15 percent of the work force, will be lost in the near future as computerization spreads.[23] Robots, computers, CAD-CAM systems, and allied technologies are eating into the jobs previously done by factory workers, seemingly ushering in an era of widespread unemployment.

Alarming as this sounds, it is nothing new. In the last hundred years or so, two-thirds of the pre-1880 jobs have been lost to mechanization, yet at the same time total employment has greatly expanded.[24] This has happened because the bulk of job creation, especially in recent decades, has not been in enterprises that make physical products, but in those that supply services. In 1948, manufacturing, mining, construction, and farming contributed 46 percent of America's gross national product (GNP). By 1993 these industries' share had fallen to 32.4 percent, while the portion taken by services went from 54 to 67.5 percent.[25] Similar changes can be seen in the proportion of the work force employed in these sectors. In 1948, 20.9 million workers were employed in construction, farming, raw materials extraction, and manufacturing, while 27.2 million worked in the service sector. By 1990, after 74.5 million people had been added to the work force, these figures were 28.2 million and 84.4 million, respectively. Manufacturing jobs did not decline in absolute terms, but they did fall sharply relative to services. Between 1948 and 1990, the manufacturing work force gained 3.6 million workers, while in the same period the service sector grew by 57.2 million workers, a ratio of 1 to more than 16.[26]

Even these figures may underestimate the extent to which services have become numerically dominant in the American economy. Many of the jobs found in the manufacturing sector are actually service occupations. Independent consulting engineers provide a service, and are reckoned as service workers, but should they perform similar duties as members of the engineering staff of a manufacturing enterprise they would appear as manufacturing employees. The number of workers performing service-type functions in the manufacturing sector is large and growing; according to one estimate, in 1979, 28 percent of America's manufacturing work force was composed of workers not directly involved in production, an increase of 10 percentage points in thirty years.[27]

Many things fall under the category of services, and not all of the jobs in this sector of the economy are equally desirable. Shoeshine boys perform a service, just as neurosurgeons do. What many of these varied services have in common, however, is that they incorporate a personal element. This, in fact, is often what makes them desired by consumers who already have a surfeit of manufactured goods. For many people, the most sophisticated machine is no substitute for personal contact. This fact was underscored by the failed attempt of New York City's Citibank to require customers to use automatic teller machines when their transactions amounted to less than $5,000. The howls of protest from the bank's clientele soon led to the scrapping of this policy.

The desire for services shows no signs of abating, and this sector will generate many future jobs. Still, relying on services to drive the economy and provide jobs seems to violate common sense. The basis of human survival is food, clothing, shelter, and the implements necessary to sustain life. How can the majority of the work force be employed in occupations that directly contribute nothing to production? In fact, much economic progress can be interpreted as an increasing ability to reduce the number of workers employed in producing the necessities of life. Two hundred years ago, about 70 percent of the population of Europe earned their livelihood by growing food, just as is the case in China or India today. With the rise of industrialization and the provision of such things as irrigation pumps, mechanized farm implements, and chemical fertilizers, the labor requirements of agriculture steadily dropped, until today less than 5 percent of the work force in most industrial countries is engaged in farming. There is no reason that the same thing cannot happen in manufacturing.

The supplantation of agriculture by industry was closely tied to the rapid progress of technology during the nineteenth and twentieth centuries. In the case of the service sector, however, modern technologies are often conspicuous by their absence. This is a major reason for the slow growth in productivity that has been characteristic of the American economy in recent years. While manufacturing employees have made much use of machines that embody increasingly sophisticated technologies, many service occupations have been only lightly touched by them. As a result, the desire for more services can be met only by employing more workers. According to the projections of a recent government report, the occupations showing the most substantial employment growth will be salespersons, waiters and waitresses, nurses, janitors, and cashiers.[28]

Perhaps the clearest example of the labor-absorbing quality of services can be found in education. Despite efforts to use computers, audio-visual aids, and machine-scored exams, teaching is conducted much as it was in the time of Socrates: through lectures and discussions, both of which require some degree of personal contact. Much of the appeal of a service has to do with the human contact it provides. As a result, the more "productive" a service job is, the less satisfying it may be for the consumer. A teacher could greatly increase his or her productivity by lecturing to a class of 500 rather than 25, but most students would find so "productive" a class much less enjoyable, and the lack of personal attention would probably result in lowered levels of learning for many of them.

To be sure, a good many services have experienced sizeable productivity gains through the introduction of new technologies. The classic example of this can be found in the telephone industry. If telephone calls were still manually switched as they were in the 1920s, today's level of telephone traffic would require no fewer than 50 million operators. Although the impact may not be as profound, it is likely that many of today's services will also experience technologically induced productivity gains and corresponding drops in employment. Peter Drucker has estimated that by the end of the century up to 15 million office jobs may be rendered redundant by electronic mail, word processing, automated banking, and the like.[29] But does this mean that there will be a net loss of 15 million jobs? Here, again, the example of the telephone industry may be relevant. Some actual, and millions of potential, jobs were lost due to the introduction of automatic telephone switching, but total employment in the telephone industry continued to rise because increased levels of demand stimulated the industry's expansion. Equally important, the expansion of telephone communications has acted as a powerful stimulant to the economy as a whole, leading to the creation of many jobs that would not otherwise exist. The job-creating qualities of a new technology are usually hard to predict in advance, but they are there all the same.

SOFTENING THE BLOW

It is not the intention here to replace an immobilizing pessimism with a naive optimism. Technological changes do destroy certain jobs by increasing productivity and rendering certain skills obsolete. On balance, technological advance expands employment opportunities by stimulating particular industries and the economy as a whole, but this fact may not help the formerly well-paid auto worker whose job is now being done by a robot welder. Although qualified workers will be needed to program and repair the robot, such jobs are not likely to go to the former auto worker. They may go to his son or daughter, but for a mature worker with obsolete skills, the future is bleak unless training programs and some form of income supplement can help him through a difficult period of transition. A general optimism regarding the consequences of technological change for employment should not obscure the fact that adjustments will not be smooth and automatic. There will always be a need for purposive action if technological change is not to result in a great number of individual tragedies alongside a general economic advance.

In fact, there have been few effective programs to mitigate the inevitable destruction of some jobs by technological advance. When the necessary adjustments have been made, they have been done on an *ad hoc* basis. A few well-organized workers have been able to nullify the job-destroying effects of technological advance by striking advantageous bargains with their employers. For example, locomotive firemen were left with little to do when the diesel-electric replaced the steam locomotive, but they were able to negotiate contracts that preserved their jobs.

In England until recently, powerful trade unions in the newspaper industry were able to block the setting of type by electronic means. Printers in the United States were not as successful in hanging on to their jobs, but they often were able to get something in return for jobs lost to new technologies. In the last twenty years, half of the jobs in America's newspaper composing rooms have been eliminated by computerized typesetting, but some displaced workers were retrained to perform new jobs. Many workers who were terminated received substantial monetary compensation. Union workers received an average severance payment of $14,000, while nonunion workers received an average of $7,500. Some workers received as much as $85,000 in compensation for the loss of their jobs.[30] It must be emphasized that these cases are not typical; most holders of jobs lost to technological change have not fared nearly as well as locomotive firemen and newspaper printers. Nevertheless, these examples do show that job loss as a result of technological advance need not be totally traumatic.

At the same time it can be easily argued that make-work schemes and generous severance allowances do not address the real problem of technologically induced job loss. Most people seem to have a real need to work, and to do work that has some intrinsic meaning. At the same time, as was noted above, technological change has not eliminated the need for work. The problem, then, lies in putting the two together by preparing workers whose jobs have been eliminated by technological change to take on jobs for which a real need exists. In an era of rapid technological change there is a fundamental need for retraining programs that help displaced workers learn new skills and prepare them for new jobs.

Unfortunately, current training programs in the United States have not been up to the task. Although the federal government has a number of job training programs in place, they are still restricted in scope and scale. One successful program that was initiated by the Job Training Partnership Act of 1982 has served only 5 percent of workers eligible for it.[31] In part this is due to the chronic underfunding of such programs, resulting in their inability to meet demand for them. At the same time, many workers fail to take advantage of them because these programs do not take into account the skills and competencies of the workers. Many of the jobs eliminated by technological advance have been held by unskilled workers with limited education. A basic remedial program may thus be an essential component of a successful retraining program. And, sad to say, many young people are entering the labor force with serious deficiencies in essential skills. According to a test given in 1985 to Americans in their early twenties, 6 percent could not read at a fourth-grade level; 20 percent could not read at an eighth-grade level; nearly 40 percent were below an eleventh-grade level; 43 percent had trouble following directions on a map; and 20 percent could not decipher a bus schedule.[32] Workers operating at these low levels are especially likely to be victimized by technological change, and it will be a major challenge to help them develop the skills that allow them to adjust to a technologically dynamic economy.

Many retraining programs also suffer from an inadequate understanding of actual employment needs. Some retraining programs take workers who have lost their jobs because of technological change, and then train them for nonexistent jobs. Particularly sad cases can be cited of factory workers who were displaced by robots and then trained to be robot technicians. The demand for this type of work is limited so all that often happens is that retraining converts a jobless unskilled worker into a skilled worker still without a job.[33]

In the United States there have been no coordinated efforts to deal with technologically induced unemployment. A patchwork of programs exist to provide labor market information, set labor standards, train workers, stimulate the creation of new jobs, and dole out unemployment benefits, but these are not systematically integrated. As one government report sums up, "Compared with policies in most European countries and Japan, U.S. labor market policy is reactive and uncoordinated, and it is not linked to other, industry-oriented programs for structural adjustment in the national economy."[34]

BENEFITS, BUT DISRUPTION TOO

All in all, the fear that technological advance will lead to widespread unemployment seems unwarranted. Technological change does eliminate specific jobs, but it does not eliminate work itself. While some occupations are rendered redundant by new products and processes, others are being created. To be sure, this is cold comfort to mature workers who have been replaced by new machines. They may find work in America's expanding service sector or even in businesses that have been created by recent technological advances, but in all likelihood they will suffer a cut in pay.[35] Even those who land good jobs may do so at the cost of considerable dislocation. Technological change often results in the rise of some geographical regions and the decline of others. Eighty years ago the rise of the automobile industry stimulated the rapid development of the Detroit area. In recent years the development of the semiconductor industry has produced substantial growth in Northern California's Silicon Valley. Taking advantage of new job opportunities may require pulling up stakes and moving to an unfamiliar part of the country, leaving friends, family, and community behind.

There is no getting around the fact that technological change can be a painful, disruptive process. Occupations rise and fall, and with them the fortunes of many individual people. There are spectacular winners and pathetic losers. Special efforts must be made if the latter are not to be ruined by the advance of technology. Technological change may destroy a career in which one has invested his or her whole life. It would be irresponsible and cruel to turn a blind eye to the needs of workers displaced by technological advance.

At the same time, it would be foolish to attempt to limit the development and spread of new technologies for fear of massive job losses. Far from being the product of technological advance, unemployment is much more likely to occur during periods of technological stagnation. One British labor historian has estimated that during the late seventeenth century half the population of England were paupers because there was not enough work for them to do.[36] This situation began to change only when a series of technological changes helped to produce the Industrial Revolution. After two centuries of technological advance significant numbers of people are still dogged by poverty and unemployment, but there is no denying that things are much better than they once were. Certainly our situation stands in stark contrast to the more technologically backward parts of the world, where vast unemployment and the poverty that accompanies it are the order of the day.

This, however, is not the end of the matter. Although technological advance has expanded employment opportunities for millions of people, it still can be argued that it has led to the degradation of the work they do. People may have jobs, but they are dreary, alienating, and unrewarding. In the next chapter we will try to determine if this is really the case.

Questions for Discussion

1. List a few jobs that have been rendered obsolete by technological change. Are some of these jobs so unpleasant that they are better done by machines? On the whole, have more "good" jobs or "bad" jobs been eliminated by technological advance?
2. What new products and services have been created by technological advance during the past twenty years or so? To what extent are they "essential" parts of contemporary life?
3. Is there anything problematic about an economy that centers on the production of services instead of the production of goods? Is a country with such an economy in an inherently dangerous position when it confronts international economic and political competition?
4. Should federal and local governments come to the aid of workers who have lost their jobs because of technological changes? Is this a legitimate role of government? If so, what sort of policies and programs might be formulated to meet workers' needs?

Notes

1. Sar A. Levitan and William B. Johnston, *Work is Here to Stay, Alas* (Salt Lake City: Olympus, 1973), p. 160.
2. Ian Reinecke, *Electronic Illusions: A Skeptic's View of Our High-Tech Future* (Harmondsworth, England: Penguin, 1984), p. 93.
3. David Treadwell and Tom Redburn, "Workplace: Site of Latest Revolution," *Los Angeles Times* (24 April 1983), part I: 12.
4. Allan Jalon, "Synthesizers: Sour Sound to Musicians," *Los Angeles Times* (6 December 1985): 1.
5. Neal Templin, "A Decisive Response to Crisis Brought Ford Enhanced Productivity," *The Wall Street Journal* (15 December 1992): 1.
6. Walter Buckingham, *Automation: Its Impact on Business and People* (New York: New American Library, 1961), p. 107.

7. Melvin Kranzberg and Joseph Gies, *By the Sweat of Thy Brow: Work in the Western World* (New York: G.P. Putnam's Sons, 1975), p. 177.
8. Bureau of Labor Statistics, "Industrial and Occupational Manpower Requirements, 1964–75," in Howard R. Bowen and Garth L. Mangum (eds.), *Automation and Economic Progress* (Englewood Cliffs, N.J.: Prentice-Hall, 1966), p. 67.
9. Reineke, op. cit., p. 49.
10. Malcolm I. Thomis, *The Town Labourer and the Industrial Revolution* (New York: Barnes and Noble, 1974), pp. 88–105.
11. Malcolm I. Thomis, *The Luddites: Machine-Breaking in Regency England* (New York: Schocken, 1972), pp. 14–15.
12. Witold Rybczynski, *Taming the Tiger: The Struggle to Control Technology* (New York: Viking/Penguin, 1985), p. 103.
13. Cited in Judy Hushbeck, "The Impact of Technology on Consciousness: Some Viewpoints on the American Case, 1875–1930," *Social Science* 49, 1 (Winter 1974): 20.
14. Marvin Cetron and Thomas O'Toole, "Careers with a Future: Where the Jobs Will Be in the 1990s," in Edward Cornish (ed.), *Careers Tomorrow: The Outlook for Work in a Changing World* (Bethesda, Md.: World Future Society, 1983), p. 12.
15. Ibid., pp. 11–15.
16. Ernest Braun and Stuart McDonald, *Revolution in Miniature: The History and Impact of Semiconductor Electronics* (Cambridge: Cambridge University Press, 1978), p. 57.
17. James Fallows, "America's Changing Economic Landscape," *The Atlantic* 255, 3 (March 1985): 54.
18. Barry Jones, *Sleepers, Wake! Technology and the Future of Work* (Melbourne: Oxford University Press, 1982), p. 33.
19. Gregory Stricharchuk and Ralph E. Winter, "Second Thoughts," *The Wall Street Journal* (16 September 1985): 14C.
20. Marjory Blumenthal and Jim Dray, "The Automated Factory: Vision and Reality," *Technology Review* 88, 1 (January 1985): 29–30.
21. Ibid., p. 29.
22. Harley Shaiken, "The Automated Factory: The View from the Shop Floor," *Technology Review* 88, 1 (January 1985): 18.
23. Treadwell and Redburn, op. cit.
24. Max Geldens, "Towards Fuller Employment," *The Economist* (28 July 1984): 19–22.
25. The statistics for 1948 can be found in Eli Ginzberg and George J. Vojta, "The Service Sector of the U.S. Economy," *Scientific American* 244, 3 (March 1981): 48; those for 1993 are calculated from figures that appear in *Statistical Abstract of the United States* (Washington, D.C.: U.S. Government Printing Office, 1993), p. 443.
26. The 1948 figures are from Ginzberg and Vojta, op. cit., p. 49; those for 1990 appear in *Statistical Abstract*, op. cit., p. 410.
27. "Will Services Be the Great Job Spinner?" *The Economist* (3 January 1981): 57.
28. U.S. Bureau of Labor Statistics, *Projections 2000* (Washington, D.C.: U.S. Government Printing Office, 1988), p. 57.
29. Peter Drucker, *The Next American Workforce: Demographics and U.S. Economic Policy* (Washington, D.C.: National Council for Economic Development, 1981).
30. James N. Dertouzous and Timothy H. Quinn, *Bargaining Responses to the Technological Revolution: The Case of the Newspaper Industry* (Santa Monica, Calif.: The Rand Corporation, 1985).
31. *Automation of America's Offices* (Washington, D.C.: Office of Technology Assessment, 1986).

32. "Many Do Poorly on Literacy Tests," *Los Angeles Times* (25 September 1986) part I: 21.
33. Dale D. Buss, "Retraining of Workers for Automated Plants Gets Off to Slow Start," *The Wall Street Journal* (13 April 1983): 22.
34. *Computerized Manufacturing Automation: Employment, Education, and the Workplace* (Washington, D.C.: Office of Technology Assessment, 1984), p. 29.
35. Paul O. Flaim and Ellen Sehgal, "Displaced Workers of 1979–1983: How Have They Fared?" *Monthly Labor Review* 108, 6 (June 1985): 3–17.
36. Thomis, *Town Labourer*, op. cit., p. 147.

Chapter 10 Technological Change and Life on the Job

The word "manufacture" literally means the production of something by hand, and for most of human history, it was just that. Goods were produced in small numbers by individual craftsmen who relied on little more than their own skills and a few tools. This situation began to change markedly in the middle of the eighteenth century as industrialization began to transform the way that things were made. To be sure, this was a revolution that was a long time in the making; people did not pick up their newspapers one day in 1750 and read "Industrial Revolution Breaks Out in England!" Many of the changes that produced an industrial society were small and slow to develop. Others were more dramatic, but even these required numerous incremental changes before they were effective. But taken together these changes utterly transformed the way things were made, and with them the way people worked. This process continues today, with technological advance constantly changing the nature of work. In this chapter we will consider the historical relationships between work and industrialization, as well as where they seem to be headed today.

INDUSTRIAL PRODUCTION

An economy based on industrial production has a number of special characteristics. In the first place, it requires large amounts of energy. Before industrialization, almost all of the energy used came from organic sources: human and animal muscle power and the burning of wood and other plant products. Civilization rested on the physical labor of peasant farmers and artisans (and in many places, slaves), whose efforts produced the surplus that supported the activities of small numbers of artists, priests, and government officials. The use of new sources of energy allowed vast increases in production while at the same time lightening some of the physical burdens of work.

The first new energy source for the emerging industrial economy was water power. Water wheels date back to antiquity, and were widely employed during the Middle Ages; over five thousand of them could be found in England according to the eleventh-century *Domesday Book*. But these were simple devices that produced little power. With the mechanization of key industries such as spinning, weaving, and flour milling, larger, more elaborate wheels began to be used in great numbers. To an increasing degree they were built according to systematic, empirically derived principles, making them significantly more efficient.[1]

Water wheels continued to be important industrial power sources well into the nineteenth century, at which time they slowly began to be supplanted by steam engines. Originally built to pump water out of mines, during the late eighteenth century the steam engine was brought to a higher level of efficiency by James Watt's invention of the separate condenser. Of equal importance, Watt devised new linkages and gearing systems so that the up-and-down motion of a piston could be translated into rotary motion, which was essential for the powering of many types of industrial machinery.

During the Industrial Revolution steam engines and water wheels animated a vast number of new machines. Among the most important of these were machine tools that could produce large numbers of identical parts. The possibilities of this system were first indicated in the early nineteenth century when the Franco-British engineer Marc Brunel teamed up with Henry Maudsley, a pioneer designer and builder of machine tools, to produce large numbers of pulley blocks for the British Navy. After they installed forty-four machines at the Portsmouth (England) Navy Yard, their staff of ten workers was able to produce 160,000 blocks a year, a larger output than the one that had been achieved by 110 men using traditional techniques.[2]

The installation of power-driven machinery was complemented by the establishment of a new setting for productive work: the factory. Instead of the craftsman's shop with its half-dozen or so workers, the dominant manufacturing enterprise was the large factory in which hundreds and even thousands of people performed the tasks that large-scale production required. As industrialization took hold, the size of manufacturing establishments grew rapidly. In 1870 one of the largest industrial enterprises in the United States, the McCormick reaper plant in Chicago, employed no more than 500 workers. Thirty years later, more than a thousand factories had between 500 and 1,000 workers, and over four hundred had more than 1,000.[3] Three steel plants and a locomotive works each had more than 8,000 workers.

With the spread of the factory system, most manufacturing workers no longer worked as self-employed artisans. They now worked for a business firm, and were dependent on it for the tools they used, the facilities they worked in, and of course the money they took home. Independent workers ceased to be the foundation of the economy. The typical worker was now a hired member of a firm, a wage-earner rather than an independent producer. Industrialization had created a society of employees.

The supervision and coordination of large numbers of employees in turn required entirely new methods of organization. Not only were far more people working in a single enterprise, they were also socially and psychologically separate from their employer. Hired through the workings of an impersonal labor market, workers labored in jobs detached from their other social roles and responsibilities. As a result, the face-to-face contact and personal attachments found in the craftsmen's shop gave way to impersonal managerial methods. Hierarchical command structures, written rules, strict job definitions, the precise scheduling of work, and bureaucratic procedures in general became typical ways of organizing production.

MACHINE-PACED LABOR

The spread of industrial technologies helped to make work a more regular and precisely scheduled process. New power technologies were of particular importance in making work follow strict temporal rhythms. Industrial processes that made use of steam engines and water wheels were obviously unsuited to irregularly paced methods of working. Since they were expensive pieces of capital equipment, these power sources and the machines they ran had to be operated constantly if they were to be used efficiently; sitting idle they produced no income. This meant that not only were the hours of work more precisely scheduled, but work had to be performed at all hours. Shift work and labor at unsociable hours made their appearance, a situation that stood in marked contrast to the life of the craftsman working under guild regulations that usually prohibited working at night or under any form of artificial illumination.

Many of the new industries that emerged in the nineteenth century gave further impetus to precisely scheduled, clock-regulated work patterns. This was especially evident in the operation of one of the era's key industries, the railroad. The size of this new enterprise, the multiplicity of tasks required, and above all the need for adhering to precise schedules made the railroad a key instrument for the temporal regularization of work. The emphasis on scheduling and the efficient use of time also had an effect on the larger economy and society. As Henry David Thoreau said of the railroads of his era:[4]

> They come and go with such regularity and precision, and their whistles can be heard so far, that farmers set their clocks by them, and thus one well-regulated institution regulates a whole country. Have not men improved somewhat in punctuality since the railroad was invented? Do they not talk and think faster in the depot than they did in the stage office?

During the railroad's earliest years railwaymen were required to carry "good watches and correct them daily."[5] From that time onward, the conductor's pocketwatch, accurate to a few seconds a day, became emblematic of his occupation. On a larger scale, the demands of railroads led to the establishment in 1883 of Standard Time and the division of the United States into four time zones. No longer would each community observe a unique time that was based on a noontime determined by the highest point of the sun's passage. A new technological system had produced the artificial method of marking time that we take so much for granted today.

The railroad was not the only industry that required high standards of temporal regularity in the workplace. In factories of every sort, workers had to accommodate themselves to the demands of the machinery, and in many cases the machines were harsh taskmasters. And not only was the pace more frenetic and unvarying, the work itself was often highly unsatisfying. For many of the new industrial occupations, the required work often called for only the most rudimentary skills. Many machines were designed so that the skill required inhered in the machine and not in the worker who tended it. Although the traditional labors of artisans and farmers

Machine-paced labor: dressing chickens. (Cary Wolinsky/Stock, Boston.)

also had their periods of drudgery and monotony, the factories that sprang up in the nineteenth and twentieth centuries seemed to usher in a new era of deadening labor—repetitious, narrowly specialized, radically simplified, and paced by the machine.[6]

The employment of large numbers of workers, each performing specialized tasks that had to be closely scheduled and coordinated in accordance with the needs of machinery, necessarily resulted in the loss of worker autonomy. Even Friedrich Engels, Karl Marx's collaborator and a strong advocate of the rights of working people, was convinced that industrial technologies resulted in a regime that was "more despotic than the small capitalist who employs workers ever has been." Engels stressed that industrial technology required strict managerial authority:[7]

> If man, by dint of his knowledge and inventive genius has subdued the forces of nature, the latter avenge themselves upon him by subjecting him, in so far as he employs them, to a veritable despotism independent of all social organization. Wanting to abolish authority in large-scale industry is tantamount to wanting to abolish industry itself, to destroy the power loom in order to return to the spinning wheel.

Industrial technology thus stands indicted as the destroyer of long-established ways of working that had allowed workers to enjoy at least some measure of independence. Machinery and the accompanying factory system increased production, but only at the cost of a cold, regimented on-the-job existence. Many would consider it a poor bargain. But the story is not yet complete; before we blame the advance of technology for the degradation of work, we must consider how these technologies came into being and whose interests they served.

IS TECHNOLOGY TO BLAME?

Industrialization resulted in rapid and sustained economic growth and a massive increase in consumer goods. But at the same time, for many people it meant a thoroughly unpleasant work environment. Were industrial technologies in themselves the cause of the long working hours, monotonous routines, and the general degradation of labor often found in industrial economies? Here we again meet a fundamental question about the nature of technology: does it determine a basic human process such as work, or is technology itself shaped by economic and social relationships?

The history of industrial technologies seems to indicate that similar technologies can support a variety of working arrangements. The use of steam power did not always require massive factories under centralized direction. In the Coventry ribbon industry, weaving was done on power looms situated in blocks of independent workshops that were arranged so that they could make use of a single steam engine.[8] By the mid-nineteenth century, these cottage factories contained over a thousand power looms, as many as could be found in the conventional factories throughout the city.[9]

The Coventry ribbon-weaving industry is not an isolated example. In Sheffield and Birmingham, cities at the very heart of England's industrial revolution, local artisans could rent workspaces supplied with transmission belts connected to a centralized steam engine; "Power to Let" was a sign commonly seen in Birmingham during this period.[10] Fundamental mechanical innovations of the early nineteenth century, such as the Jacquard loom (which introduced the use of punch cards for the guidance of an industrial operation), were used not in large factories, but in small workshops that were often part of the workman's home.

By themselves, these examples do not prove much. As we have seen, a fundamental feature of technological advance is that innovations begin to be used on a small scale and are steadily expanded as experience is gained with them. Still, the late-nineteenth and early-twentieth-century deployment of machines in factories of ever-increasing size was not simply the result of increasing technological sophistication; putting workers into large factories may have been motivated by a need to control and discipline workers.

Employers had long contended with strong-willed, independent workers under a variety of working arrangements. Before the Industrial Revolution one such method was the domestic (or "putting out") system of manufacture. An employer supplied workers with tools and raw materials that were to be used in the workers' own homes. The employer then picked up the finished product several weeks later. This system gave the workers a good deal of autonomy and flexibility—too much, as far as many employers were concerned. The lack of direct supervision often resulted in shoddy products and a slow pace of production. Even worse, some workers bilked their employers by selling the raw materials, using an inferior substitute, and pocketing the difference.

Given these managerial problems, it is easy to see the appeal that the factory had for many employers. Within the walls of the factory the workers were required to put in regular hours while the pace of their work was controlled by foremen who closely supervised their activities. Although perhaps overstated, one scholar's summation captures an essential reason for the development of the factory:[11]

> It was purely for purposes of discipline, so that workers could be effectively controlled under the supervision of foremen. Under one roof, or within a narrow compass, they could be started to work at sunrise and kept going until sunset, barring periods for rest and refreshment. They could be kept working six days a week. And under the penalty of loss of all employment they could be kept going almost throughout the year.

The development of the factory made close managerial control possible. Even so, the closest management by itself cannot ensure regular work efforts. Unless one supervisor can be assigned to each worker, the latter will always find ways to escape scrutiny and evade work if so inclined. In extreme cases, disgruntled workers can even resort to sabotage, a word that derives from the throwing of wooden shoes—sabots—into the machinery. In a setting where workers are employees and have no direct stake in the success of their enterprise, such inclinations are likely to be widespread. In working environments like these, machinery takes on an importance that goes beyond the improvement of productive processes. Machine-based technologies can ensure that work is steadily performed and that it is performed in accordance with the requirements of the management. As one critic of management-inspired industrial technologies put it:[12]

> Machinery offers to management the opportunity to do by wholly mechanical means that which it had previously attempted to do by organizational and disciplinary means. The fact that many machines may be paced and controlled according to centralized decisions, and that these controls may thus be in the hands of management, removed from the site of production to the office—these technical possibilities are of just as great interest to management as the fact that the machine multiplies the productivity of labor.

Machines can be complex and hard to manage, but people are even more difficult. Many employers eagerly sought machine technologies as a way of counteracting the vagaries of human motivation and performance. Machines provided a model of reliable performance, as well as a way of coaxing it out of their workers. The factory system that began to emerge in the late eighteenth century was not exclusively the consequence of the technologies employed; at least as important as the operational requirements of machine-based technologies were the desires and intentions of management.

INDUSTRIAL TECHNOLOGY AND THE DIVISION OF LABOR

A major characteristic of an industrial society is an extensive division of labor. Instead of a single craftsperson performing all of the tasks involved in making a product, production is broken into small segments, each one performed by a different worker. This reduces production costs, as it is not necessary to pay top wages

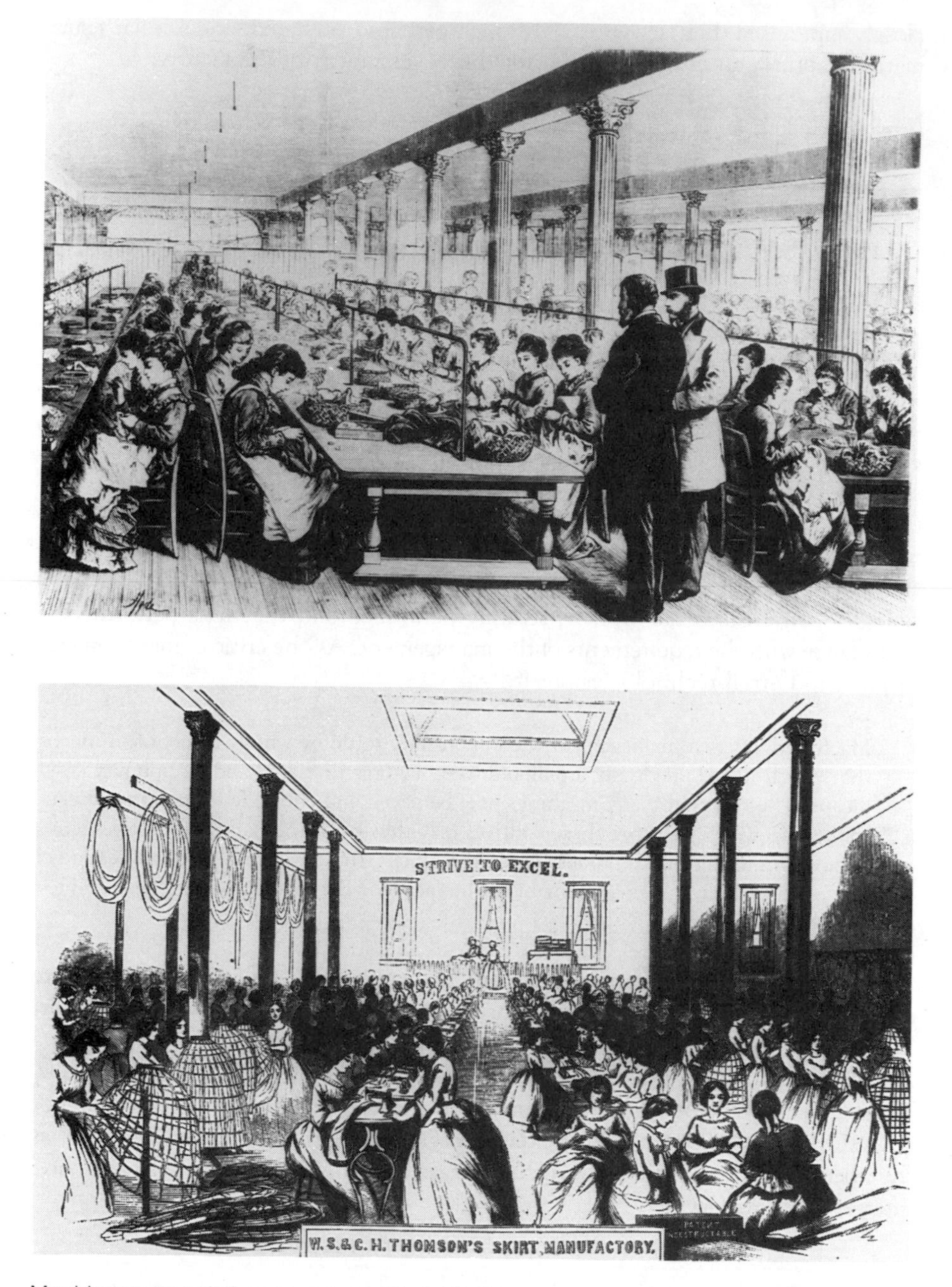

Machinery was not always a prerequisite for factory production. Compare the general layout of two clothing factories, one operating before the invention of the sewing machine, the other afterwards. (By permission of Smithsonian Institution Press from *Engines of Change: The American Industrial Revolution, 1790–1860* by Brooke Hindle and Steven Lubar. Figures 12–4 and 12–5, pp. 210 and 211. © Smithsonian Institution, Washington, D.C., 1986.)

to a highly skilled worker to do everything; some of the work can be done by a person with lesser skills at lower wages.

A productive system based on division of labor requires managerial control, but equally important, it makes such control easier to exert. As long as tasks are kept simple so that workers can be readily hired and fired, there will be little need to depend on workers with all-around skills. This can benefit management because the special abilities of skilled workers generate an independent spirit and a fair amount of economic leverage, as John Delorean's description of the behavior of skilled toolmakers in the automobile industry illustrates:[13]

> These fine old tool makers worked hard and were very proud of their craft. They kept their big tool boxes right underneath their work benches. If you looked at them the wrong way or dealt with them in any manner other than a man-to-man, professional fashion, they would simply reach under their work benches without saying a word, throw their tools into the big box, lock it up and leave. That was it. Each guy figured, "What the hell, I don't have to take this from anybody. I'm a pro. I know this business and I am not depending on you or anyone else."

The ultimate marriage of machine technology and the division of labor is the assembly line. Instead of using a variety of skills and working at their own pace, workers perform specialized and repetitious tasks that are dictated by the tempo of the moving line. The development of this system of production is attributed to Henry Ford, and rightfully so.[14] After taking his inspiration from the meat-packing industry (which of course used a "disassembly" line), Ford and his associates realized productive gains that had scarcely been imagined in the past. Before Ford's introduction of the moving line in 1913, automobiles were produced in small batches by workers with a multiplicity of skills. The work was undoubtedly more satisfying, but the product was high in price and necessarily limited to the rich. The assembly line allowed dramatic price cuts for two reasons: it drastically cut the time needed for manufacture and it allowed the substitution of low-wage, unskilled labor for the high-priced labor of all-around craftsmen.

Ford's description of work in his factory captures the essence of this new form of manufacture:[15]

> . . . one man is now able to do somewhat more than four did only a comparatively few years ago. That line established the efficiency of the method and we now use it everywhere. The assembling of the motor, formerly done by one man, is now divided into eighty-four operations—those men do the work that three times their number formerly did. . . . In the chassis assembling [room] are forty-five separate operations or stations. The first men fasten four mudguard brackets to the chassis frame; the motor arrives on the tenth operation and so on in detail. Some men do only one or two small operations, others do more. The man who places a part does not fasten it—the part may not be fully in place until after several operations later. The man who puts in a bolt does not put on the nut; the man who puts on the nut does not tighten it.

The extreme division of labor found on the assembly line was paralleled by an unrelenting tempo of work, and the absence of any control over it by the workers. Until recent efforts at "job enlargement" began to change the nature of automobile

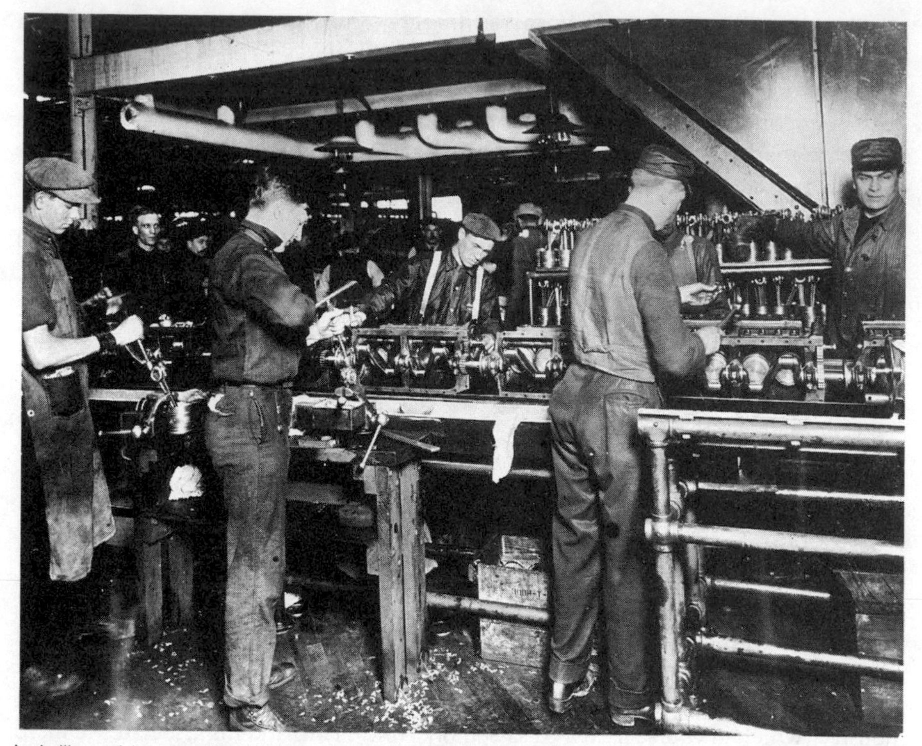

Installing pistons on the Highland Park, Michigan, Ford assembly line, 1914. (From the collections of Henry Ford Museum and Greenfield Village.)

production, manufacturing methods followed the Ford pattern closely. With the typical automobile assembly line churning out fifty to sixty vehicles per hour, each worker had but a few seconds to complete an operation such as mounting a shock absorber or installing a windshield. When problems arose, the worker had to make the best of a bad situation, even if it meant installing something improperly. Under these circumstances work can be little more than a rather unpleasant way of earning a living. Extracting any sense of personal fulfillment from the work performed is just about impossible.[16]

SCIENTIFIC MANAGEMENT ONCE AGAIN

In addition to controlling the pace of work through the use of the assembly line, managers also attempted to regulate work by using elaborate systems that removed all vestiges of worker control over work processes. The most significant of these was Frederick Taylor's Scientific Management, which was briefly discussed in Chapter 1. The important thing to reiterate here is that Scientific Management stipulated the precise scheduling and organizing of work activities, and that these procedures were never to be left to the workers' discretion. As Taylor explained his system, "Perhaps the most prominent single element in modern scientific management is the task idea. The work of every workman is fully planned out by the management at least one day in advance, and each man receives in most cases complete written instructions, describing in detail the task which he is to accomplish, as well as the means to be used in doing the work. . . . This task specifies not only what is to be done, but how it is to be done and the exact time allowed for doing it."[17]

The production of these instructions was the work of a small group of technical specialists who were not directly involved with the work being performed, for as Taylor stipulated, "All possible brain work should be removed from the shop and centered in the planning or laying-out department."[18] Efficient work required unfettered control by specially trained managers, who were to be the repository of all knowledge regarding the work being done: "The managers assume . . . the burden of gathering together all of the traditional knowledge which in the past has been possessed by the workmen and then of classifying, tabulating, and reducing this knowledge to rules, laws, and formulae."[19]

Many would see Taylorism as a perversion of scientific means of inquiry and a travesty of engineering practice. But repellent as it may seem to us today, the spirit of Scientific Management still lives in technologically advanced societies. The explosion of knowledge that is the driving force of our sophisticated economy can fragment the population into groups of specialized experts. This may be a natural outcome of the growth of knowledge, but it takes on a pernicious quality when specialized knowledge is the exclusive possession of one group of people who attempt to use it as a means of dominating others. Under these circumstances science and technology do not simply supply detached knowledge that is used for universal human betterment. They are used to strengthen the position of a dominant group by removing the control over work processes from the hands of those doing the actual work. Technological advance may represent an advance in human knowledge and productive capacity in the aggregate, but for many individual workers just the opposite may occur. "Improved" productive technologies may result in an actual decline of worker skill levels, making workers little more than flesh-and-blood robots.

INDUSTRIAL WORK AND RECENT TECHNOLOGICAL DEVELOPMENTS

Many social critics have decried the human consequences of technologies and managerial strategies that take the skill out of human labor. Others have taken a different tack, arguing that an industrial system that "de-skills" workers is doomed to failure. A great deal of contemporary industrial work, according to this perspective, requires substantial amounts of independence if it is to be done effectively. A modern factory and its constituent departments cannot be expected to hum along indefinitely like a well-oiled machine, nor can all of the jobs be precisely defined and completely subject to control from above. Breakdowns, disruptions, and normal deterioration are inevitable, and they all require prompt attention and often a fair amount of innovative ability on the part of the workers.[20]

Even routine tasks often require knowledge of special quirks in the machines and work processes. This knowledge can be gained only through intimate acquaintance with the work itself; engineers cannot plan everything in advance.[21] If workers are reduced to totally mindless operators, they will be incapable of rising to the challenges of keeping the machinery going. Effective managers realize that the destruction of worker skills through mechanization and rigid centralized controls is self-defeating, for an ignorant and hostile work force means production losses as well as high costs for maintenance and direct supervision.[22]

A precisely programmed production process that makes use of specialized machinery while sharply separating the tasks of those who plan and those who work is possible only when the product is a standardized one. This system generated a great outpouring of mass-produced goods, but it may not be appropriate to today's economy and society. With improvements in the overall standard of living, consumers are less willing to buy the same things that their neighbors do. Instead, they have an appetite for goods with distinctive qualities. These goods cannot be produced by traditional manufacturing processes. As Robert Reich has argued, rigid production technologies were appropriate only when business enterprises were "engaged in high-volume, standardized production, entailing a large number of repetitive tasks with clearly defined goals, in which almost all contingencies could be identified and covered."[23] In contrast, the production of nonstandardized items requires much more flexibility, making it difficult to replace human skills with machines and machinelike processes. At the same time, the continual modification and customization of products and procedures require the ongoing efforts of a variety of workers—designers, toolmakers, computer programmers, systems analysts, setup personnel, maintenance and repair staff, construction workers, and machinists, as well as personnel involved in marketing, distribution, sales, and service. All of these occupations require a fair degree of skill and on-the-job autonomy.

The production of standard products for a mass market creates a fertile ground for the extreme division of labor and the de-skilling of work that was typical of many businesses during the nineteenth century and a good part of the twentieth. But when consumers continually demand new, nonstandard products, existing ways of producing things no longer make sense from an economic standpoint. Firms that cling to the old routines not only oppress their work force; they run the risk of eventual bankruptcy.

TECHNOLOGICAL CHANGE AND WHITE-COLLAR WORK

At this point it can be fairly argued that a concentration on industrial work is misplaced. The number of workers in manufacturing enterprises has steadily declined in relative (although not absolute) terms. Less than 30 percent of the labor force currently is employed in the manufacturing sector. Work on the assembly line, often considered the epitome of industrial work, in reality occupies only 1 out of 150 workers.[24] Due to fundamental changes in the economy, the typical worker no longer toils on some factory floor.

As the number of manufacturing jobs relative to the total labor force has declined, more and more people are now found in white-collar occupations. More workers perform clerical work than any other type of occupation; today, the typical workplace is not the factory but the office. Some analysts have seen this as an indication that the general occupational level has been improved. Moreover, it has been argued that technological change has in many cases increased skill requirements and allowed clerical workers to take on broader responsibilities. In particular, the introduction of the computer is said to be making clerical work a more skilled activity than ever before.[25]

This is undoubtedly true of some jobs, but contrary examples can easily be found, as when general-purpose secretaries are supplanted by specialized typists who spend all day in front of word-processing equipment.[26] At the same time, computer-based technologies can make the work environment more unpleasant by giving management greater opportunities for snooping on workers and constantly monitoring their performance. Some managers seem to believe that adequate performances can best be insured by electronically looking over their employees' shoulders. As one manager puts it, "I'm a great one for believing that people really will do what's *in*spected and not what's *ex*pected."[27] Many word-processing systems include a means of counting keystrokes in order to determine a typist's productivity. Telephone operators are monitored in order to determine average working time and to record how long it takes an operator to answer a call.[28] Monitoring technologies can also be used for jobs outside the white-collar sector. In order to save fuel, microprocessor-equipped "black boxes" have been installed in trucks in order to determine if drivers have exceeded sixty miles an hour. A clothing factory has a system that allows a worker's pace to be monitored to a thousandth of a minute; comparisons of a worker's performance with the factory standard determine the worker's pay.[29] To some critics, therefore, "Computers are the electronic equivalent of an assembly line to mechanically pace and control workers rather than a tool to enhance a job."[30]

Computers and other electronic information systems can affect more than rank-and-file workers; managerial personnel may also feel their impact. Computers can give top management access to up-to-the-minute information that formerly was available only to lower-level management. For example, top management can now quickly find out the daily production output of a particular plant or division, where in the past they had to be satisfied with monthly or quarterly reports. This removes an important source of power that the lower-level managers formerly possessed. At the same time, these managers may be less willing to take risks or make decisions, for they know that they are much more vulnerable to second-guessing by their superiors.[31] Perhaps this is the reason that one study found that of all employees, managers were the most likely to report that computers posed a threat in the workplace.[32]

The experiences of different industries and organizations, as well as those of the workers employed by them, vary immensely, making it impossible to come to a summary judgment of the consequences of technological change for white-collar work. It is also important to keep in mind that personal characteristics of workers, such as their gender and ethnicity, affect the way in which technological changes have altered their jobs. There is some evidence that office automation has been more beneficial to male rather than to female workers. In the insurance industry, for example, high-level positions in the management of electronic data processing systems have become almost exclusively male preserves, while routine clerical work has become an increasingly female occupation. A similar pattern has been observed in the Internal Revenue Service, where computerization reduced the number of middle level jobs, thereby blocking advancement opportunities for a largely female clerical staff.[33] These examples indicate that when a labor market is already segmented along gender or ethnic lines, new technologies can reflect or even reinforce existing divisions. One group may benefit from new ways of doing things, while another group may find themselves in the same, or possibly worse, circumstances.[34]

The differential effects of new workplace technologies can also be seen in the way new computer and communications technologies have been used for work done at home instead of in an office. Ironically, these new technologies allow a return to a mode of employment that the Industrial Revolution had largely displaced: work in the home. As we have seen, this system was riddled with problems of motivation and discipline that were ultimately overcome by the development of the centralized factory and its regimen of regular supervision. Now, with personal computers and modern communications equipment, it is possible for workers to go about their tasks while far from the central office. Their employers need not worry about personally supervising their work performance because it can be electronically monitored.

For some workers (journalists and computer programmers, for example) electronic homework can be a boon, for it allows a way to reconcile the conflicting demands of work and family. Other workers, however, are less fortunate. Home workers doing routine typing and data entry may have to work long hours for low pay and no fringe benefits.[35] Concerns about the re-emergence of exploited domestic laborers led the AFL-CIO to pass a resolution at its 1983 convention calling for a complete ban on "electronic homework." As one spokesman put it, "If history is any guide, we can say with certainty that abuse of electronic homeworkers is inevitable. . . . An early ban would try to prevent a repeat of past experiences in a new guise."[36]

Is history necessarily a reliable guide in this instance? As organized labor's official stance implies, the problem lies not with the new technology itself, but with a form of work organization that maintains wide differences in power between workers and management. Under these circumstances, home workers can be socially isolated, lacking in basic work amenities, and poorly paid.[37] Abundant possibilities for exploitation exist when workers are isolated in their individual homes, are dependent on their employers for their livelihoods, and can be easily replaced. But not all work in the "electronic cottage" conforms to this model. Electronic homework need not be confined to such things as routine word-processing tasks. Jobs requiring a high degree of skill and expertise can also be done in the home. It may even be possible for many specialized services to be supplied by homeworkers who are not employees of the firm, but work on a contract basis, perhaps as a member of a new kind of autonomous cooperative. At the same time, an employee who works far from a firm's central office may not have to be electronically supervised as long as he or she "has bought in, feels the task is important, and has a piece of the action."[38] This is more likely to happen when employees occupy upper-level positions within an organization. For these workers, new technologies can make their work more autonomous, convenient, and fulfilling. For those well down the ladder, the result may be the opposite. Again, we can see that new office technologies do not affect everyone in the same way, and that they can maintain existing patterns of control and subordination.

SMART TECHNOLOGIES AND DUMB JOBS?

As we have seen, optimistic students of industrial change are fond of pointing out that as increasingly sophisticated machines take the place of manual labor, a growing portion of the industrial labor force will be found in jobs that require high degrees of skill. Instead of monotonously tightening nuts and bolts, increasing numbers of factory workers will be engaged in the monitoring of automated processes and taking corrective actions when problems arise. Other workers will be charged with the challenging task of repairing and maintaining the complex machinery that advanced factories require. In this view, industrial skill levels are going up, and the unskilled operative is becoming an anachronism in the modern manufacturing enterprise.

It is undeniable that these trends can be found in many industrial establishments, but to less optimistic observers, predictions of the radical upgrading of factory work are at best premature. Plenty of jobs still require only the most minimal levels of skill and training. The task of monitoring industrial processes can be just as dull as the most routine assembly operation, and in any event, many of these functions can be performed by sensors connected to computers. Even maintenance and repair jobs may represent only a slight improvement in required skill levels. Technological advances can lower maintenance requirements, and when repairs need to be done, they can often be accomplished by simply disconnecting a faulty module and replacing it with a new one. The task of diagnosis, which often requires the greatest amount of the repairperson's skills, has itself been simplified through the development of special test circuits and computer-based "expert systems." To be sure, the need for repair skills will always be present, and it is unlikely that technological developments will allow the elimination of repair personnel. Still, any dilution of the repairperson's skill will have significant consequences for the overall skill levels of the labor force, for these workers currently comprise the largest single group of skilled workers.[39]

Moreover, even if the optimists are largely correct, their predictions only hold for industries that make extensive use of advanced technologies. High-technology factories and offices may be the scene of an expansion of worker skills, but these are not the workplaces of the majority of employees. Nor will new high-technology industries generate many future jobs; according to one projection, only 6 percent of the new jobs created during the 1990s will be in these industries.[40] Many of the jobs characteristic of these industries (systems analyst, laser technician, circuit designer, and the like) will expand in number, but relatively few people hold these positions now, so even a large-scale expansion will not result in the creation of many jobs.

Instead, the bulk of new jobs are likely to require rather modest skills. The U.S. Labor Department's tabulation of the occupations expected to undergo the greatest growth from 1986 to 2000 listed (in descending order): salespersons, waiters and waitresses, registered nurses, janitors and cleaners, general managers and top executives, cashiers, truck drivers, office clerks, food counter workers, nursing aides and orderlies, and secretaries.[41] With the possible exception of managers and executives, and nurses (none of which has a necessary connection with technological sophistication), all of these jobs are traditional occupations that will remain largely unaffected by technological advance. Those looking for great improvements in work skills as a consequence of technological change are likely to be disappointed.

Basic organizational patterns are the most important determinant of the skills used by workers. When there is a strong division between those who manage and those who work, and when managers view their workers as hostile and unreliable, workers will be treated as replaceable parts. Their skill level will be low, and they will be controlled by centralized supervision and machine-paced work processes.[42] By contrast, when a more democratic and egalitarian order prevails, technologies that require the initiative and commitment of workers are more likely to be selected, and workers will be given a chance to develop their skills and take their places as valued members of the organization.

New technologies can be developed to increase levels of skill, or they can be used to diminish them. Like all matters regarding technology, choices have to be made. These choices emerge in an environment where the intentions and relative power of employees, managers, and the owners of enterprises are usually the most important determinants of technological change and its consequences for the worker.

Questions for Discussion

1. Do you agree that many early industrial technologies were used as ways of controlling labor? What sort of historical research could be done in order to test this proposition?
2. In recent years, automobile manufacturers have begun to develop some alternatives to assembly-line production. What has motivated them to do so? What sort of new manufacturing technologies might they use? What problems are they likely to encounter in this venture?
3. Can you see any indications that consumers are showing a preference for nonstandardized products? How will changed consumer preferences affect workplace technologies?
4. According to some predictions, a growing number of jobs will be held by "telecommuters"—employees who work at home while using computers to receive, process, and transmit information. What sort of jobs could be done in this way? Would you like to work as a telecommuter? Why?

Notes

1. See Terry Reynolds, *Stronger Than a Hundred Men* (Baltimore: The Johns Hopkins University Press, 1983).

2. Melvin Kranzberg and Joseph Gies, *By the Sweat of Thy Brow: Work in the Western World* (New York: G.P. Putnam's Sons, 1975), p. 112.
3. Daniel Nelson, *Managers and Workers: Origins of the New Factory System in the United States, 1880–1920* (Madison: University of Wisconsin Press, 1975), pp. 4–7.
4. Quoted in Stewart Holbrook, *The Story of American Railroads* (New York: Crown, 1947), p. 15.
5. Walter Licht, *Working for the Railroad: The Organization of Work in the 19th Century* (Princeton, N.J.: Princeton University Press, 1983), p. 88.
6. Daniel T. Rodgers, *The Work Ethic in Industrial America* (Chicago: University of Chicago Press, 1978), p. 67.
7. Frederick Engels, "On Authority", in Karl Marx and Frederick Engels: *Selected Works* (Moscow: Foreign Languages Publishing House, 1962), p. 637.
8. Malcolm I. Thomis, *The Town Labourer* (New York: Barnes and Noble, 1974), pp. 108–109.
9. Witold Rybczynski, *Taming the Tiger: The Struggle to Control Technology* (New York: Viking/Penguin, 1985), pp. 46–47.
10. Charles F. Sabel, *Work and Politics: The Division of Labor in Industry* (Cambridge: Cambridge University Press, 1982), p. 41.
11. N. S. B. Gras, *Industrial Evolution*, quoted in Sidney Pollard, *The Genesis of Modern Management: A Study of the Industrial Revolution in Great Britain* (Harmondsworth, England: Penguin, 1968), pp. 22–23.
12. Harry Braverman, *Labor and Monopoly Capital: The Degradation of Work in the Twentieth Century* (New York: Monthly Review Press, 1974), p. 195.
13. J. Patrick Wright, *On a Clear Day You Can See General Motors* (New York: Avon Books, 1979), p. 94.
14. David A. Hounshell, *From the American System to Mass Production, 1800–1932: The Development of Manufacturing Technology in the United States* (Baltimore: The Johns Hopkins University Press, 1984), pp. 217–261.
15. Quoted in Alfred D. Chandler, *Giant Enterprise: Ford, General Motors, and the Automobile Industry* (New York: Harcourt, Brace and World, 1964), pp. 39–40.
16. See Charles Rumford Walker and Robert H. Guest, *The Man on the Assembly Line* (Cambridge, Mass.: Harvard University Press, 1953).
17. Quoted in Braverman, op. cit., p. 118.
18. Ibid., p. 113.
19. Ibid., p. 112.
20. Larry Hirschhorn, *Beyond Mechanization: Work and Technology in a Post-Industrial Age* (Cambridge, Mass.: The MIT Press, 1984), pp. 61–86.
21. Ibid., p. 157.
22. Sabel, op. cit., p. 74.
23. Robert B. Reich, *The Next American Frontier* (Harmondsworth, England; Penguin, 1984), p. 81.
24. Sar A. Levitan and William B. Johnston, *Work is Here to Stay, Alas* (Salt Lake City: Olympus, 1973), pp. 99–100.
25. Jon Shepard, *Automation and Alienation: A Study of Office and Factory Workers* (Cambridge, Mass.: The MIT Press, 1971); Barbara Gutek, Tora K. Bikson, and Don Mankin, "Individual and Organizational Consequences of Computer-Based Office Information Technology," in Stuart Oscamp (ed.), *Applied Social Psychology Annual, No. 5* (Beverly Hills: Sage, 1984), pp. 231–254.
26. Jonathan Schlefer, "Office Automation and Bureaucracy," *Technology Review* 86, 5 (July 1983): 36.

27. Michael W. Miller, "Computers Keep an Eye on Workers and See If They Perform Well," *The Wall Street Journal* (3 June 1985): 1.
28. Rybczynski, op. cit., p. 65.
29. Miller, op. cit., p. 15.
30. Quoted in Cathy Trost, "The Price of Progress," *The Wall Street Journal* (16 September 1985): 34C.
31. Shoshana Zuboff, "New Worlds of Computer-Mediated Work" *Harvard Business Review* 60, 5 (September–October 1982): 148.
32. Barbara A. Gutek and Tora K. Bikson, "Differential Experiences of Men and Women in Computerized Offices," *Sex Roles* 13, 3–4 (1985): 133.
33. Roslyn L. Feldberg and Evelyn Nakano Glenn, "Technology and Women's Degradation: Effects of Office Automation on Women Clerical Workers," in Joan Rothschild (ed.), *Machina ex Dea: Feminist Perspectives on Technology* (New York: Pergamon Press, 1983), pp. 59–78.
34. Gutek and Bikson, op. cit.
35. Philip Mattera, "Home Computer Sweatshops," *The Nation* 236, 13 (2 April 1983): 390–392.
36. "AFL-CIO: Ban on Electronic Homework," *Profiles* 2, 9 (May 1985): 37.
37. Barbara A. Gutek, "Women in Clerical Work," in Ann Stromberg and Shirley Harkess (eds.), *Women Working: Theories and Facts in Perspective*, 2d ed. (Palo Alto, Calif.: Mayfield, 1988), pp. 236–237.
38. Gil Gordon, "Micro Futures," *Profiles* 2, 9 (May 1985): 28.
39. Levitan and Johnson, op. cit., p. 90.
40. Bryant Robey and Cheryl Russell, "A Portrait of the American Worker," *American Demographics* 6, 3 (March 1984): 16–21.
41. U.S. Bureau of Labor Statistics, *Projections 2000* (Washington, D.C.: U.S. Government Printing Office, 1988), p. 57.
42. Louis E. Davis and James C. Taylor, "Technology, Organization and Job Structure," in Robert Dubin (ed.), *Handbook of Work, Organization and Society* (Chicago: Rand-McNally, 1976), p. 412.

PART FOUR

COMMUNICATION

Chapter 11 Printing

When I needed a permanent means of recording and subsequently disseminating my ideas about how technology is connected with social change, I naturally thought about writing a book. If I weren't so verbose, I could have written a magazine article. There are, of course, other ways of retaining and transmitting information and ideas. They can be stored in a person's memory and then transmitted orally; indeed, that was the way things were done for most of human history, and still are done in many nonliterate societies. Perhaps sometime in the future, books, newspapers, and magazines will be largely supplanted by videotapes, optical discs, and computer networks. But as of now, nothing equals the printed word as a rapid, cheap, and portable means of storing and transmitting information. Moreover, the importance of printing goes well beyond the storage and transmission of information. The social effects of printing have been immense, making printing one of the most significant technological innovations of all time.

The development of a written language is one of the hallmarks of civilization. While a spoken language is a basic element of all human societies, not all of them developed a way of writing in their languages. Of the approximately 5,000 languages that are spoken today, only a very small percentage appear in written form, and in many cases these written languages have been recent inventions, stimulated by missionaries or other contacts with the outside world. One interesting example of an effort to create a written language appeared in the 1820s, when a Cherokee Indian named Sequoyah single-handedly constructed a written language so that his people could more effectively resist the encroachments of white settlers. A written language was an essential technology, for as Sequoyah had observed, "Much that red men know, they forget; they have no way to preserve it. White men make what they know fast on paper like catching a wild animal and taming it."[1] After much laborious effort, Sequoyah succeeded in constructing a system of writing that could be mastered in a few weeks. Unfortunately for Sequoyah's people, the new written language by itself could not stem the advance of the white man into Cherokee territory.

The writing systems of the great languages of the world were also constructed to serve practical purposes. In ancient Egypt, Babylonia, and China, writing was used to keep accounts and to facilitate communication between government officials. Writing also served a spiritual purpose in many cases, for it allowed the recording and diffusion of concepts that were central to the local religion. Since church and state were tightly interwoven in these ancient states, writing met the needs of the political elite in their dual role as priests and secular officials. As with many technologies, writing owed its development to the needs of one group of people and was used to maintain and extend their influence over others. For example, the priests of Egypt during the days of the pharaohs were able to refer to written calendars that allowed them to predict the cyclical flooding of the Nile, thereby demonstrating their supposed indispensability.

THE PRINTING REVOLUTION

Other than the maintenance of a social and political hierarchy, the effects of a written language were minimal when the written language was understood only by a small segment of the population. Few of the common people had the time and inclination to become literate, and even if they were able to read, the great cost of producing written documents kept these works out of their reach. Written materials were rare and costly because they had to be laboriously drafted and copied by hand. Documents central to a civilization, such as the Bible of medieval Christianity, were scarce and inaccessible, for monks and other scribes had to labor for many months to produce a single volume. Books were rare and precious; two of the largest book respositories in England, the abbey libraries of Canterbury and Bury, each held 2,000 books, while the library of Cambridge University had only 300.[2]

In the absence of widespread literacy, the ability to memorize was developed to a level scarcely imaginable today. People made use of elaborate mnemonic devices in order to recall important facts and concepts. This facility made it possible to preserve laws, ceremonies, and stories in the minds of individuals. It was not unusual to find people who could flawlessly recite sacred texts and heroic sagas that might today occupy hundreds of pages of printed text.

All of this began to change dramatically with the introduction of printing with movable type. As with many other technological innovations, printing originated in the Orient. In the fourth century A.D., the Chinese were copying stone inscriptions through the use of paper rubbings. By the seventh century, Chinese printers were using carved wooden blocks to produce books on agriculture and medicine. Chinese accomplishments using this method were prodigious; during a twelve-year period that began in 971, printers in the city of Chengdu published a 5,048-volume compilation of Buddhist scriptures that required the carving of 130,000 blocks. A significant advance came in the middle of the eleventh century when a Chinese named Bi Sheng invented a system of printing that used individual pieces of clay type set in a wax matrix. Two hundred years later another Chinese, Wang Zhen, invented a rotating type frame that allowed the typesetter to easily pick out the required pieces. The Chinese also experimented with metal type, although it was their Korean neighbors who achieved success in this endeavor during the fourteenth century.[3]

Printing developed more slowly in Europe. By 1400 playing cards and pictures of saints were being printed from wooden blocks in Northern Italy. This method was not suitable for printing the many pages that a single book required. Efficient printing was possible only when a page was set with individual pieces of type, which could be reused after a sufficient number of pages had been printed. As we have just seen, the Chinese had already learned how to do this, but the basic idea had to be independently discovered in the West.

Mass-produced communication: a sixteenth-century print shop. (The Mansell Collection, London.)

Although there are other European claimants to this invention, Johann Gutenberg (1400?–1468?) is usually given credit for devising a workable system of printing with separate pieces of type. In order to put the idea into practice, Gutenberg also had to develop a number of new techniques, such a method to uniformly cast pieces of type. This was done by punching the letter into a copper matrix; the resulting mold was then used for casting the actual pieces of type. Setting these individual pieces of type in a straight line also presented a problem. Gutenberg solved it by designing a mold that kept each piece of type within acceptable tolerances, and by equipping each piece of type with a ridge on one side and a corresponding groove on the other. In this way pieces of type could interlock with one another when they were set in a frame.[4]

Important as the invention of movable type was, the printing of books in large quantities required more than simply composing a page from individual letters. As is always the case with major technological changes, a whole system of interrelated elements had to be developed. Obviously, some sort of a press was required; here the ancient wine-press provided a useful model. New kinds of ink, compounded from lampblack and linseed oil, were also necessary. The production of durable type required considerable experimentation with different alloys; a mixture of tin, lead, and antimony proved to be the winning combination. And to realize the economies offered by printing, costly parchment was replaced by paper, a Chinese invention that had passed through the Islamic world, and began to be produced in Europe early in the fourteenth century.

In a world that produced its goods through individual craftsmanship, the first books printed from movable type were harbingers of the age of mass production.

While the hand-copied and illuminated books of the past were works of art available to only a few, printed books, tracts, and pamphlets were standardized products that reached a large and growing audience. According to one estimate, more than 10 million copies of 40,000 different titles were produced during the fifty years following Gutenberg's invention. Had they been copied by hand, the task would have taken a thousand years.[5] During the sixteenth century, as many as two hundred million books were printed in Europe.[6]

PRINTING AND THE EXPANSION OF KNOWLEDGE

Printing and the consequent expansion of the number of books in circulation did much to transform European society. By systematizing knowledge and making it readily available, printing helped to promote the rational modes of thought that characterize a modern society. Information flowed much more readily, as everything from treatises on accounting to printed musical scores received a wide distribution. With so much information circulating, it was no longer necessary to simply accept the judgment of a few authorities, and a more critical spirit emerged. Printing also allowed the widespread distribution of maps and accounts of explorations. As a result of this, new geographical discoveries could be widely publicized, and errors and inconsistencies could be made evident through the cross-checking of existing maps and printed descriptions.[7] The clearer definition of space made possible by printed maps gave a great impetus to voyages of discovery that opened the world to European civilization and vice versa. Better maps made journeys less hazardous, while the ongoing definition of the world's spatial contours motivated explorers and colonists to fill in the remaining blank spaces on the maps.[8]

While explorers were making fundamental geographical discoveries, others were gaining new insights into how the world worked. The great advances in scientific knowledge that began to emerge in the fifteenth and sixteenth centuries were pushed forward by printing. As with the progress of geographical exploration, the advancement of science was to a substantial degree the result of the spread of information through print media. Science requires accurate data, but before the age of printing, information was scattered, quirky, and often highly inaccurate. To be sure, much of the information disseminated by printing was bogus, but because information reached a much greater audience, it was subject to informed criticism, making for greater accuracy. In short, printing allowed the establishment of an intellectual feedback system that helped to drive out errors.[9]

More was involved than the transmission of factual information. Human knowledge was significantly advanced by a greater exposure to different theories, perceptions, and interpretations. As Elizabeth Eisenstein has summarized the process:[10]

> Much as maps from different regions and epochs were brought into contact in the course of preparing editions of atlases, so too were technical texts brought together in certain physicians' and astronomers' libraries. Contradictions became more visible, divergent traditions more difficult to reconcile. . . . Not only was confidence in old theories weakened, but an enriched reading matter also encouraged the development

> of new intellectual combinations and permutations. . . . Once old texts came together within the same study, diverse systems of ideas and special disciplines could be combined. Increased output . . . in short, created conditions that favored new combinations of old ideas at first and then, later on, the creation of entirely new systems of thought.

The spread of information through printing allowed major advances in scientific understanding. Everything from astronomical tables to accurate anatomical illustrations became grist for the scientific mill. Theories could be readily checked against a growing mass of information. Equally important, printing helped to change the nature of learning. Scientific investigators were freed from the drudgery of copying tables and texts. Also, as literacy spread and books were much more readily available, students of all ages did not need to spend vast amounts of time on rote memorization and the learning of elaborate mnemonic techniques. Freed from these tasks, advanced students and natural philosophers could devote their energies to "solving brain teasers, designing ingenious experiments and new instruments, or even chasing butterflies and collecting bugs if they wished."[11] The pursuit of knowledge could take on a playful quality, and play can be a key ingredient in the development of scientific understanding.

PRINTING AND THE RISE OF PROTESTANTISM

Just as printing helped to alter people's views of the physical and natural world, the new medium played a key role in transforming conceptions of the spiritual world. It is certainly no coincidence that the Protestant Reformation began less than two generations after Gutenberg's invention. Before then, unorthodox interpretations of prevailing Christian religious beliefs had sprung up from time to time, but they had remained confined to particular locations. Although several instances of religious dissent may have occurred simultaneously, they were not linked together into an integrated attack on the authority of the centralized Church. The circulation of printed religious books and tracts created a more unified challenge to the dominant Catholic order, and helped to turn a number of separate revolts into the Protestant Reformation.

Literacy and Protestantism reinforced each other; while Protestantism benefited from the expansion of the printed word, the religious beliefs of Protestant reformers did much to encourage printing. One of the central tenets of Protestantism was that individuals should study the Bible and interpret its message unencumbered by the guidance of priests, bishops, and popes. The production of printed Bibles carried the word of God directly to the faithful, while at the same time a great outpouring of religious tracts spread the words of religious reformers throughout Christendom. For good reason, Martin Luther praised printing as "God's highest and extremest act of grace, whereby the business of the Gospel is driven forward."[12]

Luther later lost his enthusiasm for popular Bible reading when it became apparent that the unguided reading of the Bible could result in "dangerous" beliefs that contradicted his own.[13] But by then individual reading of the Bible had become an integral part of Protestant life. Sixteenth-century English clergymen were told to "comfort, exhort, and admonish every man to read the Bible in Latin or English, as the very word of God and the spiritual food of man's soul, whereby they may better know their duties to God, to their sovereign lord the king, and their neighbor."[14]

The Catholic Church also used printing as a weapon in its battles against the Protestant "heresy." But the cultural order it attempted to maintain was fundamentally antagonistic to the use of printed sources and the spread of literacy. The Church was unwilling to allow laypeople to pursue Biblical scholarship, and it was deeply suspicious of popular Bible reading and interpretation. Catholicism was still rooted in the medieval past, and it was unprepared for the challenge presented by print technology.

If the Bible was to be made accessible to the Protestant faithful it had to be printed in a language they understood, rather than in the Latin that had served as the universal language of the medieval clergy. While the Catholic Church strongly resisted the production of Bibles in modern European languages, Protestantism made it an imperative.[15] In Protestant lands the translation and printing of Bibles in vernacular languages spread the Gospel, but equally important, it stimulated nationalism. A Europe that had been united under the cultural domination of the Catholic Church gave way to a patchwork of separate states, each with its own established church and a growing literature printed in the local language. Printing helped to destroy the unity of the medieval world, while within the borders of discrete national entities, a considerable amount of homogenization took place. Printing was of great importance in standardizing their languages, while dialects unsupported by a printed literature slowly faded away, supplanted by a standardized form of French, German, or English.

PRINTING, LITERACY, AND SOCIAL CHANGE

Motivated by the religious requirement of Bible reading, Protestants became literate to an extent not previously found in European civilization. Throughout the Middle Ages the great bulk of the population could not read. Even rich and powerful people were illiterate. Charlemagne, who reigned as King of the Franks from 768 to 814, and after 800 was the emperor of much of present-day Western Europe, was an intelligent and learned man, but he needed to have books and documents read aloud to him, and he relied on scribes to write down his words.

The only people who found literacy to be a necessity were members of the Christian clergy. (The connection between literacy and a religious vocation can be seen in the linguistic relationship between the word "clerk" and "clergy.") The ability to read was often taken as sufficient proof that a person was a member of the clergy, and therefore not subject to civil law. This could be a great advantage, for a person claiming "benefit of clergy" could escape corporal punishment or even execution if convicted of a crime. The fifty-first psalm of the Old Testament was known as the "neck verse," for a person able to read it would be spared the gallows.[16] As might be expected, many illiterates tried to escape punishment by memorizing the psalm and then pretending to read it at the appropriate moment. This resulted in some farcical scenes in which a person attempted to demonstrate his literacy by "reading" from an upside-down Bible!

Protestantism, with its emphasis on Bible reading, created strong incentives for laypeople to learn to read. This skill could of course be useful in other endeavors. A person who could read and write was able to keep accurate records and accounts, learn about new ways of doing things, understand the law, enter into contracts, communicate with large numbers of people, and make use of banks. Thus, the often noted connection between Protestantism and rapid economic growth in early modern Europe may have been in part due to the superior levels of literacy in Protestant lands.[17]

As we have seen, the rise of Protestantism created a fertile environment for the development of the printing industry. At the same time, urbanization and the spread of capitalism also made for a society that was more open to—actually hungry for—the new ideas that books could disseminate. On a more utilitarian level, the development of commerce and industry generated a need for written records and accounts. In later centuries, the desire for literacy was further stimulated by the growing concerns with social equality. In particular, the greater accessibility of education that culminated with the introduction of mass education in the nineteenth century produced a literate public and expanded the market for the printed word.[18]

All of these historical forces produced an environment that was ripe for the large-scale production of printed works. Had they not been present, the invention of movable type would have been of much more limited significance. After all, printing had first emerged in East Asia, yet social conditions limited its influence in that region. Korean printers had independently invented movable type, but few books were printed by this means in Korea until the end of the nineteenth century. Korean scholars even invented a phonetic alphabet known as *Hangul* in the fifteenth century, which greatly facilitated the use of movable type. But Korea's traditional aristocracy clung to the use of Chinese characters for the written rendition of Korean words, for their ability to understand the difficult-to-learn characters set them apart from the common folk. Although a mass readership, literate in *Hangul*, could have provided a large audience for printed works, no self-respecting writer was willing to use the "inferior" script. Consequently, printing with movable type had minimal effects on Korean society.[19]

In Europe, social and cultural conditions created a strong demand for the products of print technology. At the same time, the technological advances that made the mass production of printed works possible did much to create these conditions. As with all major technological advances, we find a reciprocal, reinforcing relationship between technological change on the one hand, and social change on the other. Printing with movable type did not effect a sudden revolution in Western society; even as late as the middle of the eighteenth century, 40 percent of English men and more than 60 percent of English women were illiterate.[20] Still, working in conjunction with the great changes that were sweeping over Western civilization, the development of print technology produced a transformation that has only begun in many parts of the world today.

PSYCHOLOGICAL EFFECTS OF PRINTING

The effects of printing on religious, economic, and political life seem reasonably evident. A consideration of the consequences of reading for individual attitudes and ways of thinking, however, is a more speculative venture. The Protestant Reformation is an identifiable historical event, and its connections with printing seem clear. It is much more difficult to demonstrate that printing fundamentally changed basic ways of thinking, or that it altered how we understand our place in the world.

Still, we might reasonably speculate that printing resulted in a greater sense of one's separateness from the rest of society. Reading is usually a solitary activity. Novels, with their typical emphasis on one key figure, have been the central literary product of the last two centuries. It is therefore likely that a culture based on print is more individualistic than one that relies on collective means of communication, such as Sunday sermons or old-fashioned political oratory.

Some fascinating possibilities of this sort have been suggested by Marshall McLuhan, for whom media such as print or television had consequences that far outweigh the overt messages that they carry.[21] Printed books fundamentally changed civilization not because of the information they transmitted; the greatest consequences of printing lay in the different modes of thought and perception that it fostered. In McLuhan's analysis, each medium of communication engenders a distinctive way of looking at the world; as his famous aphorism has it, "the medium is the message." The reading of the printed word makes readers think in sequence, just as a sentence is sequentially read from left to right. Reading also produces an egocentric view of the world, for the reader's involvement with the printed word is solitary and private. For McLuhan, the intellectual and social consequences of print were immense, encompassing such diverse things as the Newtonian view of the universe, the employment of fixed prices for business transactions, and the "detribalization" of society.

It is impossible to summarize McLuhan's ideas in a few sentences. The intrepid reader is advised to turn to McLuhan's writings after first being warned that their author can be cryptic, elliptical, hyperbolic, and at times simply maddening. Nonetheless, some taste of his thinking can be obtained by looking at how McLuhan has interpreted changed painting styles as a reflection of printing, or what he has termed "The Gutenberg Galaxy." Reading, as McLuhan points out, inculcates a single "point of view," the fixed position of the reader.[22] In similar fashion, the visual arts that developed in the post-printing era began to depict scenes as they might be viewed from a single point in space. Given this central vantage point, the scene appears to have a three-dimensional quality due to the artist's employment of the principles of perspective. The employment of perspective is not a natural or necessary part of pictorial art. As any art student knows, to paint or draw something in perspective requires the application of special procedures and a good deal of practice. These artistic techniques were not employed in medieval painting, which was characterized by a visual flatness. Figures were not presented according to the rules of perspective, but in accordance with the importance of the subjects; in a typical Medieval painting, Jesus, an archbishop, and the faithful appear in the same plane, but they are not all of the same size, for their size is commensurate with their importance in the spiritual hierarchy. Until the advent of printing there was no reason to depict subjects according to rules of perspective, for printing and widespread reading had not conditioned people to see the world as it might appear from a single vantage point. The rules of perspective were developed and used only after printing produced an individualistic perception of the world.

NEWSPAPERS

For several centuries after the invention of movable type the most important product of the printing press was the book. Beginning in the eighteenth century a new kind of publication appeared that was to have equally revolutionary consequences for society. That publication was the newspaper. The transmission of written news has a long history, originating with the *acta diurnia* that was posted in the Roman Forum, and continuing with the printed broadsheets that circulated from late medieval times onward. The first regularly appearing newspapers emerged in the seventeenth century, and by the beginning of the next century, newspapers were established institutions in a number of European cities. Colonial America followed suit, with the appearance of the *The Boston News-Letter* in 1704 marking the beginning of newspaper publishing in what was to be the United States.

These early newspapers were unimpressive, usually consisting of no more than four pages of hard-to-read type. Their circulation was also limited, being confined to the small number of people who were both literate and able to pay for them. This situation changed dramatically during the nineteenth century, when the newspaper became a mass-produced item, a product of Industrial Revolution technology.

One of the distinguishing characteristics of the Industrial Revolution was the use of new sources of energy, with steam power taking on a growing importance. The first steam-powered printing press was invented by Koenig and Bauer in 1812, and was put to use by the *Times* of London in 1814. The use of steam power for this purpose necessitated a fundamental redesign of the printing process. Instead of manually pressing a sheet of paper against a flat bed containing the type, the job was done by a steam-powered rotating cylinder, while another cylinder did the inking. It ran off 1,000 copies of the newspaper each hour, one-third the time that the hand press required.[23]

Other refinements ensued. Printing on a continuously running roll of paper (instead of feeding single sheets of paper) greatly speeded up the process, as did the development of presses that printed on both sides of a sheet and automatically cut individual pages. Improved inking processes contributed to the speed of production, while the use of woodpulp for paper also lowered production costs.

Improvements to the rotary press concept followed, most notably Richard Hoe's use of another rotating cylinder instead of a flat horizontal bed to hold the type. First employed in 1847, improved variants of the rotary press could produce as many as 20,000 impressions per hour by the beginning of the Civil War.[24] Printing was further accelerated by the invention of the stereotype plate, which allowed the easy production of duplicate plates for use on multiple presses. The stereotype plate was made by pressing a paper mat onto a frame of set type. After

A mid-nineteenth century Hoe rotary press. (Division of Graphic Arts, National Museum of American History, Smithsonian Institution.)

the mat was dried and shaped into a half of a cylinder, molten metal was forced into it, forming the actual printing plate. The other side of the rotating cylinder held type for another page. The first of these presses contained two cylinders, each with four plates, so an eight-page newspaper could be printed with each turn of the cylinders.[25] By 1890 improved rotary presses were capable of running off 48,000 twelve-page papers each hour.[26]

While the presses turned out newspapers with increasing rapidity, the need to set type by hand restricted the rate of production. A great deal of inventive energy therefore was expended on the development of automatic typesetting. Even Mark Twain was involved as a financial backer for such a venture; it unfortunately failed, dashing the great author's dreams of making a vast fortune.[27] Success instead came to a German immigrant, Ottmar Mergenthaler, whose Linotype machine was the first effective typesetting device. First used in 1886, by the end of the century 3,000 of these machines were in use around the world.[28]

The development of the mass-circulation newspaper was further stimulated by a host of nineteenth-century inventions that allowed the rapid gathering and dissemination of news. Railroads and steamships conveyed reporters to scenes of newsworthy events. The telegraph quickly relayed their stories back to the newsroom, and had the incidental effect of making stories less verbose because the senders of telegraphic messages were charged by the word.[29] The news-gathering power of the telegraph was enhanced by the laying of trans-Atlantic cables that greatly speeded up communications between Europe and the United States. The telephone and, by the end of the century, the radio further assisted reporters and the newspapers for which they worked.

CIRCULATION WARS AND THE SHAPING OF PUBLIC OPINION

As with other mass-production technologies, the improvements in newspaper printing dramatically lowered production costs. Before the 1830s, newspapers sold for six cents a copy, which in those days could buy a quarter pound of bacon or a pint of whiskey.[30] As production costs fell, the first American penny paper, the *New York Sun*, made its appearance in 1837. Aiming for a mass audience, the *Sun* offered sensational stories, such as a report on the observance of life on the moon through the use of a powerful telescope located in South America. Other newspapers followed suit, and by the second half of the nineteenth century the cheap mass-circulation daily newspaper was a fixture in every city and town of Europe and North America. The expansion of newspaper readership occurred at a rapid rate. During the early decades of the nineteenth century *The Times* had led all English newspapers with 10,000 readers,[31] while the largest New York newspapers usually printed editions of no more than 2,500.[32] By 1870, 2.6 million newspapers were printed each day in the United States, and by the end of the century that figure had risen to 15 million.[33]

Unfortunately, the spread of these newspapers was often stimulated by journalism of the worst sort. Newspapers built up circulation by running stories that stressed scandal, crime, and bizarre occurrences. The debasement of news reporting accompanied the increase in newspaper readership; the "yellow journalism" of that era, in the words of Edwin and Michael Emery, "turned the high drama of life into a cheap melodrama and led to stories being twisted into the form best suited for sales by the howling newsboy."[34] We can still see this sort of thing in the headlines of the tabloids conveniently situated next to supermarket checkout lines: "UFO Piloted by Elvis Lands on Madonna's Lawn!"

Both in the past and today, reportage of this sort demonstrated many newspapers' low regard for the taste and intelligence of their readers. Beyond this, the mass-produced newspaper became a force in its own right, and its editorial policies did much to shape public opinion. This can be seen in the way the popular press helped to push America into a war with Spain in 1898. To be sure, other forces were at work; having conquered the frontier of its own territory, the United States was seeking new triumphs of Manifest Destiny. Still, without the galvanization of public opinion by newspapers such as William Randolph Hearst's *New York Journal*, it is possible that the war could have been averted.

Hearst was engaged in a fierce struggle with Joseph Pulitzer's *New York World*, and inflamed coverage of a rebellion against Spanish rule in Cuba was a surefire way to sell papers. Through the use of sensational reporting, both papers enlarged their circulations to over a million each, a record at that time.[35] Reports of Spanish atrocities in putting down the rebellion, while having some foundation, were wildly exaggerated and given extensive coverage. Reports such as one that appeared in the *World* left little to the imagination: "Blood on the roadsides, blood in the fields, blood on the doorsteps, blood, blood, blood!"[36] Photographs were beginning to appear in newspapers at this time, and the stories were accompanied by many lurid examples, some of them phony.

From the beginning of the revolt to America's going to war with Spain, hardly a day went by without a story about Cuba in one of the New York newspapers.[37] And when the American battleship *Maine* blew up in Havana harbor, many newspapers were quick to attribute it to a Spanish mine (to this day the cause of the ship's sinking remains unknown). Even though disputes with Spain over a variety of issues were almost fully resolved by this time, a significant portion of the popular press had whipped up a war hysteria that could not be extinguished. The brief war that ensued concluded with the United States assuming the role of an imperial power, with colonial holdings that stretched from the Philippines to Puerto Rico.

It would be an exaggeration to say that the Spanish-American War, or any other major historical event, was directly caused by the reportage of circulation-hungry mass newspapers. The remnants of Spain's American empire were a convenient target for turn-of-the-century America's expansionist ambitions. Under these circumstances, overheated news reporting reinforced opinions and ambitions that were already evident. Still, there can be no denying that the journalistic policies of Hearst and Pulitzer made a peaceful resolution of disputes between the U.S. and Spain far more difficult.

It would be even farther from the truth to attribute such events as the Spanish-American War to technological advances in printing. To be sure, the mass-circulation dailies depended on improved methods of newspaper production, but much more was involved than such innovations as the steam-powered rotary press. The spread of education and literacy, spurred by the expansion of public schooling, created a large potential readership for these publications. The expansion of industry and commerce was accompanied by the emergence of a growing advertising industry that provided vast revenues for the newspapers. Urbanization and immigration produced large concentrations of population from which a mass readership could be drawn. Finally, a more democratic social order generated an environment in which the "common man" gained in political and economic importance; as the first of the mass media, newspapers were a natural outgrowth of "mass society."

All of these trends reinforced one another. New printing technologies contributed to literacy and the rise of a mass readership, commercial expansion, and perhaps even the expansion of democracy. At the same time, the emergence of new printing technologies was stimulated by all of these economic, social, political, and cultural factors. As we have already seen, Gutenberg's invention of movable type was accompanied by changes on many fronts. The same thing happened during the second half of the nineteenth century as a series of technological innovations made possible the publication of mass-circulation newspapers. In neither case did technological developments by themselves produce the epochal changes with which they are associated. The development of these technologies was part of a larger process of change. Yet at the same time, these technologies were major contributors to this process. From movable type to the steam-powered rotary press, the technology of printing exemplifies the complex, mutually reinforcing interactions between technological and social change.

Questions for Discussion

1. Do statements that appear in print seem more true than those that are presented through some other medium? Why?
2. One of the major trends in the world today is the spread of literacy. In many parts of the world, only a small percentage of the population could read a generation ago; today, the majority of the population is literate. What sort of

social, political, and economic effects will this produce? Will all of them be good?

3. Some social critics are of the opinion that the spread of the electronic media is destroying literacy. Standardized test scores of reading and writing ability have in fact gone down in recent years. Are the new media the main cause? If so, is the slow destruction of the printed media by the electronic media necessarily a bad thing?
4. Lurid newspaper stories helped to fan a conflict between Spain and America that ultimately led to war. The Vietnam war, another Asian conflict involving America, met with a great deal of opposition, due in part to the continual presentation of that war on television. Marshall McLuhan would have thought this significant. Can you imagine why? Can you think of any fundamental differences between newspaper accounts of a war and those presented on television?

Notes

1. Quoted in Maurice N. Richter, Jr., *Technology and Social Complexity* (Albany: State University of New York Press, 1982), pp. 16–17.
2. James Thorpe, *The Gutenberg Bible: Landmark in Learning* (San Marino, Calif.: The Huntington Library, 1975), p. 3.
3. Xing Runchuan, "The Invention and Development of Printing and Its Dissemination Abroad," in *Ancient China's Technology and Science* (Beijing: Foreign Languages Press, 1983), pp. 383–391.
4. D. S. L. Cardwell, *Turning Points in Western Technology: A Study of Technology, Science and History* (New York: Science History Publications, 1972), pp. 20–24.
5. Thorpe, op. cit., p. 4.
6. Ferdinand Braudel, *Capitalism and Material Life, 1400–1800* (New York: Harper & Row, 1975), p. 298.
7. Elizabeth Eisenstein, *The Printing Revolution in Early Modern Europe* (New York: Cambridge University Press, 1984), pp. 195–199.
8. Lewis Mumford, *Technics and Civilization* (New York: Harcourt, Brace and World, 1934), pp. 21–22.
9. Eisenstein, op. cit., p. 75.
10. Ibid., pp. 43–44.
11. Ibid., p. 239.
12. Ibid., p. 147.
13. Richard Gawthorp and Gerald Strauss, "Protestantism and Literacy in Early Modern Germany," *Past and Present* 104 (August 1984): 6.
14. Quoted in David Cressy, *Literacy and the Social Order: Reading and Writing in Tudor and Stuart England* (New York: Cambridge University Press, 1980), p. 3.
15. Eisenstein, op. cit., p. 170.
16. Cressy, op. cit., p. 16.
17. A. L. Stinchcombe, "Social Structure and the Invention of Organizational Forms," in Tom Burns (ed.), *Industrial Man* (Harmondsworth, England: Penguin, 1969), pp. 163–165.
18. Scott Cook, "Technology and Social Change: The Myth of Movable Type," *The Weaver* 4, 1 (Fall 1985): 5.

19. Kichung Kim, "*Hyol-ui Nu*: Korea's First 'New' Novel," *Korean Culture* 6, 4 (December 1985): 41–45.
20. Cressy, op. cit., p. 176.
21. See Herbert Marshall McLuhan, *The Gutenberg Galaxy: The Making of Typographic Man* (Toronto: University of Toronto Press, 1962); and *Understanding Media: The Extensions of Man* (New York: New American Library, 1964).
22. McLuhan, *The Gutenberg Galaxy*, op. cit., p. 111.
23. Robert W. Desmond, *The Information Process: World News Reporting to the Twentieth Century* (Iowa City: University of Iowa Press, 1978), pp. 62–63.
24. Edwin Emery and Michael Emery, *The Press and America: An Interpretive History of Mass Media*, 4th ed. (Englewood Cliffs, N.J.: Prentice-Hall, 1978), p. 144.
25. Desmond, op. cit., p. 295.
26. Emery and Emery, op. cit., p. 234.
27. John F. Kasson, *Civilizing the Machine: Technology and Republican Values in America, 1776–1900* (New York: Penguin Books, 1977), pp. 203–204.
28. Desmond, op. cit., p. 298.
29. John Tebbel, *The Compact History of the American Newspaper* (New York: Hawthorne Books, 1969), p. 121.
30. Emery and Emery, op. cit., p. 116.
31. Desmond, op. cit., p. 77.
32. Emery and Emery, op. cit., p. 108.
33. Edwin Emery, *The Press and America*, 3d ed. (Englewood Cliffs, N.J.: Prentice-Hall, 1972), p. 285.
34. Emery and Emery, op. cit., p. 244.
35. Desmond, op. cit., p. 388.
36. Quoted in Emery, op. cit., p. 365.
37. Emery and Emery, op. cit., p. 250.

Chapter 12 The Electronic Media

During the 450 years that followed the invention of movable type in the West, the printed word was the dominant form of mass communication. Toward the end of the last century new forms of communication began to emerge as scientists and engineers in several countries learned how to send and receive signals through empty space. The first product of their efforts was radio. The experience gained with radio subsequently supplied the foundation for an even more far-reaching medium: television. The consequences of these electronic media have been at least as great as those of printing; indeed, we are still trying to comprehend the effects the electronic media are having on our world. Some of these effects will be explored in later sections of this chapter, but first we will briefly survey the major technological changes that made electronic media possible.

THE INVENTION OF RADIO

The telegraph was the first device to make use of electricity for sending and receiving signals. Invented by Samuel F. B. Morse in 1844, the telegraph used electrical pulses to transmit combinations of dots and dashes that represented individual letters—the Morse Code. Telegraph communication took on a vital importance as it tied together far-flung communities, facilitated the running of railroads, and brought news from far-off places. Still, the telegraph had its limitations. Not only were large numbers of trained operators required to encode and decode messages, large amounts of capital were need to string up telegraph wires between communication points. And it could not be used when the installation of wires was physically impossible or the costs were prohibitive, as in the case of ships at sea or geographically remote places.

It might be thought that the limitations of telegraphic communication created the motivation to develop a method of electrical communication that did not require wires. And it is certainly true that much of radio's development was "pulled" along by the demand for better methods of communication. But that came later; the first successful attempt at sending and receiving wireless communications had nothing to do with practical concerns. As we have already seen, radio transmission and reception was first used by Heinrich Hertz as a way of empirically testing the theoretical formulations of James Clerk Maxwell. Hertz produced radio waves with an oscillator, a device that rapidly generated electrical pulses. To receive or "detect" the radio waves produced by these pulses, Hertz employed a piece of wire with small balls at each end, bent into a near-circle. When this was placed in the vicinity of the oscillator, a spark jumped between the two balls in conjunction with the sparking of the oscillator. Through a series of experiments using these devices, he was able to confirm Maxwell's calculations that electrical waves traveled at the speed of light. Additionally, he showed that these waves propagated themselves in accordance with the laws of optics that had first been formulated for visible light.

Hertz's concerns were entirely in the realm of scientific inquiry, but others saw in his apparatus an embryonic technology that might realize the dream of wireless communication. Many experimenters tackled the numerous technical problems that kept radio from being a practical possibility. Although Guglielmo Marconi is often hailed as the inventor of radio, as with most significant inventions there are many others whose contributions were equally important, such as Oliver Lodge, who developed the method of tuning a receiver so it could receive signals of a specific wave length.

The basic apparatus used by early radio experimenters was crude. A spark generated the radio waves, while the signal was received by a "coherer," a tube containing metal filings. A radio signal reaching the coherer would cause the filings to stick together and allow the flow of current that activated a recording device. A "trembler" would then shake up the tube and prepare it for another signal. As can be imagined, these rather primitive devices set sharp limits to the quality of radio transmission and reception. At best they could be used for sending telegraphic messages; voice communication was impossible.

The transmission and reception of normal sounds required the replacement of the spark-and-coherer system with one that used continuous waves to carry the signal. First used successfully by Ronald Fessenden, this technique was augmented by his application of the heterodyne principle. Fessenden's heterodyne system was based on the operation of two separate alternating currents, one generated by the transmitter, and the other by the receiver. Oscillating waves transmitted at a certain frequency mixed with the receiver's lower-frequency waves to produce a wave that could be easily received. This system allowed the transmission and reception not just of dots and dashes, but of a whole range of audible signals.

Important as the heterodyne principle was, an even greater breakthrough came with the invention of better equipment for the transmission and reception of radio waves. These devices were based on the principle of using a small flow of current to govern a larger current. Thus, the small current produced by a radio wave could be used to regulate the stronger current that drove the speaker of a radio. The first of these devices was the diode, invented by John Ambrose Fleming in 1904. It established the principle of controlling current electronically, but it had its limitations. A much more workable device was the triode, an invention of Lee de Forest. By using a separate grid to switch current, the triode served as an excellent basis for a radio receiver, as well as for its amplifier.[1]

THE ORIGINS OF COMMERCIAL RADIO

While technical development proceeded at a good clip during these early years, few thought of radio as a means of providing entertainment and information for a large audience. Radio was seen simply as a wireless telegraph; for many years its main use was for ship-to-shore communications. Radio was also used to transmit messages across oceans, thereby obviating the need for undersea cables, but high cost and unreliability plagued these efforts. Military forces also made extensive use of radio communications, and the First World War stimulated a good deal of technical progress, especially in the development of mass-production techniques for the manufacture of radio components.

During the 1920s interest in radio mounted as large numbers of amateur radio operators constructed their own equipment, sent out messages, and attempted to pick up the signals of others. Their chatter cluttered up the airwaves, and with a mischievous intent similar to that of today's computer hackers, a few even sent out bogus orders to naval vessels. But others had more serious aims. Throughout the United States, amateur radio operators, many of whom had developed their skills during World War I, began to broadcast news, weather bulletins, musical recordings, and even live musical performances. In a manner resembling the early days of the personal computer, enthusiastic amateurs made significant contributions to the new technology by constructing their own apparatus, finding new applications for radio communication, and in general exploring the potentials of the new medium.

At this time business enterprises began to take notice of radio's commercial possibilities. Realizing that regular programming could stimulate a sizeable market for its radio sets, in 1920 the Westinghouse Electric and Manufacturing Company engaged one of its employees, himself a dedicated radio amateur, to construct a broadcasting station atop one of the buildings of its Pittsburgh headquarters. On November 2, radio station KDKA went "on the air," giving up-to-the-minute results of the 1920 Presidential election to an audience of several hundred. Within a matter of months Westinghouse and General Electric had set up radio stations in a number of cities. Other entrepreneurs followed suit, so by 1922 more than 500 stations were transmitting music, sporting events, speeches, and news programs. Commercial broadcasting took on a growing importance as radio began to reach the general public and not just a relatively small group of technically minded amateurs.

At first, many of these programs were low-budget endeavors; phonograph records required only small initial costs, and most live performers were willing to appear free of charge. This did not last for long; the American Society of Composers, Authors, and Publishers (ASCAP) began to demand payment for the broadcast of recorded music, and in 1923 it backed this demand with a successful lawsuit. In similar fashion, live performers started to expect payment for their services. Other countries, most notably Great Britain, had by this time established a national radio system through which the expenses of broadcasting were defrayed by licensing fees paid by owners of radio sets. But in the United States free enterprise was the order of the day, and radio broadcasting was sustained by companies willing to sponsor programs in return for the airing of their commercials. A host of new programs made their appearance over the airwaves, and in the years to follow, listening to radio programs became an integral part of American life. At the same time, however, listeners were subject to a steady barrage of commercial "messages," exhorting them to buy the sponsors' products and promising them that all of their problems could be solved by using the proper mouthwash or driving the right car. As had happened before with other advances in the ability to communicate, impressive technological achievements were put into the service of the mindless and the mercenary.

THE RISE OF TELEVISION

For all of radio's accomplishments, only a portion of the electronic media's potential had been demonstrated. Radio could only reach one of the senses; if sight could be added to sound, the appeal of electronic communications could be greatly enhanced. The dream had long been there; the idea of electrically transmitting and receiving visual images existed before even radio was a practical reality. In 1879, readers of the English magazine *Punch* were presented with an illustration of a couple watching a tennis match being shown on a large screen over their fireplace. Three years later, a French artist drew equally prescient illustrations that depicted people viewing television screens filled with pictures of distant lectures, wars, and even girlie shows.[2] And in 1907 in the pages of *Scientific American*, the term "television" appeared for the first time in print.

The realization of these early visions was long in coming. Early efforts to transmit pictures electronically depended on a mechanical device, the Nipkow disk. First demonstrated in 1884, it employed a disk perforated by holes arranged in a spiral that was interposed between an object and a screen composed of selenium cells, which had the property of activating an electrical current when light fell upon them. As the disc rotated, pinpoints of light moved across the screen with a scanning motion. Unfortunately, Nipkow was unable to reconstruct the image on a screen through the use of a separate synchronized disk.[3]

The Nipkow disk and similar mechanical scanning devices, such as one using a revolving drum fitted with small mirrors, were the basis of subsequent attempts to develop a practical system of television. In 1924 a Bell Laboratories team sent broadcasts by wire over considerable distances, including one by Secretary of Commerce Herbert Hoover that traveled from Washington to New York. Similar experiments were undertaken at General Electric's Schenectady laboratories. Meanwhile, commencing in 1929, the British Broadcasting Corporation employed another mechanical system devised by John Baird to send out half-hour programs five days a week.

The mechanical system had many drawbacks, not the least of which was that it required extremely high levels of illumination. Attention therefore turned to the development of all-electronic methods of transmission and reception. These efforts were boosted by a series of fundamental discoveries and inventions that began to occur at the end of the nineteenth century. Most important of these was the cathode ray tube. This device was based on a principle, discovered by Ferdinand Braun in 1897, that a beam of electrons moving along a closed tube could be deflected by a magnet. In Russia, Boris Rosing used this tube for television reception by placing a screen of photoelectric cells in the tube, and activating them with moving electrons.

This was only a laboratory demonstration and not a complete system of sending and receiving television images. A crucial step was taken by Vladimir Zworykin, a

Russian émigré who had served as an assistant to Rosing while a student at the Institute of Technology in St. Petersburg. After coming to the United States he was employed by Westinghouse and then by the Radio Corporation of America (RCA), where he worked on television in addition to a number of other projects. In 1928 he produced the first workable electronic television camera, which he called the "iconoscope." It employed a screen made up of thousands of individual elements that took on an electrical charge when they were struck by light reflected from the object whose image was to be transmitted. The screen was rapidly scanned by a beam from an electron gun, which resulted in an electrical discharge that was amplified and transmitted to the receiver. The receiver then reconstructed these electrical discharges into points of light on the television screen.

Zworykin's iconoscope was a major accomplishment, but as often happens in the history of technology, parallel inventions were being made elsewhere. While Zworykin was engaged in the early phases of his research, an Idaho farmboy named Philo Farnsworth astonished his high school science teacher by presenting him with plans for an apparently workable television system. A few years later Farnsworth succeeded in getting financial backing for his research, and by the late 1920s he had produced a workable, if crude, system of electronic television. But after an expenditure of a million dollars by 1938, Farnsworth had not produced a commercially viable method of transmitting and receiving television. Even so, Farnsworth held a large number of key patents on devices and processes of considerable importance, which had been upheld only after a good deal of litigation with Zworykin and his backers.

In 1939 the Radio Corporation of America (RCA) began the regular broadcast of television programs to a few thousand receivers in New York City. A number of drawbacks were still evident; most of the early receiving sets had five-inch screens, and their dim pictures meant that they had to be viewed in darkness. Many technical problems remained to be solved in such areas as the design of antennas and the relaying of signals over large distances. Still, television had become a practical possibility.

THE FEDERAL GOVERNMENT STEPS IN

The development of radio and television in the United States was largely the work of private enterprise. Even so, the activities of the federal government gave a substantial boost to the electronic media. Military contracts stimulated a great deal of useful research. National security was also invoked when the government played the leading role in the creation of a unified radio industry. In the years after World War I, government officials, especially those in the Navy, were increasingly concerned about the monopoly that the British Marconi Company had over international radio communication. After some prodding by the Navy, the General Electric Company bought a controlling interest in Marconi's American subsidiary and transferred it to the newly formed RCA. There still remained the problem of RCA's gaining access to the numerous patents that covered various aspects of radio production and broadcasting. This was ultimately resolved by giving the two other major patent holders, Westinghouse and American Telephone and Telegraph, a financial stake in RCA in return for the use of their patents. The creation of a radio monopoly was thus the result of an action undertaken at the behest of the U.S. government.

In addition to these activities, the federal government used its regulatory power to insure an orderly environment for broadcasting. For clear reception it is necessary for each broadcaster to remain on a single frequency (when you "tune in" to a radio station you are selecting a specific frequency, such as 790 kHz or 89 MHz). Should a number of radio stations broadcast their programs on the same frequency, the result is a chaos of conflicting sounds. In order to address this problem, in 1912 the Department of Commerce and Labor began to require the licensing of all broadcasters. When a license was awarded, it stipulated the exact frequency on which the station could broadcast, as well as its times of operation. But this authority was successfully challenged in court, and by the mid-1920s a large number of broadcasters were transmitting all over the frequency spectrum. The airwaves had become a cacophony of interfering signals.

To remedy this situation, Congress created the Federal Radio Commission in 1927, giving it broad powers to issue licenses and assign frequencies. In 1934 government oversight of communications media was consolidated through the creation of the Federal Communications Commission (FCC). The Commission was (and is) made up of seven members appointed by the President, each serving a seven-year term. The licensing and frequency allocation powers of the FCC occasionally worked to the detriment of the very few stations whose licenses were revoked because their broadcasts were deemed not to be in the public interest. The strict assignation of frequencies also frustrated would-be broadcasters who were unable to gain licenses. But for the industry as a whole, the Federal Radio Commission and its successor the FCC were essential elements in the system of radio broadcasting. Without Federal regulation, broadcasters could not be assured of their own special frequencies, and the air would be filled with signals that constantly intruded on one another. Most radio entrepreneurs probably believed in free enterprise, but they realized that in the absence of regulation their pursuit of individual self-interest would result in the destruction of their industry.

The issue of federal regulation once again came into prominence when television became ripe for commercial development in the late 1930s. One of the major issues confronting the industry was the setting of technical standards. Each one of the major companies involved in television research employed a different system for transmitting and receiving televised images. At the same time, one of these companies, RCA, had made the most progress in putting together a total television system. This meant that the standards employed by RCA would have to be used if the FCC decided to permit the immediate broadcast of commercial television programs. The other manufacturers knew that they lagged behind RCA, but they felt that their own technologies had a great deal of promise. They certainly were not eager to adopt RCA's standards and pay royalties for its patents.

Faced with these contentious issues, the Commission gave the go-ahead for commercial television broadcasting in 1940, only to rescind its decision less than a year later. A special engineering committee was then set up to consider the whole issue of standards. By the middle of 1941 these had been successfully resolved, although America's entry into World War II put a halt to commercial television. When TV went into its period of rapid growth in the late 1940s and early 1950s, some of its success could be attributed to the uniform technical standards that were developed under FCC sponsorship.[4] Television did not suffer from the incompatible equipment and processes that fragmented the market during the early years of the personal computer.

PROBLEMS OF REGULATION

The efforts of the FCC to promote uniform technical standards were essential to the growth of the electronic media industry. The consequences of the FCC's other activities, however, are more controversial. According to the 1934 Communications Act, the FCC is supposed to grant and renew broadcasting licenses on the basis of "the public interest, convenience and necessity." In fact, the FCC has been quite reluctant to influence the radio and television industries through the exercise of its licensing powers. Very few licenses have not been renewed; as long as a station has met some minimal level of community service it is deemed to have met the letter, if not the spirit, of the regulations. An FCC chairman might indict American television as "a vast wasteland," as Newton Minnow did in 1961, but other than sponsoring the development of noncommercial television, his agency did little to upgrade the quality of TV programming.

There have been times when the ineffectiveness of FCC regulation could be attributed to out-and-out malfeasance. One particularly notorious case involved the agency's decision to support a system of color television that worked to the advantage of RCA, and to the disadvantage of CBS. Seven months after this decision, the chairman of the FCC resigned his post and took a high executive position with NBC, at that time an affiliate of RCA.[5] Other instances of apparent wrongdoing could be found, such as the widespread practice of FCC commissioners being paid for travel expenses when they addressed meetings of broadcasters' associations. In one case, a commissioner went on a week-long yacht trip to a Caribbean island at the expense of a station owner whose case was being reviewed by the agency. The commissioner saw nothing improper in this; neither did his superiors, for he was subsequently elevated to the chairmanship of the FCC.[6]

These may have been extreme cases; even critics of the FCC do not generally believe that the agency has been riddled with corruption and conflicts of interest. But many critics do believe that it has failed to serve the public interest because it has been "captured" by the broadcast industry. This is a common feature of government regulation; the Interstate Commerce Commission often champions the cause of the trucking industry, as the now-defunct Atomic Energy Commission did for the nuclear power industry. Many regulatory agencies owe their creation to the demands of the industry that they regulate, as when in the 1930s the airlines promoted the idea of a new regulatory agency with the intent of using it for their own purposes.[7] This is precisely what happened in the case of the regulation of the radio industry, for it was the demands of broadcasters that led to the establishment of the FCC in 1934.[8]

Regulatory agencies such as the FCC are supposed to be independent of other branches of government and therefore insulated from political pressure. At the same time, however, their insulation often results in a lack of political support, so the agencies find it in the very industries that they are regulating.[9] Moreover, regulatory agencies often lack the expertise and information necessary to act independently. Compounding a potential conflict of interest, there is often close personal contact between the personnel of these agencies and representatives of the industry that they are supposed to be regulating. And, as in the case of the FCC commissioner who took a job with NBC, they are often alert to the possibility of going over to the other side.

Occasionally, a regulatory issue will leap into political prominence and disrupt the comfortable realtionship between a regulatory agency and the industry that it is regulating. This is what happened when the accident at the Three Mile Island nuclear power plant put the Atomic Energy Commission's role in regulating nuclear power into the political spotlight. But this is rare; for the most part regulatory agencies view the industry that they regulate as a client to be served, sometimes to the detriment of the public interest.

The FCC seems to conform to this general process, for it has failed to take strong and independent positions vis-à-vis the radio and television industry.[10] Hampered by lack of sufficient information and expertise, the FCC usually has agreed to the initiatives of industry representatives. And tempted by the opportunity to take lucrative positions in the industry, FCC commissioners have generally opted for the status quo, thereby preserving the primary purpose of broadcasting as being little more than a way of making large sums of money. This proclivity for maintaining existing arrangements has also manifested itself in technological matters. The FCC did not enthusiasticly support innovations such as FM broadcasting, UHF, and pay and cable TV. This conservatism was financially advantageous for established broadcasters, but it has been detrimental to technological innovation.[11]

THE SOCIAL AND PSYCHOLOGICAL CONSEQUENCES OF TELEVISION

Some indication of the pervasive influence of television can be gained by looking at some basic statistics. In 1992 there were 192 million television sets in American homes, and more than 98 percent of all households owned at least one television set.[12] More and more time is being devoted to watching television. In 1950 Americans spent an average of 4.6 hours per day watching TV. By 1970 the figure had jumped to 5.9 hours, and by the end of the decade the figure had increased to 6.5 hours. In 1988 American televisions were turned on for an average of 7.1 hours each day.[13] There has never been a year in which the average number of hours spent watching TV has declined to a significant degree. Only work and sleep take up more time than watching television, and children spend more hours in front of a television set than they do in school.

These figures have to be treated with some caution, however. Just because a television set is turned on doesn't mean that it is being actively watched. Research conducted on people's actual viewing habits reveals that attention is often divided. For about 20 percent of the time that the television is turned on there is no one in the room. During an equal amount of time the TV room is occupied, but nobody is watching the set. As one researcher summarized his findings, children "eat, drink, dress and undress, play [and] fight . . . in front of the set." And adults "eat, drink, sleep, play, argue, fight, and occasionally make love" in front of the flickering screen.[14] In many households, a switched-on television is as much a part of the domestic environment as the wallpaper, and often receives about as much attention.

Given the many complaints about the vapidness of most television programming, it might be thought that better-educated people spend less time watching television. But in truth, their viewing habits are little different from those of other Americans. They are more inclined to watch public television, but their tastes run to about the same mixture of comedy, adventure, and public affairs programming.[15] People of all socioeconomic groups seem generally satisfied with television programming. Most viewers are annoyed by frequent commercials, and about one-half express concerns about the effects of TV on children, yet on the whole there seems to be no widespread concern about television's impact on modern life.[16]

VIOLENCE ON TELEVISION AND ITS CONSEQUENCES

Is this complacency justified? Television has been indicted for fostering violent behavior, giving a distorted view of the world, and altering the substance of politics, to name but a few of its alleged effects. Do these criticisms have substance, or are they little more than the natterings of a few intellectuals? Does television simply provide harmless entertainment and useful information, or is it a pernicious influence on our society?

One of the most hotly debated issues concerning television's influence has centered on the medium's role in stimulating violent and aggressive behavior. There can be little doubt violence is a TV staple. By the time he or she is eighteen, the average young viewer will have seen 18,000 murders on television. Eighty percent of the programs have some violence, and eight violent episodes occur each hour. Nearly seven out of ten leading characters are involved in violence, and between one and two out of every ten are involved in killing.[17] With all of this mayhem appearing on TV, it is understandable that critics have indicted television for contributing to the rampant violence that plagues American society.

Particular attention has been given to the connection between TV viewing and aggressive behavior in young children, and a large number of studies have been devoted to this topic. Some of these studies have been based on laboratory experiments that allow the researcher to eliminate extraneous factors in order to determine the relationship between exposure to a filmed act of aggression and subsequent behavior. In a typical experiment, one group of children is exposed to a film depicting violence while a control group is not. The subsequent behavior of the two groups is then compared. A classic experiment of this sort was conducted in the 1960s. One group of preschool children was shown a film in which an actor physically and verbally assaulted Bobo the Clown, a large, inflated plastic doll. The control group saw nothing. The children were then allowed to play in a room containing a Bobo doll and other toys. The results were unequivocal; the children who had seen the simulated aggression displayed higher levels of aggressive behavior toward the hapless toy, in some instances even directly imitating the acts that had appeared in the film, such as yelling out "Lickit! Stickit" while pounding on the toy with the handle of a mallet.[18]

Many studies of this sort have been conducted. Some have found indirect as well as direct consequences of exposure to filmed violence. For example, children are more likely to play with toy guns after seeing filmed aggressive acts, even if gunplay did not appear in the film. All in all, these experimental studies are nearly unanimous in finding a connection between seeing acts of violence and subsequently engaging in aggressive acts:[19]

> The studies differ greatly in the degree to which the indicator of agressiveness employed equals violence that might be performed in day-to-day life, but a good many of them do represent valid indicators of willingness to inflict discomfort or pain on another human being, and across the total array the results are similar. The findings of experiments using different stimuli, different sets of instructions and varied samples of children and young adults are remarkably consistent.

Still, laboratory experiments such as these can always be criticized because of their artificiality. In their attempt to reduce or eliminate extraneous influences, experimenters may create an environment that bears little resemblance to reality. A film clip presenting violent behavior is not a television program, in which violent acts are presented in conjunction with a variety of other messages. Experiments also deal only with immediate cause-and-effect relationships. They therefore do not take long-term, cumulative effects into account. Finally, in these experiments the subjects may think that aggressive acts are permitted, and even expected. For all these reasons, the real-world applicability of these experiments is highly questionable.

Other researchers have attempted to circumvent the inherent limitations of laboratory experiments by conducting more "naturalistic" studies. In these, one group of children views "normal" violent programming, while another group does not. Their subsequent behavior in everyday settings is then directly observed, or is reported by parents or teachers. In contrast to the laboratory experiments, these studies when taken as a whole are inconclusive. Some studies show a positive relationship between viewing violent programming and subsequent aggressive behavior, while others do not. In some cases, aggressive behavior even diminishes after a violent program has been watched.[20] It has therefore been hypothesized that in some instances the viewing of violent acts on TV may serve as a "catharsis" that allows the harmless release of violent emotions.

A third type of research examines actual viewing habits to see if there is an association (or "correlation") between viewing violence on television and aggressive or violent behavior. These studies generally use interviews and self-reports as their primary source of information about viewing habits and behavior. For the most part, these studies do show a positive correlation between watching violent TV programs and aggressive behavior. Children and adolescents who prefer violent TV programs tend to be more aggressive in their behavior.[21] There is a basic problem with studies of this sort, however, for correlation is not the same thing as causation. The *correlation* of A with B does not necessarily mean that A *causes* B. B could cause A, or a third factor or factors could be the cause of both of them. Accordingly, a child who was already violence-prone might seek out violent programs. It is equally possible that an impoverished, culturally deprived background may be the cause of both the viewing of violent programming and aggressive behavior. In any event, a review of the existing research does not indicate that watching violence on TV *causes* aggressive behavior.[22]

A final item that provides some insight into the possible connection between televised and real violence comes from the Republic of South Africa. Due to long-standing conflicts between speakers of English and speakers of Afrikaans over the control of the media, South Africa had no television until 1974. From 1945 until 1974, a period when the U.S. murder rate rose by 93 percent, the murder rate for white South Africans actually declined by 7 percent. To be sure, more separated the United States and South Africa than the availability of television. Even so, potentially significant variables such as economic growth, urbanization, alcohol consumption, capital punishment, and the availability of firearms do not explain the difference in murder rates.[23] It is, of course, quite possible that differences other than the availability of TV explain the divergent murder rates of the two countries. Still, they are striking; if nothing else, the experiences of South Africa before and after the introduction of television merit further study.

It should also be stressed that the conclusions drawn from all of these different studies cannot be applied to the behavior of specific individuals. For most children, watching violent acts on television may not lead to aggressive or violent behavior, but for some children it may. TV affects different people in different ways. Hard-and-fast generalizations are difficult to come by, for as Wilbur Schramm has noted, "For *some* children under *some* conditions, some television is harmful. For *other* children under the same conditions, or for the same child under *other* conditions, it may be beneficial. For *most* children, under *most* conditions, *most* television is neither particularly harmful nor particularly beneficial."[24]

Like any voluntary act, violent behavior is a product of a complex set of motivations and inhibitions. All of us probably have had moments when we wanted to commit a violent act. We usually haven't done so for three basic reasons: (1) we have learned that such actions are likely to result in retaliation; (2) we know that they usually do not solve the problem; and (3) we have internalized a code of behavior that discourages such acts. Most likely, the operation of all of these prevents us from giving vent to our violent urges.

Television has the capacity to alter all of these inhibiting factors. In some televised depictions of violence, the retaliation for the violent act is delayed and muted; in many cases, a criminal may receive his or her just deserts only at the end of the program. To make things worse, the second inhibiting factor, a belief that violence does not solve most problems, is not always supported by television programming. Illegal and violent acts are sometimes used to attain socially approved goals.[25] Finally, there is the problem of television's contribution toward desensitizing people, especially children, to violence, thereby attenuating the development of a personal code of ethics that discourages violent acts. Such long-term effects of televised violence are difficult to prove, but there is some evidence that exposure to televised violence dampens a child's emotional response to the witnessing of violent acts, and diminishes their concern about such acts.[26]

It cannot be said with certainty, however, that these are the primary messages actually received by viewers of televised violence. TV is one influence among many, and powerful as the medium is, it does not automatically override the influence of family, school, church, and peers. Research into this area faces the formidable challenge of separating the effects of television from other influences, and up to now it has not been successful in this endeavor. The present state of knowledge in this area makes us reluctant to say that television violence has been a direct cause of aggressive behavior, but subsequent research may eventually prove that there is a connection. If nothing else, the uncertainties surrounding this subject show how difficult it is to determine the consequences of a particular technology with any degree of precision.

TELEVISION, INFORMATION, AND NEWS

Until recent times most people were blissfully ignorant of the world around them. Travelers might bring stories of distant places, and the literate few could read of them. Information traveled very slowly; the Battle of New Orleans was fought two weeks after a treaty formally ended the War of 1812, for the combatants were unaware of its signing. During the nineteenth century the railroad, the telegraph, and the appearance of popular newspapers brought the world closer, but coverage was still slow, spotty, and often inaccurate. Past the borders of one's own community was a world dimly preceived by most people. All that has changed today. According to one tabulation, the United States can boast 17,000 newspapers, 12,000 periodicals, and more than 400 million radios to go with the aforementioned 192 million television sets.[27] Instantaneous communications tell us of wars in the Middle East, famines in Africa, and political realignments in Europe. Although we do not quite live in a "global village," modern communications technologies have extended our vision to a degree scarcely imaginable not long ago.

Television has been a major, perhaps *the* major element in the communications revolution. For many people, television news is synonymous with news in general. In 1960, survey respondents indicated for the first time that TV had replaced newspapers as a their prime source of news,[28] and it has held that position ever since then. In addition to its primacy as a news source, television also tops all other media in regard to perceived credibility, fairness and relative absence of bias.[29] For a majority of the public, TV appears to be the most accurate and objective source of news.

At the same time, however, TV is not completely dominant. According to one survey, during one two-week period more than one-half of the respondents did not watch any national news programs.[30] The average nightly audience for network news was 34 million in 1977, an impressive figure, but one that pales against 104 million daily readers of newspapers, and 47 million weekly readers of *Time*,

Newsweek, and *U.S. News and World Report*.[31] Moreover, television is much less important as a source of local news than newspapers. Only for national and international events is television the prime source of news.[32]

It is not certain that the viewing of television news has contributed to a greater awareness and understanding of the world. One survey of teenagers and adults found that except for those who watched no television news at all, a knowledge of public figures was not connected with the extent to which local and national TV news was watched.[33] Another study found no differences in knowledge of presidential candidates' stands on issues between those who watched evening network news and those who did not, although exposure to televised political advertisements did result in viewers' being more aware of the candidates' positions.[34] One study even found that the ability of voters to give cogent reasons for liking or disliking candidates was negatively related to exposure to television news, while it was positively related to exposure to news presented in newspapers.[35]

The news presented on television is fundamentally different from the news that appears in newspapers. In part, it is a matter of coverage; all of the verbal information presented in a thirty-minute news program would fit on one page of a newspaper. The way the news is covered in the two media is also strikingly different. Newspaper reporting tends to be detached and impersonal, whereas TV reporting is more like storytelling. A news report is presented in narrative form, with a theme being developed through the use of sound and pictures as well as verbiage. A great deal of television news is presented as a kind of entertainment, narrated by newscasters who are selected primarily on the basis of their physical attractiveness. The stories are short and fragmentary, with very little contextual information. News of this sort is quickly forgotten; one study found that more than twenty percent of the viewers of TV news could not recall a single item of news an hour after having seen a news broadcast. Another study found that the average viewer retained only 20 percent of the information presented in a simulated news story.[36]

By contrast, newspaper stories are written so that the key item can be quickly grasped. Background and analysis then follow for the reader who wants to continue with the story.[37] Viewers of TV news and casual newspaper readers may have equally superficial understandings of the events in the news, but at least the latter can extend their knowledge. Television news gives few such opportunities.

TELEVISION AND POLITICS

Numerous analysts have charged that television has fundamentally altered the political process. TV, they assert, has reduced political campaigns to trivial "media events," selected the sensational over the substantive, undermined political parties, greatly inflated the cost of political campaigns, and made "image" the primary criterion of a candidate's appeal. Are these accusations true?

There can be little question that the politics of the television age is different from the politics of earlier times. In 1948 Harry Truman could wage a successful presidential campaign from the back of a railroad car. Today, televised news and political commercials have taken the place of the whistle-stop tour. On occasion, candidates have successfully contended for office without the aid of television advertising; in 1974 a Democrat won Gerald Ford's vacant congressional seat in a special election despite being too impecunious to pay for television commercials. But incidents such as this are highly atypical. Televised political advertising has become an inescapable part of electioneering.

Television advertising is expensive; a minute of prime time can cost over $200,000. And candidates buy a lot of minutes; in the run-up to the 1992 presidential election, the Bush, Clinton, and Perot campaigns spent $42.5 million, $32.1 million, and $39.5 million, respectively, on television commercials. Congressional and Senate campaigns, as well as those for other elected offices, have also made extensive use of television advertising. The typical campaign now budgets at least one-third of its campaign funds for television commercials.

As television has greatly increased the costs of conducting campaigns, the need for large campaign contributions has mounted. This in turn has increased the danger of successful candidates being beholden to the special interests that supply a significant portion of these contributions. As some cynics have suggested, we may end up with "the best elected officials that money can buy." Corruption has always been a part of political life, but the demands of television campaigning have created something new, a scramble for campaign funds that is perfectly legal, but nonetheless a threat to democratic political institutions.

Television has become a major force in American political life. Even so, there are limits to its influence. Televised political advertising isn't always decisive, and money doesn't always win elections. In most elections the majority of voters have made up their minds about who will get their vote well before the campaign gets under way. For these voters, televised advertisements help to crystalize their choices and reinforce their existing beliefs, but do not produce any fundamental alterations.[38] At the same time, however, there are many elections where the outcome hinges on the choices made by voters who make their decisions late in the political campaign. Sometimes they are a significant portion of the electorate; in the 1980 presidential campaign (which offered an unusually clear choice of ideologies and policies), 35 percent of the voters did not decide who would get their vote until the last week of the campaign, and 10 percent were undecided right up to the day of the election.[39] Many of these voters are not greatly concerned with political affairs. They have not closely followed the campaign, and their interest emerges only during the last few days of the campaign. For this reason a flurry of televised political advertisements appears during the last week of a campaign.[40] Television can therefore be decisive when an election hinges on the choices made by this group of voters.

In most cases, the voters who have made up their minds long before the election takes place are the ones most strongly influenced by party loyalties. This makes them fairly impervious to televised appeals to vote for another party's candidate. Yet party loyalties have been eroding in recent years. Only 65 percent of the electorate identify themselves as Democrats or Republicans, and most of them are not strongly committed to their party.[41] Television has been responsible for a good deal of the erosion of party loyalties because a candidate can use the medium to appeal directly to the electorate. Also, presidential and other candidates for national office are no

Even when he is out on a jog, the president cannot escape media attention. (AP/Wide World Photos.)

longer selected by party bosses in smoke-filled rooms. They are often the products of state primaries, in which televised appearances and political commercials can be the most important part of the campaign.

At first glance, the ability of candidates to use television in order to directly appeal to the electorate seems like a good thing, an improvement over the days of party bosses and political machines. In reality, electronically transmitted communications from candidates to voters are usually devoid of substance. Numerous critics of television have pointed out that the effectiveness of a TV presentation hinges on its visual appeal, with content a secondary consideration. Accordingly, televised political appeals have little room for ideas, policies, and well-articulated political positions; what really matters is the kind of personality and presence that a candidate projects on the screen. At the same time, television often conveys the idea that problems can be quickly resolved—requiring less than thirty minutes in the case of a particular show, and thirty seconds for individual commercials. It is no wonder, then, that the electorate is often attracted to candidates that offer simple and quick solutions. And these "solutions" are of course neatly packaged for television; instead of lengthy discussions of a candidate's stand on particular issues, the typical political commercial consists of thirty seconds of visual images and a few empty slogans.

The pervasive influence of television on political life should be of concern to anyone concerned about the fate of our political institutions. As some critics have argued, the greatest threat to democracy may not come from the assaults of hostile nations, but from the trivialization of the political process that occurs when television dictates the basic mode of discourse and comprehension.[42]

TELEVISION AND THOUGHT

For some students of the media, the effects of television go well beyond the ones just described. The most notable of these has been Marshall McLuhan, whom we met briefly in the preceding chapter. In McLuhan's schema, television is a "cool" medium: the televised image is indistinct and requires that viewers "fill in the blanks" in a way that involves all of the senses. Viewing television is not a single-minded, linear process, as reading is. Exposure to television in turn leads to a mode of perception that affects many other aspects of life. For McLuhan, a television-based culture is less concerned with sequence, and more with complete, all-at-once involvement. In his view, these changes in perception are transforming every aspect of our lives; McLuhan attributed everything from a desire for small cars to changes in church liturgy to the influence of television. Examples such as these seem far-fetched, and some of his predictions have not been borne out, such as his prophecy that baseball will necessarily decline because it is a "linear" game. McLuhan's focus on the influence of different types of media opens up all kinds of fascinating possibilities, but his conclusions lack clear proof, and in fact the basic thesis is virtually unprovable.

Other effects of television on mental processes are more concrete. It is likely that TV helps young children to understand how actions take place in sequence, and that it also helps them to see how a three-dimensional space is rendered in two dimensions.[43] For very young children, watching television requires new perceptual skills such as understanding that an object viewed from different perspectives is still the same object. This perceptual skill is not inherent, and watching TV may assist in its development.[44] So regardless of content, *something* is being learned when young children begin to watch TV.

Children's ability to visualize may be improved by watching television, but at the same time other elements of imagination may be depressed. Experimental studies have shown that children supply more imaginative continuations to interrupted stories that they read than they do to televised stories. In similar fashion, young children who are heavy watchers of television engage in less imaginative play than other children.[45] There is also a fair amount of evidence that television watching may contribute to underdeveloped reading skills in children. The relationship is not a perfectly direct one; it may emerge only when a certain threshold of TV watching is reached, and it seems to be affected by the kind of programming selected.[46] Still, it seems evident that for many children television is a negative influence in this regard. Finally, one might also speculate that the lack of concentration and the short attention spans that seem so common today are at least partially the result of growing up with television programs that are constantly punctuated by commercial messages and other distractions.

TELEVISION'S PLACE IN MODERN SOCIETY

In assessing television's consequences on society we are left with many unanswered questions. As with every other major technological change, the effects of television have been so diffuse and so intertwined with other changes that it is impossible to come to an exact reckoning of TV's effects. It is also important to avoid taking too narrow a view of television and its consequences. Television did not come into our world as an alien invader. The invention and diffusion of television, like those of books and radio before it, were part of larger social, economic, and political changes. Television is only the latest of a series of technologies that have been designed to expand our ability to communicate. As society has expanded in size, complexity, and mobility, the need to tie it together through improved communications media has increased apace. As we have seen, long before the ability to transmit and receive sound and pictures over great distances had been attained, some visionaries were describing the future technologies that would accomplish this. The need for electronic communication systems had been firmly established long before there was the capability of developing them.

Still, as we have seen in earlier chapters, the need for something does not ensure that it will be produced; somebody has to be willing to pay for its development. In the case of emerging radio and television technologies, there was no lack of financial support. During its early years, radio was supported because military, governmental, and business organizations perceived its relevance to their needs. Later, the emerging commercial possibilities of radio and television were eagerly exploited, and the market opened up by broadcasting further stimulated technological development.

Television and radio also have become an essential feature of modern life because they have helped to mitigate the disruption of stable community ties that has been a characteristic of our era. Economic and social developments have produced a great deal of mobility and general rootlessness. In response, television and radio have contributed to the construction of a new common culture that has served as at least a partial substitute for local ties. The loosening of these ties has also resulted in a greater degree of privatization in our lives; in a rapidly changing and often bewildering world, we retreat to the stability and security of our homes. Privacy, however, is often accompanied by loneliness. For many people, the electronic media fill a void by bringing into their private environments information and entertainment that helps to mitigate their aloneness.[47] Even so, there is ample room for debate about whether these benefits outweigh the negative consequences of the electronic media.

Questions for Discussion

1. In many countries, a considerable amount of television and radio broadcasting is done by government organizations such as the British Broadcasting Corporation. In the United States, most broadcasting is done by commercial stations. What are the advantages and disadvantages of the two systems?
2. Are you satisfied with the quality of radio and television programming? What changes would you like to see? Should the FCC actively promote these changes? How might it do so?

3. Do you agree with this chapter's general conclusions about television's influence on violent behavior? Do you think that there is too much violence on TV? Should television be more closely regulated so less violence appears on the screen?
4. What medium do you rely on most for coverage of current events: newspapers, radio, or television? How does your reliance on this medium affect the information and analysis that you receive?
5. One of the most exciting technological developments on the horizon is high-definition television (HTV), which produces televised images of great sharpness and clarity. American companies have been reluctant to invest in this new technology; as a result, both inside and outside the industry there has been a call for government sponsorship (i.e., money) of HTV. Would this be a proper role for government? How might HTV influence the development of future technologies? Would a failure to develop HTV affect America's industrial competitiveness?

Notes

1. For extensive coverage of the early history of radio, see Hugh G. J. Aitken, *Syntony and Spark: The Origins of Radio* (Princeton, N.J.: Princeton University Press, 1985) and *The Continuous Wave: Technology and American Radio, 1900–1932* (Princeton, N.J.: Princeton University Press, 1985).
2. Eric Barnouw, *Tube of Plenty: The Evolution of American Television* (New York: Oxford University Press, 1975), pp. 4–5.
3. S. Handel, *The Electronic Revolution* (Harmondsworth, England: Penguin, 1967), pp. 129–130.
4. W. Rupert Maclaurin, *Invention and Innovation in the Radio Industry* (New York: Macmillan, 1949), pp. 225–240.
5. Barnouw, op. cit., p. 100.
6. Ibid., pp. 200–201.
7. David Nachmias and David H. Rosenbloom, *Bureaucratic Government USA* (New York: St. Martin's Press, 1980), p. 23.
8. Kenneth J. Meier, *Politics and the Bureaucracy: Policymaking in the Fourth Branch of Government* (North Scituate, Mass.: Duxbury, 1979), p. 72.
9. B. Guy Peters, *The Politics of Bureaucracy: A Comparative Perspective* (New York: Longman, 1978), p. 120.
10. Barry Cole and Mal Oettinger, *Reluctant Regulators: The FCC and the Broadcast Audience* (Reading, Mass.: Addison-Wesley, 1978).
11. George Comstock, *Television in America* (Beverly Hills: Sage, 1980), p. 15.
12. United States Department of Commerce, Bureau of the Census, *Statistical Abstract of the United States 1984* (Washington, D.C.: U.S. Government Printing Office, 1985), p. 737, and *Statistical Abstract, 1990*, p. 550.
13. *Statistical Abstract, 1984*, p. 542, and *Statistical Abstract, 1989*, p. 544.
14. Comstock, op. cit., p. 29.
15. Gary A. Steiner, *The People Look at Television* (New York: Alfred A. Knopf, 1963).
16. Comstock, op. cit., pp. 40–41.

17. George Gerbner and Larry Gross, "Living with Television: The Violence Profile," *Journal of Communication* 26, 2 (Spring 1976): 187.
18. Albert Bandura, Dorothea Ross, and Sheila A. Ross, "Transmission of Agression through Imitation of Aggressive Models," *Journal of Abnormal and Social Psychology* 63, 3 (1961): 575–582.
19. George Comstock et al., *Television and Human Behavior* (New York: Columbia University Press, 1978) p. 229.
20. Jonathan L. Freedman, "Effects of Television Violence on Aggressiveness," *Psychological Bulletin* 96, 2 (September 1984): 234–235.
21. Ibid., pp. 236–237.
22. Ibid., pp. 237–243.
23. Brandon S. Centerwall, "Television and Violent Crime," *The Public Interest* no. 111 (Spring 1993).
24. Wilbur Schramm, Jack Lyle, and Edwin B. Parker, *Television in the Lives of Our Children* (Stanford, Calif.: Stanford University Press, 1961), p. 1.
25. Comstock, op. cit., p. 83.
26. Comstock et al., op. cit., p. 238.
27. Neil Postman, "Of Learning, Luddites, and Life" *Technos* 2, 4 (Winter 1993): 26.
28. Barnouw, op. cit., p. 314.
29. Comstock, op. cit., p. 47.
30. John P. Robinson, "The Audience for National TV News Programs," *Public Opinion Quarterly* 35, 3 (Fall 1971): 403–405.
31. Comstock, op. cit., p. 48.
32. Comstock et al., op. cit., p. 138.
33. Ibid., pp. 170–171.
34. Ibid., p. 332.
35. Comstock, op. cit., p. 63.
36. Neil Postman, *Amusing Ourselves to Death: Public Discourse in the Age of Show Business* (New York: Viking/Penguin, 1985), p. 152.
37. Melvin L. DeFleur and Everette E. Dennis, *Understanding Mass Communication*, 2d ed. (Boston: Houghton Mifflin, 1985), p. 450.
38. Comstock et al., op. cit., pp. 340–341.
39. Thomas E. Patterson, "Voter's Control of Information," *Society* 22, 4 (May–June 1985): 56.
40. L. Patrick Devlin, "Campaign Commercials," *Society* 22, 4 (May–June 1985): 45.
41. Patterson, op. cit., pp. 52–53.
42. Postman, op. cit., pp. 125–141; Jerry Mander, *Four Arguments for the Elimination of Television* (New York: Quill, 1978).
43. Patricia Marks Greenfield, *Mind and Media: The Effects of Television, Video Games, and Computers* (Cambridge, Mass.: Harvard University Press, 1984), p. 33.
44. Ibid., p. 17.
45. Ibid., pp. 88–89.
46. Johannes W. J. Beentjes and Tom H. A. Van der Voort, "Television's Impact on Children's Reading Skills: A Review of Research" *Reading Research Quarterly* 23, 4 (Fall 1988).
47. Raymond Williams, *The Long Revolution* (New York: Penguin Books, 1965), pp. 26–27.

PART FIVE

THE TOOLS OF DESTRUCTION

Chapter 13 Weapons and Their Consequences

For most of human existence, the baser instincts of humanity were checked by the limitations of the weapons that could be used. Clubs and stone knives could be used at close range, and rocks and other missiles could be hurled at a target a few dozen yards away. There are still places in the world where warfare is conducted as it was thousands of years ago. Unfortunately, however, human ingenuity has not always been confined to activities that make our lives better. On many occasions our most inventive minds have used their talents to find new ways of wreaking havoc on other human beings. Warfare presents the dark side of technological progress; although technological advances have bettered many aspects of human life, they have also led to a terrifying expansion in the ability to kill and destroy.

MILITARY TECHNOLOGY IN THE ANCIENT WORLD

The technological changes that marked the beginning of civilization's development were from the start used for martial purposes. The production of metals, first bronze and then iron, resulted in more lethal weapons, as well as armor to protect against them. The use of metal was not the only source of improved weaponry. Around four thousand years ago a more powerful bow was created by laminating it with sinew on one side and with horn on the other. This combination of compression and tension greatly expanded the range at which an enemy could be assaulted. In the ancient Middle East archers often shot their arrows from a horsedrawn chariot, which was invented at about the same time, giving invading armies great mobility and striking power. So equipped, the armies of Assyria and Persia successfully invaded their neighbors' territories and established extensive empires.

Effective as they were, military forces based on the chariot and bow did not conquer everything in their path. Horses and chariots operated with great effectiveness in the open expanses of Central Asia and the Middle East, but they were far less useful in mountainous areas, as the Persians found to their sorrow when they invaded Greece. Equally important, horses had to be fed, and in the absense of adequate forage, an army using large numbers of horses quickly outran its supply lines. The armies of ancient Greece made some use of horses, but on the whole they relied on the foot soldier, armed with a spear and protected by a shield, a helmet and possibly a breastplate. Although Greek military technology was less impressive than that of Persia, this deficiency was compensated by tight organization and discipline. Arrayed in a phalanx, a mass formation up to eight deep, Greek warriors were able to resist the horse-mounted invaders from the East, who, for all their equipment and skills, could not effect a decisive breakthrough.

The Romans brought a greater degree of mobility to infantry combat by deploying their soldiers in smaller units of 120 men each. Although armaments were little changed from the time of the Greeks, the Romans carved out a vast empire by employing superior tactics and, above all, because they could rely on the greater discipline and coordination of their legions. Mobility, after all, is no advantage if it leads to the dispersal of troops and the lack of unified action. The genius of Rome was at least as much organizational as it was technological, and it gave the Romans a military superiority that lasted for centuries.

While the weapons of the infantry soldier changed little in the ancient world, significant strides were made in the development of siege machinery. The most fearful weapons of antiquity were catapults, first used in Sicily during the fourth century B.C. They were then employed with considerable success by Philip of Macedon, the father of Alexander the Great, and played an important role in many of the battles waged by the Greeks and Romans, as well as by their enemies. It is also significant that siege machinery was the creation of some of the first specialists in technology; indeed, the word "engineer" is derived from the Latin *ingenium*, an ingenious device used for siege warfare.[1]

Catapults hurled rocks or large arrows by releasing the stored-up energy of a torsion spring, made of animal sinew and human hair. Hair was much prized as a military supply; there is an ancient report of thousands of pounds of hair being sent as a present from one king to another.[2] The most powerful catapults were capable of hurling a bolt or a small stone a distance of 800 yards (although 400 yards was a more normal range), while a sixty-pound rock thrown a distance of 150 yards produced the best effect when fortifications were the target. A smaller catapult (known as the *onager*) used by the Romans for battlefield service was probably capable of flinging an eight-pound projectile up to 500 yards.[3] The main limitation of all catapults was that hair and sinew stretch when wet, so in damp climates the springs lost their resiliency and therefore much of their power. Despite this limitation, these devices played an important role in siege warfare. Each Roman legion was supported by 55 catapults, and 300 catapults were deployed by Rome during the siege of Jerusalem in 70 A.D. Even after the invention of gunpowder and cannon, many decades were to pass before the destructive power of these weapons was surpassed.

The use of such siege machinery conferred great strength on offensively minded empire-builders like the Romans. The crude fortifications of the "barbarians" usually provided only scant defenses against invading armies armed with these early missile launchers. Conversely, Roman military superiority was less evident when Rome itself was put on the defensive during the barbarian invasions. Roman defensive technology was inferior to the offensive weaponry of the time; this incapacity, when coupled with the many internal problems besetting Rome, resulted in the eventual disintegration of one of the world's greatest empires.

MILITARY TECHNOLOGY AND THE FEUDAL ORDER

Military technologies made few advances during the centuries following the fall of Rome. But in the sixth century a new technology emerged that fundamentally changed the nature of warfare, and with it the social structure of Europe. This innovation was not itself a weapon, but it made existing weapons far more effective. The device was the stirrup. Originating in the Middle East or possibly China, the stirrup added a new dimension to combat. Horses had long been used in battle; as we have seen, many battles were won through the use of horse-drawn chariots from which arrows could be shot and spears thrown. But a warrior mounted directly on a horse was precariously perched. Although cavalry charges were sometimes employed in battle, in most cases when it came time to fight, the warriors dismounted and fought on foot.

The stirrup greatly amplified the destructive potential of a warrior by fixing him firmly to his horse. The horse was no longer merely a means of conveyance; its power could now be used to augment the striking power of a spear or lance. As we shall see, the horse-mounted knight was by no means invincible, but his determined charge could be devastating. The stirrup made the mounted knight the key figure in warfare, and medieval people were hopelessly insecure if they could not enlist his service.

The central importance of the mounted knight forced a number of economic and political changes. A horse was an expensive item, and its maintenance required considerable outlay. At the same time, the mounted knight was the product of years of specialized training, during which time he could not be engaged in productive activities. The maintenance of a horse, knight, his assistants, and their training required three to four hundred acres of land.[4] Even kings lacked the financial means and administrative capacity to raise and support an army of knights on their own. Instead, they gained the military support of mounted warriors by granting them tracts of land in return for their military support. This was the basis of the feudal order, the dominant form of political organization throughout the Middle Ages.

Feudalism was an inherently decentralized system, and armies were temporary and disorderly assemblies. Knights relied on their individual skill and courage, and were disinclined to submit to centralized discipline and strategic planning. Battles were often chaotic affairs involving a great deal of individual combat and little in the way of careful tactical planning. On occasion opposing forces might spend several days just trying to find each other so that a battle could take place. The military technologies of the Middle Ages thus led to a retreat from the principles of discipline and precise organization that had made Roman armies so formidable.

Horse-mounted combat also generated the culture of chivalry. The word itself derives from the French *chaval* or horse. The horse-mounted knight was different from ordinary soldiers by virtue of his noble birth and the long period of training that was required to develop his proficiency in combat. The chivalrous knight lived by a code that regulated certain aspects of warfare. Noncombatants—provided that they were Christians—were to be left in peace, and combatants were supposed to adhere to specific rules of battle. Above all, war was seen primarily as a contest between members of a distinctive warrior class, in which individual glory was at least as important as the attainment of military objectives. War still remained a grisly business, but as long as war was considered to be a contest between participants in a common culture, some limits were set on the manner in which it was waged, and the land was usually spared the total devastation that has been a hallmark of more modern conflicts.

NEW WEAPONS AND THE DECLINE OF FEUDALISM

The era of knightly combat began to come apart when once again new technologies changed the nature of warfare. During the early Middle Ages mounted combatants could charge into massed formations of infantrymen and devastate them in one crushing attack, but by the fourteenth century an extraordinarily simple weapon, the pike, was effectively employed to break up cavalry assaults. Used to especially good effect by the Swiss, the nineteen-foot-long pike was wielded by a solid formation of soldiers with four rows of pikemen extending their weapons to the front line. United by an iron discipline and a high morale that came from being free men in the service of their nation, the Swiss pikemen became the most feared soldiers in late medieval Europe. Hard-charging knights were dispersed by the pikemen and then assaulted by other soldiers wielding halberds, a long battle axe.[5]

Infantry soldiers truly got the upper hand when they gained the ability to attack knights from afar. The first weapon to confer this advantage was the longbow. Archers had been part of combat for thousands of years—the bow appears to have been in use 17,000 years ago[6]—but they were limited by the weapons they used. The bow was transformed into a decisive weapon simply by making it longer. At the same time, however, the longbow was a difficult weapon to use properly, for it required considerable strength and skill, and that came only with extensive practice.

First used by the Welsh during their resistence to English conquest, the longbow was adopted by King Edward I during the early fourteenth century. In skilled hands, this bow, which was nearly six feet long, could be a devastating weapon, capable of rapidly firing a dozen arrows in succession at a range of up to 650 feet.[7] Drawn back to the ear instead of the chest, as was the case with smaller bows, the longbow was capable of driving an arrow through several inches of oak, and although it could not penetrate the best plate armor, it easily pierced chain mail or the joints in plate armor.[8] Horses, having less armor protection, were especially vulnerable, and when a horse went down its rider was in serious trouble. The prowess of English archers was clearly demonstrated at the Battle of Crecy in 1346. Despite a two-to-one numerical inferiority, the English forces won a decisive victory. A rain of arrows penetrated the chain-mail armor of the French knights and wrought even greater destruction on their horses.

The mounted knight was further threatened by an eleventh-century invention, the crossbow. A mechanically sophisticated device, the crossbow used a lever or a crank-and-ratchet assembly to draw the string. The crossbow had a lot of stopping power, for it fired a bolt that weighed a half pound or more. Its range and accuracy were superior to that of the longbow, but the cumbersome process of winding it restricted its rate of fire to only two bolts per minute. It was thus most effective when the archer could find cover where he could safely reload.[9] Despite this deficiency, the crossbow was a formidable weapon that made the life of the knight all the more precarious.

One answer to these new offensive weapons was the replacement of chain mail by plate armor. This armor became increasingly thick, until by the late sixteenth century a mounted knight was weighted down with as much as a hundred pounds of armor. His horse, too, required protection, for a felled horse often meant death for its dismounted rider. The result was a crippling loss of mobility for the knight and his steed. Maneuverability and speed, hitherto the great virtues of the mounted warrior, were lost, and the knight ceased to be the mainstay of the European battlefield. Horse-mounted soldiers continued to play an important part in combat, not as lance-wielding knights, but as cavalry armed with pistols and sabres. If carefully coordinated with artillery salvos, a cavalry unit could force the enemy to maintain solid ranks, thus presenting a convenient target for artillery.[10] But the traditional knight was no longer the dominant player in the game. At best he was an auxiliary, at worst a vulnerable target.

The castle, the other great symbol of the Middle Ages, was also challenged by new military technologies. Early medieval fortifications were little more than mounds of earth surrounded by wooden walls, a far cry from the elaborate defenses constructed by the Romans. During the Crusades, Europeans learned a great deal by observing the castles constructed by their Moslem opponents, and they took much of that knowledge back home. Throughout the early Middle Ages castles stood as virtually impregnable redoubts that conferred as much security as could be hoped for during those turbulent times, for about the only way to conquer a fortified position was to starve it into submission.

Attackers attempted to breach the wall of castles and other fortifications with weapons little changed from Roman times, until, in the twelfth century, besieging armies began to employ a device known as a trebuchet. This weapon consisted of a long arm that was unequally balanced on a fulcrum. The long end, which held the projectile, was held in place by a catch, while the short end was loaded with heavy weights. When the catch was released, the long arm flew up and hurled the projectile in the direction of the enemy. These could be formidable weapons, capable of delivery a 300-pound projectile a distance of up to 300 yards.[11] Although its range was less than that of the catapults of antiquity, the greater weight of the projectile made the trebuchet a more effective siege weapon. Also, since it did not depend on torsion springs made from hair as the older missile launchers did, it was effective in all kinds of weather. Even so, improved siege engines were rarely up to the task of battering down sections of walls. Failing this, they were used in the hope of demoralizing a castle's inhabitants by lobbing incendiaries, live snakes, and dead horses over the walls. And on occasion they were used to speedily return prisoners to their place of origin over the walls.[12]

THE GUNPOWDER REVOLUTION

The crossbow, longbow, pike, and trebuchet posed a severe challenge to the knight and the feudal system that maintained him. New weapons based on gunpowder sealed his fate. Incendiary weapons had long been used in combat. Flaming arrows, vases filled with a mixture of pitch, sulphur, and boiling oil, and primitive grenades filled with naphtha were all used in battle. One of the most terrifying weapons used in antiquity was Greek fire. Although its ingredients are a matter of debate even today, there can be little doubt of its effectiveness. Like modern-day napalm, it clung to whatever it came into contact with and burned fiercely, even on water. First used in seventh-century Byzantium, it served as a decisive weapon in naval engagements and as an anti-siege weapon. At the same time, however, Greek fire was used only sparingly, in part because it was more effective as a defensive weapon than as an offensive one. No less important, the rulers of Byzantium were reluctant to make widespread use of it for fear that it would fall into enemy hands, a concern that has limited the employment of many other "secret weapons" throughout history.[13]

Early incendiary devices were generally employed as auxiliaries to armies whose main weapons continued to be swords, spears, and bows. With the invention of gunpowder, the nature of warfare underwent a profound transformation. As with many of the inventions that transformed medieval European society, gunpowder originated in the East. First used for medicinal purposes by its Chinese inventors, gunpowder became the propellant for simple rockets during the Tang Dynasty in the eighth century.[14] When first used in Europe during the early fourteenth century, gunpowder was used to fire large arrows and stone cannonballs. At first, these artillery pieces were little more than a nuisance. They had less destuctive power than a good trebuchet, and their chief advantage was that they could be built more cheaply and be more easily transported than mechanical missle launchers. Their barrels were often made of wood; when iron was used the barrel was made from parallel strips that were welded together. In neither case could the cannon withstand repeated firing, and their projectiles were stone balls that often shattered when fired.

Artillery had begun to show its mettle by 1453, when Moslem attackers successfully used cannon to batter down the walls surrounding Constantinople prior to taking the city. In the succeeding years, a number of improvements resulted in larger, more accurate and reliable weapons. Iron and bronze gun barrels were cast through the use of techniques that had been used for the production of church bells—an ironic application of a peaceful technology to a martial one. Iron cannonballs replaced stone ones. "Corned" gunpowder, which was made into coarse grains, produced a more consistent and rapidly burning charge. As a result of these innovations, by the middle of the sixteenth century the fortified castles that had long been immune to direct attack were now vulnerable.

Still, the age of the castle was by no means at an end. Military engineers strengthened castles by backing their walls with loose dirt, which absorbed much of a cannonball's force, and by constructing bastions from which their own cannon could be fired. Such fortifications provided a good deal of security until the mobile artillery developed during the late eighteenth century resulted in massed firepower, once again tipping the balance in favor of the besiegers. Until then, castles conferred a great deal of security against attack, and did much to preserve the independence of smaller states, thereby preventing the political consolidation of Europe.[15]

Smaller, more mobile artillery pieces were also used to good effect on the battlefield. Of particular importance were the innovations of Sweden's Gustavus Adolphus during the Thirty Years' War (1618–1648). Instead of deploying the large field pieces commonly used, Gustavus relied on cannon that fired a nine- or four-pound ball. Light in weight, they could be handled by a small crew and rapidly deployed during the course of battle. Artillery took on its classic role in battle: softening up the enemy's lines in preparation of a charge by the cavalry and infantry.

As we have seen, technological advances often occur through the expansion of a small-scale device or process. In the case of firearms, however, the process was reversed. Handguns were at first nothing more than small cannon, lacking even a stock, and it was not until a century after the introduction of cannon that useful hand-held firearms began to appear in battle. Still, they had many shortcomings. Large artillery pieces were aimed at a fixed target and then had their charge lighted through a touchhole. In contrast, small guns had to be used against rapidly moving targets. They could not be accurately aimed if a soldier had to shift his gaze to the weapon's touchhole. The answer to this problem was the matchlock, which used a smoldering wick held by a trigger-operated clamp to ignite the powder. Even so, firing this gun was a cumbersome process, requiring dozens of separate procedures to load, fire, and clean the gun. As many as fifteen minutes were required to load and fire one of these weapons.[16]

Despite these shortcomings, by the first quarter of the sixteenth century small arms had graduated from being battlefield auxiliaries to indispensable articles of combat.[17] Subsequent improvements made firearms all the more significant. The cumbersome matchlock eventually gave way to the wheellock and then the flintlock musket, which was capable of firing three rounds per minute.

The firepower and accuracy of these weapons greatly augmented the offensive power of armies, provided that they were used in conjunction with proper battlefield procedures. Military commanders learned to deploy their soldiers in rows, so that a soldier could fire his weapon and then retreat to the back rows where he could reload in relative safety. Still, troops could not always be depended on to use their weapons effectively. Although a musket could fire a ball several hundred yards, it was accurate only to a distance of about eighty yards. The familiar "don't fire until you see the whites of their eyes" was a wise admonition, but difficult to adhere to under combat conditions. In the heat of battle, soldiers were inclined to fire prematurely, and many were so distracted that they completely botched up the necessary routines, as one report from the American Civil War indicates:[18]

> The official report of the examination of the arms collected upon the battle-field of Gettysburg, states that "Of the whole number received, 27,574, we found at least 24,000 of these loaded; about one-half of these contained two loads each, one-fourth from three to ten loads each, and the balance one load each. In many of these guns from two to six balls have been found, with only one charge of powder. In some, the balls have been found at the bottom of the bore with the charge of powder on top of the ball. In some cases as many as six paper regulation caliber '58 cartridges have been found, the cartridges having been put in the gun without being torn or broken (preventing them from being exploded by the percussion cap). Twenty-three loads were found in one Springfield rifle-musket, each loaded in regular order. Twenty-two balls and 62 buckshot with a corresponding quantity of powder, all mixed up together, were found in one percussion smooth-bore musket.

An early seventeenth-century soldier loading his matchlock musket. (Peter Newark's Historical Pictures.)

This sort of bungling could be expected, given the panic and paralysis often experienced by men in combat. In order to counter it, successful military leaders had to institute rigid battlefield procedures that had been deeply imbued through countless parade-ground exercises. The numerous steps required for the effective use of firearms were incessantly driven into the troops until they became virtually automatic, even in the face of withering attacks from the enemy. In complete contrast to the unruly egocentricity of knightly combat, warfare had become a routinized procedure, while soldiers "became replaceable parts of a great military machine just as much as their weaponry."[19]

Earlier chapters have stressed the complementarity of organizational and technological changes, and nowhere is this better illustrated than in the military realm. A well-drilled army always has an advantage over one that is lax in its ways, and the technological development of warfare made this even more the case. The handling of firearms required a precise series of operations; incessant drill was required if soldiers were not to be totally unhinged during the course of battle. Adhering to an iron discipline created by constant drilling and blindly obedient to a rigid chain of command, European armies of the eighteenth century had become some of the most grimly efficient organizations the world had ever seen.

The creation of these armies had a significance that transcended their importance in the military sphere, for they served as the prototype of new forms of civilian organization: routinized, regimented, and hierarchical. Obedience to authority is nothing new in human societies; it is, after all, one of the most important lessons learned by young children. But this sort of authority is embodied in a particular person—first a parent, and then another authority figure. Soldiers of this era obeyed the authority of a king and his delegates, but equally important they had to submit to standard procedures drummed in through constant repetition. In time, churches, schools, and factories made abundant use of these principles, and even in our own times this mode of organization is predominant in many areas of life.

WAR AND THE CENTRALIZED STATE

Artillery and firearms expanded the scope of warfare. Battles were more complex, organized affairs, while at the same time sieges became more elaborate. Both forms of warfare intensified logistical problems; according to one calculation, maintaining an army of 50,000 required the daily provision 475 tons of food for soldiers and horses.[20] Bullets and powder also had to be readily available. This meant that an army required long supply lines, giving rise to continual strategic problems of maintaining these lines, and considerably enlarging the zone of military operations.

New forms of warfare greatly increased the cost of military operations. The feudal nobility had the means to pay for their own arms and armor, and a longbow did not entail large expenditures. In contrast, large contingents of musketeers and artillerymen required heavy financial outlays if they were to be adequately trained and equipped. In most cases only the monarch of a country could meet these demands.[21] At the same time, firearms gave centralized states a decisive advantage in expanding their territory and increasing control throughout their domain.[22] New weapons technologies thus reinforced the rising tide of nationalism, centralization, and royal absolutism that characterized European history from the sixteenth century onward.

The active involvement of centralized governments was also evident in the standardization of weaponry. Prior to the seventeenth century most soldiers did not wear uniforms, and the weapons they used were of great variety. Successful military leaders such as Gustavus Adolphus and Oliver Cromwell strove to standardize weaponry and the other accoutrements of war. Standardization, however, often acts as a brake on technological progress. Once a gun or cannon becomes an army's standard weapon, its replacement or substantial modification requires a multitude of parallel changes. Not only will the weapon's auxiliaries, such as bullets, have to be changed, but so will a host of procedures, as well as the culture that had grown up around it. Soldiers, like the rest of us, can become quite attached to doing things in a particular way, and they may be especially prone to adhere to established ways of doing things when engaged in such an inherently risky business as warfare.

Accordingly, the revolutionary changes brought on by the use of gunpowder lost momentum. During the eighteenth and early nineteenth centuries improvements in weapons were few and far between. The invention of rifled gun barrels greatly improved the accuracy and range of firearms, but they did not supplant the traditional musket. Indeed, the "Brown Bess" flintlock musket remained the standard English infantry weapon for 160 years before it was replaced by breechloading rifles (in which the bullet is loaded from behind the barrel, rather than through the muzzle) during the mid-1800s. Artillery fared no better; according to A. Rupert Hall, "The guns of Queen Victoria's wooden ships were capable of little more accurate practice than those of Drake's fleet which defeated the Armada."[23]

But within Queen Victoria's lifetime military technologies changed dramatically. The invention of bullets that expanded as they were fired made rifled firearms much easier to load, for bullets did not have to be laboriously rammed into the bore. The development of breechloading weapons made loading more rapid, and had the added benefit of allowing soldiers to remain prone while reloading instead of standing upright, where they presented a conspicuous target. Improved powders produced higher muzzle velocities, allowing the use of smaller bullets and therefore lighter firearms. These new powders also had the advantage of being smokeless, so a soldier did not give away his position when he fired. But none of these changes can be attributed to developments exclusively within the realm of weapons production; they were products of an emerging industrial order, and as such, they will be considered in greater depth in the next chapter.

TECHNOLOGICAL CHANGE AND NAVAL CULTURE IN THE ERA OF THE BATTLESHIP

While land warfare in the nineteenth century was being transformed by the use of new weapons, a similar revolution was taking place on the sea. One key element of this revolution was the use of steam engines to propel ocean-going vessels so they were no longer affected by the vagaries of the winds. Moreover, steam propulsion was the perfect complement to steel construction, for the new energy source allowed the construction of much larger and heavier ships. The ships used by England's Admiral Nelson at the beginning of the nineteenth century displaced 2,000 tons at most; by the 1860s they were up to 9,000 tons, and by the end of the century they displaced 20,000 tons.[24]

Ships of this size were able to carry a formidable array of guns, made more accurate and rapid-firing by rifling, breechloading, and improved powders and projectiles. All of these developments reinforced a trend that had begun when cannon first began to be mounted on ships. Before this time, the traditional pattern of naval combat had been based on closing with the enemy vessel (ramming it if possible), boarding it, and then subduing its crew through the naval equivalent of battlefield combat. Ship-mounted cannon made these tactics obsolete; opposing ships now attempted to sink or disable their foe through long-distance cannon fire. Countries such as Spain and the city-states of Italy that clung to the old methods of combat put themselves at a great disadvantage, and over time the naval balance of power began to shift to England and the other nations that embraced the use of shipboard artillery.[25] By the beginning of the twentieth century the steam-powered battleship with its collection of heavy ordnance had become the very embodiment of modern weaponry, the most fearsome military artifact the world had ever seen. More than simply a weapons system, it came to symbolize a nation's armed might, both for traditional naval powers like England and aspiring ones like the United States and Japan.

As with any profound technological change, the development of the battleship created numerous problems of adjustment. Modern navies became dependent on the availability of coaling stations located along their areas of operation, and this gave a strong impetus to imperial expansion, an expansion that was facilitated by growing naval power. Navy officers basked in the reflected glory of modern warships, but at the same time, the modern naval vessel forced changes in military routines that were accepted with great reluctance. In the days of sail, a seaman's life revolved around a unique set of routines that had imbued navies with a distinct culture. Steam changed many of these; many sailors were now little more than floating factory workers—stoking coal, watching gauges, and keeping the machinery in good repair.[26] The result was a serious erosion of traditional virtues, and the eventual replacement of old-guard officers with more technically minded ones. Just as the longbow and the gun destroyed the medieval knight and the culture of chivalry, advances in naval technology put an end to a way of life that had been common to generations of sailors.

WEAPONS AND THE MAKING OF THE MODERN WORLD

As a result of all these changes on land and sea, the conduct of war began to change markedly. Although the full implications of new military technologies were not always quickly grasped by military planners in Europe, they were put to devastatingly good use abroad by the European armies and navies that brought large portions of Asia and Africa under imperialist domination. The technological superiority of Western military forces was clearly demonstrated during the first Opium War (1839–1842), when the cannon of British gunboats battered down Chinese fortifications, and the superior firepower of British soldiers routed a numerically superior Chinese army. A single British ship armed with two 32-pound cannon destroyed nine war junks, five forts, a shore battery, and two military stations—all in one day.[27] China, the originator of gunpowder and many other advanced martial technologies, had fallen woefully behind the Western world, and paid the price in its subjugation by the West.

By the second half of the nineteenth century the technological gap between the Western powers and the rest of the world had grown even larger. The disparity was most marked in Africa, where breechloading repeating rifles allowed small military contingents to prevail over native troops that at best were armed with flintlock muskets, and often with only spears and shields. Numerous instances could be cited of a few hundred European soldiers utterly routing native armies that numbered in the thousands.[28] To take just one example, at the Battle of Omdurman in the Sudan a British force, assisted by a gunboat on a nearby river, confronted 40,000

A World War I machine gun crew. (Historical Pictures Service, Chicago.)

Dervishes. After five hours of combat, twenty of the British soldiers and an equal number of their Egyptian allies were killed. For the Dervishes, the battle was more costly; 11,000 of them lay dead.[29]

To this superiority in firepower the Europeans were able to add all of the other technological innovations that had transformed nineteenth-century life: steamships, telegraphs, and railroads, as well as newly developed medicines that protected troops against tropical diseases. The result was the swift conquest of large areas of the world, so that by the year 1914 more than 84 percent of the world's land mass was dominated by Europe.[30]

The most effective weapon of European imperial advance was the machine gun. Used for the first time during the American Civil War, the machine gun came into its own during the colonial wars of the late nineteenth and early twentieth centuries, when a single gun crew could cut to shreds any native force foolish enough to attack them. But little did the bearers of "the white man's burden" realize that this product of Western ingenuity would soon be used against their own kind. Smug with the belief that the machine gun took its lethal toll only when directed against "lesser breeds of men," Europeans were slow to realize that the machine gun had transformed the conditions of warfare.[31] That realization came with frightening force during the first years of World War I, as generals on both sides continued to hurl their troops against enemy lines fortified by machine gun nests. The result was an appalling slaughter, as attested by 60,000 British casualties during the first day of the Battle of the Somme.

While the machine gun was rewriting the rules of land warfare, the submarine was doing the same thing to war on the sea. Slipping beneath the ocean's surface and firing its torpedos at enemy ships blind to its presence, the submarine defied all of the conventions of naval combat, as well as the international laws that covered the attack of commercial vessels. Instead of adhering to established prize rules that stipulated that passenger and crew be allowed to board lifeboats before the sinking of their ship, submarines remained beneath the surface and summarily sunk their unsuspecting prey. The submarine was the cornerstone of Germany's naval strategy, and during the First World War their U-boats wreaked havoc on Allied shipping. But Germany's success was fleeting; unrestricted submarine warfare ultimately brought the United States over to the side of Britain and France in 1917, and decisively tipped the balance in their favor.

At the same time that improved artillery and the machine gun were bringing new horrors to the battlefield, the airplane was beginning to demonstrate what modern technologies could do to civilian populations. First confined to observation and reconnaissance, by the third year of World War I aircraft were being employed as an offensive weapon. Bombs dropped on England from German airplanes and Zeppelins killed 1,300 and injured another 3,000, as well as doing a fair amount of material damage.[32]

Later wars were to eclipse these figures by huge margins. Even before the first A-bomb was dropped, American bombing attacks on Japanese cities during World War II killed 260,000 and injured 412,000, and destroyed 40 percent of sixty-six cities' built-up areas.[33] A single incendiary attack on Tokyo destroyed more than a quarter-million buildings, killed 84,000, wounded more than 40,000 and left more than a million homeless.[34] On August 6 the first atomic bomb was dropped on Hiroshima, killing more than 50,000 people, injuring as many more, and destroying half of the city. A second A-bomb had similar results when it was detonated over Nagasaki.

Aerial bombardment also devastated much of Germany. Tens of thousands of bombs were dropped on urban industrial centers, leaving thousands dead and many more homeless. A single raid on Berlin that took place toward the end of the war may have killed as many as 25,000 civilians.[35] Still, it is not certain that the bombing of industrial centers with large civilian populations during World War II was the decisive factor in winning the war in Europe. In part, this was due to the inaccuracy of strategic bombing; only 20 percent of the 5.4 billion pounds of bombs dropped by the Allies landed within 1,000 feet of their intended target.[36] Although large portions of cities were leveled and many industrial plants were damaged or destroyed, German industry was never brought to its knees. Many factories were put quickly back into operation after apparently having been destroyed. Substitutes were found for components that could no longer be produced. German industrial production increased each year until 1944, reaching its highest level in the year when Allied bombardment was at its greatest. Despite massive bombardment in 1944 German industry produced three times as many military aircraft, five times as many armored vehicles, and eight times as many artillery weapons than it had in 1941.[37]

Equally important, civilian morale never broke down in the face of continual air raids. It may even have been the case that bombardment resulted in a stiffening of the German and Japanese resolve to continue the war. Although there is ample room for debate concerning the effects of strategic bombing, many historians are of the belief that it was not decisive in the Allies' winning of the war. The official U.S. Strategic Bombing Survey, conducted immediately after the war, seems to bear out this negative assessment; in Walter Millis' words, "It would probably be fair to summarize the net conclusion as showing that strategic bombing, when armed only with TNT and incendiary weapons, had involved much greater casualties and had produced much smaller military results than had been expected."[38]

Part of the explanation for this may lie in the nature of long-term bombing attacks. According to B. H. Liddell Hart's analysis, strategic bombing with conventional weapons puts pressure on an enemy, but does not produce the decisive result that comes with a sudden shock. Instead of producing quick, widespread results, strategic bombing, even if it results in considerable devastation, only increases pressure slowly. But, as Liddell Hart noted, "Human beings have an almost infinite power of accommodation to degradation of living conditions, so long as the process is gradual."[39]

Yet once again technology has altered the rules of war. Missiles with nuclear warheads are shock weapons of the greatest magnitude. A nuclear war would be like no other war of the past, and would result in death and destruction on a scale that can scarcely be imagined. A single one-megaton bomb would dig a thousand-foot wide crater to a depth of two hundred feet. No structures would be standing within a radius of nearly two miles, and damage would be heavy for a considerable distance beyond this. If the area had a normal urban population density, at least 200,000 people would be killed immediately, and half a million would be injured. Fires would spread through the city, killing thousands more. After this initial shock, radiation would spread through the area, killing many more in a few weeks or months. And, of course, this would not be an isolated attack. The arms race between the United States and the Soviet Union culminated in the two countries collectively amassing an arsenal of over 4,000 nuclear-tipped ballistic missiles, along with hundreds of manned bombers and cruise missiles with nuclear capabilities. If even a small number of them had been used in anger, devastation would have been immense.[40] The basic structures of society would undoubtedly have snapped under the strain, leaving civilization perched on the threshold of barbarism. Had this happened, it might have been better if mankind's long history of technological advance had never occurred.

The terrifying power of modern weaponry has eliminated completely the distinction between soldier and civilian that already began to break down early in the twentieth century. Wars of the past, whatever their horrors, at least left some room for individual valor; a nuclear war produces depersonalized, push-button destruction on a massive scale. In this way, nuclear warfare would be a culmination of an important trend of millitary technology. Increasingly sophisticated weapons have opened the psychological distance between the warrior and his victims. In the past, the inhabitants of a village might be slaughtered by troops wielding swords and axes; today the same thing can be accomplished by dropping an incendiary bomb from an altitude of 40,000 feet. The result is the same (or worse), but soldiers, sailors, and airmen, as well as the engineers who have designed their weapons and the political leaders who sent them into battle, are removed from the fray. As Charles Lindberg reflected after flying a World War II bombing sortie, "You press a button and death flies down . . . How can there be writhing, mangled bodies? How can this air around you be filled with unseen projectiles? It is like listening to a radio account of a battle on the other side of the earth. It is too far away, too separated to hold reality. . . . In modern war one kills at a distance, and in so doing he does not realize that he is killing."[41]

Not all military men have been bothered by the moral consequences of this separation, but they have often been irritated by the erosion of martial virtues resulting from the development of improved weapons. As one Spartan king lamented when he witnessed the demonstration of an early catapult, "O Hercules, the valor of man is at an end!"[42] During the Middle Ages, military leaders were often infuriated by the insidious way in which crossbowmen and musketeers could pick off their soldiers from afar. Their anger was reflected in their harsh treatment of captured soldiers using these weapons, which often extended to cutting off their hands and having their eyes put out.

Needless to say, punitive measures and an occasional disdain for new weapons did not stem the advance of military technology. We continue to struggle today to find ways to limit the consequences of military technology change; some of these efforts are explored in the next chapter. But before taking up this subject, we should know something about the causes of technological advance in weaponry. This will be the theme of the first part of the next chapter.

Questions for Discussion

1. Can any equivalents of medieval knights be found today? In what ways do they resemble medieval warriors? How do the technologies they use shape their activities and attitudes?
2. The use of gunpowder changed the nature of warfare and played a significant role in the transformation of European society. Yet in China, its land of origin, gunpowder did not have this effect. Why do you think this was so? Can you think of any possible differences between medieval Europe and traditional China that might have been responsible for their different experiences?
3. To make effective use of firearms, soldiers had to be well-drilled and subjected to tight discipline. How have military models of organization influenced nonmilitary social institutions? What are the advantages and disadvantages of military styles of organization when they are applied elsewhere?
4. For all of its horrors, war has historically provided opportunities for people to demonstrate some real virtues, such as resourcefulness, initiative, and courage. Have modern military technologies made these virtues irrelevant? If so, does the loss of these virtues make war more absurd than ever before?

Notes

1. James K. Finch, *Engineering and Western Civilization* (New York: McGraw-Hill, 1951), p. 22.

2. William Reid, *The Lore of Arms: A Concise History of Weaponry* (New York: Facts on File, 1984), p. 15.
3. E. W. Marsden, *Greek and Roman Artillery: Historical Development* (Oxford: Clarendon Press, 1969), pp. 86–91.
4. Robert Laffont, *The Ancient Art of Warfare*, vol. 1 (Greenwich, Conn.: New York Graphic Society, 1968), pp. 121, 128–129.
5. See Charles William Chadwick Oman, *A History of the Art of War: The Middle Ages from the Fourth to the Fourteenth Century* (London: Methuen, 1898), pp. 73–115.
6. Reid, op. cit., p. 6.
7. Laffont, op. cit., p. 240.
8. Richard A. Preston, Sydney F. Wise, and Herman O. Werner, *A History of Warfare and Its Interrelationships with Modern Society* (New York: Frederick A. Praeger, 1956), p. 85.
9. Vernard Foley, George Palmer, and Werner Soedel, "The Crossbow," *Scientific American* 252, 1 (January 1985): 104–110.
10. Laffont, op. cit., p. 444.
11. Lynn White, Jr., *Medieval Technology and Social Change* (New York: Oxford University Press, 1966), p. 102.
12. Martin van Creveld, *Technology and War: From 2000 B.C. to the Present* (New York: The Free Press, 1989).
13. Alex Roland, "Secrecy, Technology, and War: Greek Fire and the Defense of Byzantium, 678–1204," *Technology and Culture* 33, 4 (October 1992).
14. China Science and Technology Museum, *China's Ancient Technology* (Beijing: *China Reconstructs* Magazine, 1983), pp. 18–20.
15. William H. McNeill, *The Pursuit of Power: Technology, Armed Force, and Society since A.D. 1000* (Chicago: University of Chicago Press, 1982), p. 91.
16. Bernard Brodie and Fawn Brodie, *From Crossbow to H-Bomb* (Bloomington: Indiana University Press, 1973), p. 61.
17. Michael Howard, *War in European History* (London: Oxford University Press, 1976), p. 33.
18. Quoted in Walter Buehr, *Firearms* (New York: Thomas Y. Crowell, 1967) pp. 134–135.
19. McNeill, op. cit., p. 141.
20. van Creveld, op. cit., p. 107.
21. H. W. Koch, *The Rise of Modern Warfare: From the Age of Mercenaries through Napoleon* (New York: Crescent, 1982) p. 18.
22. Richard Bean, "War and the Birth of the National State," *Journal of Economic History* 33, 1 (March 1973): 203–221.
23. Quoted in Brodie and Brodie, op. cit., p. 83.
24. Howard, op. cit., p. 123.
25. Carlo M. Cipolla, *European Culture and Overseas Expansion* (Harmondsworth, England: Penguin, 1970) pp. 70–72.
26. Elting E. Morison, *From Know-How to Nowhere: The Development of American Technology* (New York: New American Library, 1977), pp. 139–152.
27. Geoffrey Parker, *The Military Revolution: Military Innovation and the Rise of the West, 1500–1800* (Cambridge: Cambridge University Press, 1988), p. 154.
28. See Daniel R. Headrick, *The Tools of Empire: Technology and European Imperialism in the Nineteenth Century* (New York: Oxford University Press, 1981), pp. 117–119.
29. Ibid., p. 118.
30. Ibid., p. 3.
31. John Ellis, *The Social History of the Machine Gun* (Baltimore: The Johns Hopkins University Press, 1975), pp. 111–147.

32. Brodie and Brodie, op. cit., p. 179.
33. Ibid., p. 224.
34. Theodore Ropp, *War in the Modern World* (New York: Collier Books, 1962), p. 379.
35. Noble Frankland, *Bomber Offensive: The Devastation of Europe* (New York: Ballantine, 1970), p. 149.
36. Paul F. Walker, "Precision-Guided Weapons," *Scientific American* 245, 2 (August 1981): 38.
37. Ian Hogg, *The Weapons that Changed the World* (New York: Arbor House, 1986), p. 144.
38. Walter Millis, *Arms and Men: A Study in American Military History* (New York: New American Library, 1956), p. 277.
39. B. H. Liddell Hart, *The Revolution in Warfare* (London: Faber and Faber, 1946), p. 25.
40. Randall Forsberg, "A Bilateral Nuclear-Weapon Freeze," *Scientific American* 247, 5 (November 1982): 61.
41. Quoted in Michael S. Sherry, *The Rise of American Air Power: The Creation of Armageddon* (New Haven, Conn.: Yale University Press, 1987), pp. 209–210.
42. Liddell Hart, op. cit., p. 30.

Chapter 14 How New Weapons Emerge—And How They May Be Contained

The last chapter outlined some of the consequences of technological change for the conduct of wars, as well as its effects on other areas. But what is the source of technological change in the military realm? Why has the development of weapons accelerated at such a frightening rate within our own lifetimes? Why have some weapons been adopted while others have languished? How have social and cultural conditions stimulated the development and use of some weapons and not others? In the first part of this chapter we will try to address some of these questions, and at the same time provide some background for the last sections, which deal with past and present efforts to limit the use of new weapons.

ACTION AND REACTION

The most obvious reason for the development of new weapons is that combatants or potential combatants continually strive to gain the upper hand over their enemies through superior weaponry. When one succeeds in doing so, its enemies are strongly motivated to develop new weapons that can neutralize this new threat. New offensive weapons are countered by new defensive weapons, which in turn stimulate the development of better offensive weapons, and so on. The process is one of constant action and reaction, as rival nations thrust and parry, each seeking to neutralize their opponent's capabilities, while at the same time augmenting their own. The development of military technologies thus becomes an endless cycle in which each new weapon stimulates the invention of another, and the capacity to kill and destroy proceeds without limit.

History supplies us with many examples of this process. We have seen how the offensive power of the mounted knight was checked by the pike and longbow. Cannon threatened the security of medieval castles, until their destructive force was temporarily checked by improved fortifications. The machine gun produced a stalemate during much of World War I, but the armored tank made the machine gun a much less potent weapon, and tipped the balance back in favor of the offense. Today, a new generation of precision weapons has greatly increased the vulnerabity of tanks, and battlefield conditions are quite different from what they had been.[1]

Useful as this action-reaction schema is, it does not tell the whole story. In the first place, it isn't always a simple matter to categorize military tactics as either offensive or defensive. When a war is being fought, the contestants do not neatly think in terms of offense and defense. Wars are won by defeating an enemy, which means offensive action, but at the same time defensive measures are necessary to protect the capacity to wage an offensive campaign. And even if a country seeks only to protect itself from an aggressor, it may still find it necessary to go on the attack, for as the old maxim has it, "the best defense is a good offense."[2]

In similar fashion, many weapons can be used for either offensive or defensive purposes. For example, the possession of a more accurate, faster-firing gun may allow an army either to rapidly advance against the enemy or to more easily defend their positions in the face of enemy attack. Also, a strong defensive capability enhances the capacity to engage in successful offensive actions. Nineteenth century imperialist powers were able to conquer less technologically advanced lands—an obviously offensive purpose—because their superior firepower allowed them to easily hold off the attacks of the natives.

The close relationship between offensive and defensive capabilities is a particularly important matter today as billions of dollars are being spent to develop an anti-ballistic missile defense system, the Strategic Defense Initiative—popularly known as "Star Wars." Should a workable system be installed—which is unlikely, given the immense technical obstacles to be overcome—this defense would allow the United States to launch its own missiles with impunity, for an enemy would be incapable of retaliation. A defensive system thus becomes the basis of an overwhelming offensive superiority.

SOCIAL STRUCTURE AND THE DEVELOPMENT OF MILITARY TECHNOLOGIES

Be they offensive or defensive, weapons are not developed and put into service simply because of their intrinsic superiority. As has been noted throughout this book, the creation and use of technologies are part of larger processes that involve existing social patterns, cultural orientations, and the motivations of individual people. The development of military technologies is no different. As Lynn White has said of the adoption of the stirrup by Medieval armies:

> [A] new device merely opens a door; it does not compel one to enter. The acceptance or rejection of an invention, or the extent to which its implications are realized if it is accepted, depend quite as much upon the conditions of a society, and upon the imagination of its leaders, as upon the nature of the technological item itself. . . . [T]he Anglo-Saxons used the stirrup, but did not comprehend it; and for this they paid a fearful price. . . . [I]t was the Franks alone . . . who fully grasped the possibilities inherent in the stirrup and created in terms of it a new type of warfare supported by a novel structure of society which we call feudalism.[3]

A military technology is part of a larger system, and the use of a weapon invariably reflects the basic features of that system along with its strengths and deficiencies. This is well-illustrated by the adoption of firearms and their supplantation of the longbow as the basic infantry weapon.[4] When viewed solely in terms of effectiveness, the longbow was by no means inferior to early arquebuses and muskets. Fired by a competent bowman, an arrow could penetrate armor just as easily as a bullet could. Its range was greater, and it could be shot more rapidly. In addition, it was less expensive, was not subject to the continual misfires that plagued early firearms, and its effectiveness was not diminished by wet weather, as was the case with early matchlocks. Why then was the bow replaced by firearms despite all their imperfections?

In the above description of the virtues of the bow, it was necessary to preface the list with "fired by a competent bowman." The bowman was of course an essential element in the technological system of medieval archery, and the level of his skills determined the bow's effectiveness. This skill could not be taken for granted; proficiency with the bow required some natural ability and a great deal of practice. In medieval England there was a strong expectation that men and boys would devote a considerable amount of time to archery practice; indeed, it amounted to the national sport of England. The longbow was thus well-suited to temporary armies made up of free men.

Not everyone was willing to submit to the rigors of archery practice. Many preferred to spend their spare time engaged in card-playing, bowling, shooting dice, and the medieval version of football. On several occasions royal edicts banned these activities, in the hope, as one Elizabethan act put it, that "archery may be revived and practiced and that kind of ancient weapon whereby our nation in times past has gotten so great honor may be kept in use."[5] As might be expected, these laws were almost impossible to enforce, and the quality of English archery underwent a steady decline.

In other nations matters took a different course. During the late fourteenth century, the king of France also promoted archery by banning all other diversions. The quality of French archery surpassed that of England, to such a degree that the ruling elite began to fear that their bowmen could pose a challenge to their rule. Consequently, mass archery was superseded by requirements that a limited number of bowmen be cultivated in each town and district, and the masses went back to their traditional pastimes.[6] In general, rulers were not inclined to support the military prowess of their citizens, which could just as easily be used in the service of rebellion. Rather, autocratic states were predisposed to build up permanent professional armies composed of soldiers lacking high levels of individual skill.

Firearms fitted perfectly into this system, for they allowed the use of poorly paid mercenary soldiers drawn from the dregs of the society. Unlike archery, the use of firearms did not require high levels of skill: "Where a few days and a good drill sergeant may suffice to train a reasonably good arquebusier, many years and a whole way of life were needed to produce a competent archer."[7] To be sure, the use of early firearms was by no means a simple matter, requiring numerous operations in precise sequence. But none of these, not even aiming and firing the weapon, required a great deal of skill. Nor was there much opportunity to instill higher levels of proficiency even if it were desired; gunpowder and bullets were expensive, and as a result practice time was sharply limited, amounting to the firing of only a few dozen rounds per year. In any event, individual skill in handling weapons was far less important than maintaining a concentrated rate of fire. Battles were won by the discharge of withering volleys, not by individual sharpshooting. What counted, therefore, was iron discipline. The main requirement for effective use of arquebuses and muskets was that soldiers have the proper procedures drummed into them through rigorous drill and that they adhere to these procedures in the heat of battle. Ranks

The effective use of muskets required constant drilling in a precise sequence of procedures. (Mary Evans Picture Library.)

of soldiers were expected to advance as a single line toward the enemy and to maintain formation despite the prospect that at least a third of them would fall victim to enemy gunfire.[8]

Mercenary soldiers of this sort were effective in batttle because they had been transformed into virtual robots, who, as Frederick the Great of Prussia wanted, feared their own officers more than they did the enemy.[9] Firearms were thus particularly well-suited to armies raised by centralized, bureaucratically organized states where discipline was much more highly valued than freedom and individual ability. Troops of archers made up of independent yeomen were appropriate to the political and social setting of late medieval England, while gun-wielding mercenary armies were well-suited to the situation prevailing in most of Europe from the sixteenth century onward.

It should also be noted that the advantage of firearms was not confined to their appropriateness to a particular mode of political organization. Technologies are sometimes embraced because of the psychological needs they meet. Although early firearms had a number of deficiencies when compared with the bow, they did have the advantage of producing a great deal of noise. Despite being inaccurate and cumbersome to use, early firearms certainly conveyed the impression that they were dangerous and terrifying weapons. And one could also speculate on the sexual connotations that have long been attached to guns; one needn't be a close follower of

Sigmund Freud to realize that handguns and cannon can easily serve as symbols of male virility and potency.[10]

As the adoption of firearms shows, the motives for using one military technology instead of another do not wholly inhere in the objective capabilities of a weapon. The fit of a military technology with established interests and ways of doing things may be of paramount importance for its adoption. At the same time, stupidity and short-sightedness can delay the adoption of a weapon, and retard its effective use when this finally occurs. Although political and military leaders might be expected to employ new military technologies that confer specific advantages, some weapons have languished due to misperceptions of their potential.

A failure of this sort was evident in the slow adoption of the machine gun by late nineteenth-century armies. This was a particularly ironic situation in the United States, for the first practical machine guns, the Gatling gun and the Maxim gun, were American inventions. But France and England were equally slow in recognizing the machine gun's potential. The problem with these weapons was that they were heavy, cumbersome devices that had to be supported by large gun carriages. Since they looked like conventional artillery pieces, they were treated as such. The leaders of the armies of late nineteenth-century England, France, and the U.S. did not grasp the unique capabilities of the machine gun, and they made no effort to develop new tactics that could take advantage of the machine gun's rapid rate of fire.[11]

In contrast, the German General Staff had a strong interest in the machine gun, for a central element of their military plans was the massive use of reservists whose marksmanship was necessarily poor. In this way the German army at this time was quite different from the small, professional armies of England, France, and the United States. Given this basic difference, the machine gun was particularly appealing to the Germans, for its firepower more than made up for the deficiencies of its operators.[12]

As we saw in the previous chapter, the capabilities of the machine gun were soon demonstrated in the early battles of World War I. Combat bogged down into trench warfare, with neither side capable of advancing in the face of machine gun fire. Decisive breakthroughs became possible only after the deployment of the armored tank, which was largely impervious to machine-gun fire. But again, military planners were slow in realizing how a new technology drastically altered the nature of warfare. Just as the machine gun caught World War I military leaders unprepared, the weapon developed in response to it, the tank, was unappreciated during the years leading up to World War II. This situation was particularly evident in England and France, where military officers tended to be recruited from the traditional elite—the figurative and in some cases literal descendents of the knights of the feudal era. A good bit of their lives still centered on the horse, and prowess on horseback was an essential requirement of aristocratic manliness. In this culture, noisy, smelly machines were no match for a noble steed. Thus, a successful demonstration of an armored car in Austria during 1910 came to naught because the vehicle had frightened the horses upon which the observing generals were mounted; worst of all, the Emperor's horse came close to bolting while he was in the saddle.[13]

Established military strategists persisted in the belief that infantry and even the cavalry were the decisive forces in battle, with the tank playing only a supportive role. Field Marshal Douglas Haig, whose tactics sent tens of thousands of men to their deaths in hopeless assaults against enemy lines protected by machine guns, was quite clear on this point: "I am all for using aeroplanes and tanks, but they are only auxiliaries to the man and the horse."[14] As a consequence of this sort of thinking, twice as much money was spent in Britain in 1935 on cavalry as on tanks.[15]

In France the situation was no better. According to the official *Directive on the Use of Tanks*, "In battle, tanks, whether singly or in groups, act within the framework of the infantry. Tanks are merely support weapons, which may be assigned to the infantry for a limited time; they considerably augment its striking force, but they do not replace it."[16] Accordingly, tanks were to be commanded by infantry officers, and were not intended to serve as independent strike weapons. Throughout the 1920s and 1930s, tanks were designed to move at the speed of the foot soldier; they were expected to rip through barbed wire and neutralize the weapons that threatened the infantry's advance, but were not to play an independent role in combat.[17]

In contrast, the army of the Soviet Union, which had been formed in the wake of the Russian Revolution and was led by an entirely different breed of officer, was far more alert to the new opportunities presented by these weapons. In similar fashion, the military forces of Nazi Germany were commanded by a new generation of officers, for Germany's defeat in 1918 had discredited the old military leadership, and the post-World War I disarmament of Germany left no entrenched military structure to resist new weapons and tactical concepts. The military staff that guided Germany's rearmament in the 1930s was highly receptive to the deployment of new weapons, as can be seen in the circulation of 30,000 copies of a translation of General J. F. C. Fuller's prescient book on tank operations. In Fuller's native Britain, only 500 copies were printed.[18] Fuller's lessons were taken to heart by the German army, which made good use of the tank in the *blitzkrieg* attacks that conquered much of Europe during the first two years of World War II.

ORGANIZATIONAL INTERESTS AND THE AIR WEAPON

Not all of the conflicts that affected the use of new weapons were based on differences of generation and social class. Rivalries between different parts of the military also shaped the development of military technologies. This was especially evident in the development of the airplane as an offensive weapon. During the 1920s and 1930s, military thinkers drawn from fledgling military air corps aggressively pressed their belief that the airplane had initiated a new era in warfare. According to these proponents of "air power," most notably General Billy Mitchell of the U.S. Army Air Corps, the airplane had rendered conventional warfare obsolete. According to this new doctrine, bombers now could wipe out ground armies and sink navies at little risk to themselves. Equally important, if horrendous in its implications, bombers could attack cities and industries with impunity, paralyzing economies and breaking the enemy's will to resist.

As we saw in the last chapter, the expectations of these prophets of air power were at best only partially realized. Although aerial bombardment produced appalling devastation, the bomber did not single-handedly win the Second World War.

The expectations that it could play a decisive wartime role rested on a number of largely untested assumptions. More ominously, the doctrine of strategic bombing also served as a rationale for advancing specific personal and organizational interests.

During the years between the two world wars, members of the U.S. Army Air Corps were eager to see their organization transformed into an independent military arm equal, of not superior, to the Army and Navy. This would require a new theory of war based on "air power"—and a weapon capable of turning theory into practice. The Air Corps found that weapon in the B–17 Flying Fortress. First flown in 1935, the four-engine B–17 was not defined as a strategic bomber, since the isolationist climate of the time was not receptive to the idea of getting involved in conflicts with far-off nations; rather, it was portrayed as a weapon to be used against a seaborne invasion. It was designed to fly at very high altitudes and have great range so it could seek out an enemy fleet far out at sea. The bomber's ability to survive the attacks of intercepting aircraft was questionable, for no escort fighters of the time could begin to match its range, but it was believed that the massed machine gun fire of a tight formation of B–17s would stave off all attackers.

None of these expectations were met. As combat experiences during World War II proved, the B–17 was just about worthless as an anti-naval weapon, and had severe operational shortcomings, most notably a vulnerability to anti-aircraft fire, and an inability to defend itself from a determined fighter attack. It flew many missions, and inflicted a good deal of damage on the enemy, but it wasn't the weapon that vindicated the extravagant claims made for air power, nor could any airplane carrying conventional bombs do so.

In one sense, the mentality of Billy Mitchell was diametrically opposed to Field Marshal Haig's way of thinking, for an air force, unlike the infantry or cavalry, represented the future and not the past. But there was still a basic similarity, for both were incapable of viewing weapons with much objectivity. Instead, personal interest and commitment to a particular service arm strongly influenced the way they looked at the utility of a particular weapons technology. The traditional military resisted new weapons and clung to the established ways of doing things, often with disastrous results. Adherents of air power, and especially of strategic bombardment, were correct in their assessment that their weapons could do an immense amount of damage, but they made the same mistake as the traditionalists in believing that their personal and organizational interests were identical with the military needs of their countries.

SOCIAL REVOLUTION AND THE ENLARGEMENT OF WAR

We have just seen how personal and organizational interests can affect the course of technological change. It is to be hoped that this discussion has illuminated a recurrent theme in this book: that technologies are shaped by the societies in which they develop, and as such they reflect culture, the distribution of power, and economic, organizational, and social relationships. At this point it may be appropriate to expand on this theme by considering how social and political change on a massive scale altered not only weapons, but the very nature of war.

During the late eighteenth and early nineteenth centuries, a series of social and political revolutions profoundly changed the nature and extent of warfare. The effects of these revolutions can clearly be seen by briefly examining warfare in the period immediately prior to this era. Throughout the eighteenth century most European wars were rather limited affairs. War was the ultimate "sport of kings," fought to advance the interests of a ruler rather than the country as a whole. Most of the populace remained aloof from these conflicts; the dirty work was done by small contingents of professional soldiers, social outcasts who were drilled to perfection by noncommissioned officers equally removed from the larger society. Commissioned officers came from the upper class of the society, and had little emotional connection with their soldiers. But these officers and their ruler valued them all the same, for they knew that well-drilled, effective fighting men could not be quickly replaced. As a result, soldiers were used with great care. When a major battle did occur, it produced frightful casualty rates that could decimate an army. This in turn could fatally weaken a country. Military and political leaders therefore preferred to avoid these confrontations, and were eager to arrange compromise peace settlements instead of engaging in go-for-broke military actions.[19]

At the same time, war was constrained by the temper of the times. The wars between Catholics and Protestants that had convulsed Europe during the previous centuries were frightful examples of what happened when wars were fought over great principles. In place of zealotry, eighteenth-century political thinkers preached a code based on reason or at least reasonableness. Wars might be an unfortunate consequence of human nature or political policy, but their worst effects could be constrained by the mutual agreement of the combatants to observe proper rules of behavior. Civilians were to be spared, hatred of the enemy kept in check, and terms of peace kept moderate.[20]

The great forces of democracy and nationalism changed all that. In an earlier time, the Swiss had shown the power of a disciplined, armed citizenry motivated by a fierce desire to fight for their homeland. The American Revolutionary War also showed the power that could be unleashed when large numbers of combatants took up arms in support of national self-determination and a more democratic form of rule.

These principles, and their military consequences, were driven to new heights by the French Revolution. This movement was energized by a desire to overthrow France's traditional ruling class and substitute democratic rule. But along with the rights engendered by democracy there came responsibilities. The people of France were no longer passive subjects of the king; they were now citizens of their nation, a nation that now faced the hostility of countries governed in the traditional manner. At the same time, the revolution had produced a zeal to extend the new order across the boundaries of France. Territorial conquest was therefore undertaken in order to extend liberty, equality, and fraternity throughout Europe. Warfare now involved a newly emergent citizenry, who fought for a cause far greater than the personal ambitions of traditional monarchs. The mercenary army had been replaced by "the nation in arms," where, as a report to the French Assembly in 1789 put it, "Every citizen must be a soldier and every soldier a citizen."[21]

Not everyone subscribed to these lofty principles. But for those who were unwilling to voluntarily take up arms in the service of their nation, there was always the draft. Partial conscription had already been employed elsewhere, but these limited efforts were completely overshadowed by the universal conscription that was proclaimed by France's revolutionary government in 1793. Conscription increased the French army to nearly 750,000 men. In turn, the increased number of soldiers changed the ways that wars were conducted. Moderation and practical restraints on warfare were swept away, and military actions became far more massive in scope. Whereas the key battles of the eighteenth century had been fought by armies rarely exceeding 80,000, in 1812 Napoleon was able to invade Russia with a force of 600,000.[22]

These huge armies were expensive to maintain. A necessary consequence of the vast expansion in the size of armies was a substantial increase in taxation. Governments have always used their power to tax, and tax revenues have been disproportionately used for military purposes. But after the French Revolution, the close alignment between the individual and the nation expanded the willingness and ability of governments to tax their citizens, while massive military operations increased their need to do so.

Large armies required huge quantities of guns and other materiel. In revolutionary France these needs were met by mobilizing the population to produce weapons and equipment. The government drafted craftsmen of every description as well as legions of workers with no prior experience in armament making. A good deal of waste resulted, and many of the muskets were of poor quality, but the job was done. Paris became a great center for weapons production, turning out 750 guns each day at a time when all of Europe never produced more than a thousand.[23]

The days of restrained, limited warfare were over. The spread of democracy and nationalism brought with it mass-based warfare. And at the same time, this new form of warfare created a receptive environment for the development of new weapons that made warfare far more destructive and terrifying than ever before. Equally important, the greatly enlarged scope of warfare stimulated the development of new production technologies that allowed the large-scale production of these weapons, as we shall see next.

INDUSTRIAL TECHNOLOGY IN THE SERVICE OF WAR

Until the nineteenth century military technology developed slowly. Science made few inputs into invention, and most work was done on a trial-and-error basis. As a result, guns were inaccurate and not very powerful because metallurgy was still in its infancy and chemistry had only recently emerged from alchemy. This situation began to change dramatically during the nineteenth century. The pace of technological change accelerated. Slower-burning smokeless powders resulted in more accurate gunnery and gave gunners a clear field of vision. The new powders also allowed the production of guns with smaller bores, making weapons lighter and allowing soldiers to carry larger loads of ammunition. The discovery of fulminates made the percussion cap possible, rendering the flintlock obsolete. Improved machining techniques stimulated the production of better weapons, such as breech-loading cannon and small arms. Accurate machine tools and the use of new alloy steels also made possible the mass production of guns with rifled barrels, which greatly increased their accuracy. By the middle of the century, these new technologies also facilitated the production of practical repeating weapons, culminating with the machine gun.

For much of the nineteenth century, most of these improvements occurred with little direct support from central governments. Invention was the work of individuals and firms that attempted to sell their devices to the armies and navies of their own nation, or failing that, to anyone who was interested. By the end of the century, some of these firms had become quite large and powerful, with steadily expanding political influence. Enterprises such as Armstrong-Vickers in Britain and Krupp in Germany employed scientists and engineers who had as their sole responsibility the development of new weaponry. Derided as "merchants of death," these firms gave a powerful impetus to the invention and diffusion of new military technologies that further increased the destructiveness of warfare.

By the twentieth century the state had ceased being a passive consumer of military technologies developed by private firms. Government sponsorship of technological innovation in the military sector resulted in the rapid advance of martial technologies. This reached new heights during World War II, when the United States government employed 30,000 scientists and engineers for the war effort.[24] Today, the military is a voracious consumer of scientific and engineering output. In the United States, by the mid-1980s, 70 percent of federal research and development funds were going to military projects,[25] and nearly one-third of the nation's total R&D expenditures were being absorbed by the defense sector.[26]

The technological advances that occurred during the last hundred and fifty years transformed armaments. But of equal or greater importance was the development of industrial technologies that allowed great increases in their production. Gunsmiths organized along traditional craft lines could never produce weapons in the quantities that became commonplace during the second half of the nineteenth century. Nor could conventional means of producing uniforms, boots, and foodstuffs meet the immense requirements of modern armies.

The expanding needs of the military were met through the development of mass production. The essence of mass production is the assembling of a product from standard parts so that no individual fitting is required. This in turn requires precision tools, assembly jigs and fixtures, accurate gauges and other measurement devices, and in general high degrees of accuracy. Mass production also requires a large market for its product; to put it another way, mass production requires mass consumption.

The military was an ideal consumer of the increased output made possible by mass production, and its needs generated a strong stimulus for the development of new production technologies. As we have seen in an earlier chapter, the manufacture of pulley blocks for the ships of the Royal Navy was one of the first instances of mass production. The technique was soon adapted to the fulfillment of other military needs, most notably the production of firearms. According to popular belief, Eli Whitney was the first to employ mass production techniques when, at the beginning of the nineteenth century, he undertook the large-scale manufacture of muskets at his factory in Mill Rock, Connecticut. In fact, Whitney never fully accomplished what he intended to do, and other enterprises have a better claim to being the initiators of mass production.[27] But whatever the source, by the middle of the nineteenth century weapons were being manufactured according to the principles of mass production, allowing huge armies to be sent into battle. At the same time, industrial technologies supplied them with mass-produced provisions, such as canned rations. Gone were the days when armies lived off the lands they marched through and fought on, often destroying more in the process than they did on the battlefield.

Large armies created immense logistical problems; the best army is worthless if it cannot be rapidly moved into a theatre of war and then be continuously supplied with arms and provisions. By the middle of the nineteenth century the railroad and the steamship made the movement of troops and the provision of their materiel a more regular and rapid process. Troops were no longer exhausted by lengthy marches, and during the course of a campaign they were much better supplied. The wounded and exhausted could be quickly evacuated and replaced by fresh troops. This in turn expanded the scope of war, as railroads allowed military leaders to relentlessly use troops and supplies in numbers that hitherto had been impossible. Improved transportation technologies thus moved the world closer to an era of total war, in which the performance of industries operating on "the home front" took on at least as great an importance as the performance of soldiers on the battlefield.[28]

Improvements in medical technologies reinforced the trend to larger armies, for far fewer soldiers were removed from combat by disease. Throughout the history of warfare many more soldiers had died of disease than from combat wounds. This situation began to change noticeably toward the end of the nineteenth century. Before 1870 for every soldier that died of wounds, five died of disease; by 1918 this ratio had been inverted.[29]

The technological achievements of the nineteenth and early twentieth centuries thus allowed a great expansion in the scope of warfare. In this way, technological development was a perfect complement to the social and political changes that put large armies at the disposal of rival states. All of this became painfully evident when Europe was engulfed by the holocaust of World War I. Enormous armies faced one another across fronts that stretched for hundreds of miles, and more soldiers were killed in a day's battle than had been killed during the duration of many eighteenth-century wars. The optimistic belief of the nineteenth century that technological advance would necessarily lead to a better life had been cruelly mocked by the second decade of the twentieth century.

Controlling Military Technologies

During the twentieth century military technologies have developed at an accelerating pace. When the century began, the battleship was the most technologically sophisticated weapon, infantrymen had only recently been armed with repeating rifles, and the airplane had not flown. Today, manned bombers fly well over twice the speed of sound, the battlefield has been transformed by electronically-guided

The launching tubes of a Poseidon submarine. The missile contained in each tube could devastate a fair-sized city. (U.S. Naval Institute.)

"smart" missiles, and our lives are threatened by intercontinental ballistic missiles (ICBMs) with nuclear warheads that could destroy our cities in a matter of minutes. On the not-too-distant horizon are laser "death rays," bacteriological weapons, enhanced radiation bombs, and many other ghastly instruments of mass murder. Is there any hope that technological progress can be arrested in the military sphere so that we will not forever live under the shadow of increasingly lethal products of human ingenuity? The following sections will provide some perspectives that help to answer this question.

HISTORICAL ATTEMPTS TO LIMIT NEW WEAPONS

The threat posed by new military technologies is in itself nothing new. Horrible as our latest weapons are, they are only the latest in a long series of "advances." On numerous occasions, people have surveyed the emergence of new weapons and fervently attempted to stop their spread. Pope Innocent II was aghast at the destructive power of the crossbow, and in 1139 the Second Lateran Council banned its use, although the ban did not extend to its use against non-Christians, so "infidels" such as the Moslems continued to be fair game.

Leonardo da Vinci kept secret the details of a submarine he designed, "on account of the evil nature of men, who would practice assassinations at the bottom of the seas by breaking the ships to their lowest parts and sinking them together with the crews who are in them."[30] In similar fashion, Niccolo Tartaglia, the first man to make a systematic study of ballistics, suppressed his own calculations because "it was a thing blameworthy, shameful and barbarous, worthy of severe punishment before God and man, to wish to bring to perfection an art damageable to one's neighbor and destructive to the human race."[31] But when the Turks threatened Tartaglia's homeland, he put aside his moral scruples, and his treatise on gunnery was quickly published.[32]

Each succeeding advance in weaponry has raised the stakes of warfare. At the same time, many of these advances have elicited a desire to limit the spread of new weapons. The percussion cap, an important, but not epochal change in armament, was viewed with great alarm by one correspondent to the *Gentleman's Magazine* in 1837: "If . . . this new system were applied to the military, war would shortly become so frightful as to exceed all bounds of imagination, and future wars would threaten, within a few years, to destroy not only armies, but civilization itself. It is to be hoped, therefore, that many men of conscience, and with a reflective turn, will militate most vehemently for the suppression of this new invention."[33]

A SUCCESSFUL EXAMPLE OF ARMS CONTROL

As things turned out, the nineteenth century had many nasty surprises that far exceeded the percussion cap. And, to their credit, political leaders made some effort to limit the spread of new weaponry. In 1899 delegates from twenty-six nations participated in a conference held at The Hague in the Netherlands that banned the use of asphyxiating gas, dumdum bullets (which exploded inside the body), and put a five-year moratorium on aerial bombardment (which in this era meant dropping projectiles from balloons).

The First World War demonstrated the terrible consequences of new military technologies, and in the early post-war period several international conferences attempted to control one of the worst of them: poison gas. All but one of the signers of the 1922 Washington Conference on the Limitation of Armaments agreed to refrain from the use of poison gas in warfare. A similar resolution was contained in the Geneva protocol of 1925, and was ultimately ratified by forty-two nations, although nineteen of them reserved the right to use gas if they were attacked by an enemy that made first use of this weapon.[34]

As with all such resolutions, there was no way of insuring compliance, for there was no supernational organization with the power to do so. Still, these resolutions struck a responsive chord, for the terrible potential of poison gas had been amply demonstrated during World War I, when hundreds of thousands of soldiers were killed or disabled by gas attacks (chlorine was first used, followed by the more lethal mustard gas). During the 1920s and 1930s, the threat of gas was quite pronounced, as military planners, novelists, and journalists presented scenarios of huge civilian populations annihilated by gas shells dropped from high-flying aircraft. During the early months of World War II the fear of gas attacks was widespread, and in Britain every man, woman, and child was supplied with a gas mask that they were supposed to carry with them at all times.

But the attacks never came. Over the course of a war made increasingly horrifying by the use of new military technologies—culminating with the dropping of the first atomic bombs—gas was never used. The declarations of the Washington Arms Conference and other such pronouncements were not the main reason, for its provisions were violated in a number of other areas. Part of the reluctance to use gas was due to inherent logistical problems. Troops had to be protected from the gas that they used, and shifts in the wind could upset a military operation. A conquered area would require detoxification before it could be occupied. But by themselves, these factors do not explain why gas was not used; any new weapon requires adjustments if it is to be successfully used, and ways could have been found to make gas a practical weapon. Yet civil and military leaders were not inclined to do so, for they had every expectation that if they used gas, their opponent would follow suit, thus nullifying whatever advantage the use of gas conferred in the first place. The nonuse of poison gas indicates that in some instances deterrence really works; a combatant may refrain from using a weapon if he is certain that his opponent will use it. The deployment of poison gas would have subjected its user to a great deal of suffering in return for a transitory advantage.

Fear of retaliation was not the only reason that gas was eschewed. It is possible that military planners would have taken the risks inherent in the use of gas if they felt a greater affinity for it. In fact, they never felt comfortable with the use of gas. Poison gas was unlike all other weapons—silent, insidious, and even dishonorable. As one World War I German commanding general put it, "I must confess that the commission for poisoning the enemy just as one poisons rats struck me as it must any straightforward soldier; it was repulsive to me."[35] At the same time, poison gas represented the sort of scientific advance that rendered obsolete many of the classic martial virtues. A technologically superior enemy with sufficient industrial capacity could triumph over superior training and leadership.[36] Some advances in military technology could be fitted into existing molds—an armored tank could be seen as a modern embodiment of the cavalry—but poison gas was something altogether different. Irrespective of their nationality, military men were disinclined to use poison gas, and their reticence helped to prevent modern warfare from being even more horrible than it actually was.

GUN CONTROL IN OLD JAPAN

The nonuse of poisonous gas is not the only example of a potentially significant military technology falling by the wayside. During the middle of the sixteenth century European traders introduced firearms into Japan, and guns rapidly diffused through the country during the succeeding decades, quickly becoming dominant weapons on Japanese battlefields. But during the early seventeenth century they began to disappear, and for 250 years Japanese soldiers reverted to swords, spears, and bows. The eclipse of firearms was not caused by a lack of technological prowess. Although pistols and muskets were foreign inventions, Japanese craftsmen quickly learned to duplicate them, making some useful improvements in the process. Not long after the introduction of firearms into their country, Japanese gunsmiths were producing weapons that were the equal of any in the world. And they produced them in large numbers; in one battle 10,000 soldiers on one side were armed with matchlock guns.[37] The Japanese had already demonstrated their technological virtuosity in building construction, metallurgy, agriculture, and a host of other areas; they certainly were not stymied by the demands of gun manufacture.

The reasons that firearms disappeared from Japan were cultural and political.[38] As had been the case in feudal Europe, the political elite of Japan were warriors, and Japanese culture was deeply suffused with a martial ethic. Even after the Tokugawa shogunate initiated a long period of peace, the nation was dominated by a warrior class, which comprised almost 8 percent of the total population. The ethos of this class centered on the military virtues of strength and courage. Their favored weapon was the sword, and only they had the right to wear and use it. Even if it was never used, the sword was the symbol of their special status, and it was much valued for its aesthetic appeal.

In contrast, firearms were foreign imports, with no connection to traditional Japanese ways. Of equal importance, the use of firearms in combat destroyed the time-honored way that the Japanese had gone about the business of warfare. War was a highly stylized affair that began with the ritual introduction of the main combatants. Its essence was hand-to-hand combat waged by heroic opponents who had ample opportunities to display their bravery and prowess. Firearms were highly disruptive to this form of warfare. As had been the case in medieval Europe, a line of soldiers armed with muskets could quickly decimate the bravest and most skillful warriors. Adding insult to injury, soldiers with firearms were social inferiors to the soldiers that they picked off with such ease and impunity.

If Japan had been threatened by foreign armies it undoubtedly would have been forced to adopt the same weaponry as its adversary. But as an island nation, Japan was able to close its doors to the outside world and avoid foreign invasion. Only in the middle of the nineteenth century did the foreign powers begin to threaten Japanese independence. When that happened, the Japanese wasted no time in arming themselves with modern weaponry.

But before this happened, Japan enjoyed two and a half peaceful centuries in which the most dangerous weapon was the samurai's sword. The government achieved this by centralizing the production of guns in one city and then strictly controlling their distribution. Purchasers of guns were required to have permission from the Commissioner of Guns, but in fact the commissioner granted these only to the government. Since the government bought very few, gun production virtually ended in Japan during the early seventeenth century, and the use of firearms became virtually unknown.

Japan's experience from the seventeenth to the early nineteenth centuries shows that a superior military technology is not always destined to be universally accepted. In the case of firearms in Japan at least, it *was* possible to "turn back the clock." At the same time, it must be remembered that the disappearance of firearms was the result of a conscious political decision, a decision that the government did not make through altruistic motives. These weapons vanished after the Tokugawa shoguns succeeded in bringing Japan under centralized control. Local power holders had been neutralized, and the shoguns naturally wanted to maintain this state of affairs. The spread of firearms would have made it much easier to mount local rebellions. At the same time, these weapons would have allowed the common folk to pose an effective challenge to the martial class that formed the core of the shogun's military power. It was therefore obviously to the advantage of the ruling elite to reverse the course of technological development by eliminating firearms from their domain.

The case of Japan during this era leaves us with a certain sense of optimism, for it shows that military technologies do not necessarily have an unstoppable momentum; the development and use of increasingly deadly weapons can be arrested. At the same time, however, it is necessary to temper this optimism with a dose of cynicism, for this case also shows that disarmament is most likely to take place when it works to the advantage of a dominant group. Today's nuclear weapons make arms control advantageous to all. But can we effectively formulate and enact arms control programs that allow us to realize our own best interests? It is to this crucial issue that we will next turn.

The Control of Nuclear Weapons

There is widespread agreement among both civilian and military leaders that an all-out nuclear war would be unwinnable. Since no combatant could expect to profit from the use of nuclear weapons, it is to the advantage of all to prevent these weapons from being used. Effective arms control agreements are not generated by one nation's willingness to make one-sided sacrifices; they are the product of the realization that all parties would gain from them. Nor is it necessary that the parties to an agreement be friends; after all, the very need to come to an agreement implies that there are basic conflicts between the two parties. Nations enter into arms-control agreements not because they are on the best of terms with the other parties, but because they have good reason to fear the consequences of unshackled military power. In sum, arms control treaties are signed when nations see them as being in their best interest, and as long as their interests are preserved, they have no reason to violate the agreements.[39]

DETERRENCE, BUT NO MORE

Despite a high level of tension between the United States and Soviet Union during much of the post-World War II era, neither side was inclined to take on the other in a full-scale war. Much of the reason for this reluctance lay in the fact that both nations had the capacity to utterly ruin each other by launching a nuclear attack in retaliation to the other's attack. This policy is known as Mutually Assured Destruction (MAD), and although not pleasant to contemplate, it had some positive results. Despite all of the insecurity it engendered, the development of nuclear military technologies helped to keep the peace, for neither nation dared attack the other for fear of retaliation.

Although the Soviet Union no longer exists, our defense strategy continues to rest on this principle. However, deterrence through Mutually Assured Destruction has its obvious perils. To have a credible capacity to retaliate is similar to keeping a gun in your house for protection. A gun is of little use unless it is loaded and readily accessible. But at the same time, this availability increases the level of danger, for the gun could easily be used by a small child, a disturbed acquaintance, or an enraged spouse. In similar fashion, nuclear weapons must be kept ready for immediate use, but this readiness increases the chances that they will be launched as a result of an accident, miscalculation, or mechanical failure.[40] Military planners are thus confronted by the inherently conflicting demands of peace and war. In peacetime, an accidental or unauthorized use of nuclear weapons is a constant peril; at the same time, an effective deterrent requires the ability to launch retaliatory attacks without excessive delay or deliberation.[41]

The maintenance of a credible yet reasonably safe deterrent is difficult in its own right, and recent technological developments have made it even harder. If one country has the capacity to destroy an enemy's retaliatory forces in a first-strike attack, the ability to deter is lost. Although it is impossible to make completely accurate predictions about the survivability of ICBM's, bombers, and missile-carrying submarines in the event of a pre-emptive attack, there can be little doubt that advances in delivery systems, such as the use of multiple warheads (MIRVs) on missiles, have increased the vulnerability of retaliatory forces.[42] Even if sufficient numbers of one's retaliatory weapons survived such an attack, the disruption of civil and military communications facilities might make the launching of a retaliatory strike difficult or even impossible.[43]

This vulnerability to a first-strike attack creates a dangerous incentive for a nation to mount an attack on an enemy before damage is suffered, or indeed, before it is entirely sure that an attack is under way. Once a potential enemy possesses a first-strike capability it instills a "use 'em or lose 'em" mentality in other nations. If these nations' retaliatory weapons are vulnerable to a pre-emptive strike there will be a strong incentive to strike first in a crisis situation; otherwise, most of these weapons could be destroyed before they were launched.

No present nuclear power has deliberately sought a first-strike capability nor used it to achieve a position of military superiority. Rather, official policy has stressed maintaining parity with potential adversaries. Even so, devising mutually acceptable arms control agreements that preserve parity is very difficult. "Parity" is an elusive concept, more so now than at any time before. In the past, it might have been possible to maintain parity by ensuring that one nation did not get ahead of others by acquiring a numerical superiority in the quantity of weapons, soldiers, and war materiel. Today the situation is different. In addition to considering sheer numbers, successful arms control agreements have to take into account the technological level of the respective military forces. These qualitative assessments are inherently more difficult than those that center on the mere counting of weapons. To cite one example, the first Strategic Arms Limitation Treaty (SALT I) between the United States and the Soviet Union resulted in restrictions on the number of missiles, but it failed to halt technological advances that produced missiles with greater accuracy. This produced a new kind of arms race, one that was qualitative instead of quantitative.[44] Further complicating matters, modern technologies are inherently dynamic, so changes in military technology can render an existing arms control agreement irrelevant. It also makes it even more difficult to negotiate a new one. As one group of scholars has pointed out, "The negotiation of arms control agreements takes time, and the development of technology can run ahead of the ability of arms control negotiators to come up with effective means of identifying, counting, and limiting weapons."[45]

This problem is illustrated by the case of cruise missiles. These weapons are small, pilotless craft that can be launched from land, submarines, or aircraft. Through the use of sophisticated navigation devices they are able to fly close to the ground, thus eluding radar detection. Although they are far slower than ballistic missiles, they can be produced in large numbers and they can be used to mount an attack against many targets simultaneously. These weapons complicate the task of arms control, for an arms control agreement requires the ability to verify how many weapons have been deployed; cruise missiles and their launching facilities are virtually unidentifiable by nonintrusive inspection measures. Additionally, it is difficult to determine through direct observation if a cruise missile is a tactical weapon with a conventional warhead or a strategic weapon with a nuclear one.[46]

THE PERILS OF PROLIFERATION

While most of the attention has been devoted to preventing a nuclear war between the superpowers, a no less pressing issue has been the slow but steady development of nuclear capabilities elsewhere. There are certainly few technological obstacles to many nations "going nuclear," for there are no real secrets about the construction of atomic weapons any more. All that is required is a supply of fissionable material, and this can be produced at a relatively low cost.[47] The present situation, where six nations (the United States, Russia, Great Britain, France, China, and India) have already tested nuclear weapons, is dangerous enough; the prospects of a world full of nuclear powers is frightening beyond description.

Still, some encouragement can be taken in other nations' willingness to cooperate in limiting the spread of nuclear weapons. The most visible manifestation of this spirit has been the Nuclear Nonproliferation Treaty (in force since 1970), through which 128 nations have pledged to refrain from acquiring or making nuclear weapons. In another effort at nonproliferation, all of the nations of Latin America except Cuba are signatories to the Treaty of Tlatelolco (in force since 1968), which mandates that their region be kept free of nuclear weapons.

Up to now, the spread of nuclear weapons has been kept in check as the majority of the nations of the world have observed the Nonproliferation Treaty. At the same time, however, a significant number of nations have not signed the treaty, including several who would have very little difficulty in "going nuclear"—South Africa, Israel, Pakistan, Brazil, Argentina, and Spain, to name the most prominent. Even the treaty's signatories have the right to withdraw on three months' notice; thus, they could easily produce nuclear weapons secretly and then begin to test them after withdrawing from the treaty. One thing that has prevented the nations that have signed the treaty from actively flouting it has been the willingness of the nuclear nations to assist the nonnuclear ones in the development of *peaceful* applications of nuclear energy. As things have turned out, this may not have produced great benefits, for nuclear power is not well-suited to the needs of many developing countries. And even "peaceful" applications of nuclear technology can often be readily applied to warlike purposes. To take only the most prominent example, Pakistan's nuclear weapons program has been based on its civilian nuclear industry.[48] It is also likely that the nonnuclear countries may be less inclined to continue practicing self-restraint if the major powers make only halting progress in reducing their own nuclear arsenals.

In sum, no issue confronting humanity is more important than preventing a nuclear war, and yet nothing is harder than controlling a dangerous arms race. The task will be a daunting one as long as individual nations are afraid of any substantial imbalance in weaponry between them and potential enemies. When one gets ahead qualitatively or quantitatively, the other side follows, and the result is an unending upward spiral. These races produce no real gains, for the advantages that come from gaining a technological lead don't last very long. The other side almost always catches up, and both nations are left with reduced levels of security.[49]

It has been argued that arms races are inevitable as long as individual countries contend with one another in a world that lacks any real controls above the level of the nation-state. From this perspective, meaningful arms controls require nothing less than the establishment of an international government.[50] It goes without saying that such a profound transformation in governance is not likely any time in the near future. At present, efforts aimed at preventing the development and spread of increasingly destructive military technologies are hampered by national rivalries, perceived verification problems, and the presence of powerful interest groups, both civilian and military, that reap substantial gains from the development and production of weaponry. Bringing a bit of sanity into a world that already spends a trillion dollars a year on armaments and has an arsenal of tens of thousands of nuclear weapons is no easy task.

Even so, from the standpoint of sheer self-interest there is every reason to make the effort. A good arms control agreement is not a giveaway to a rival nation. It can reduce the risks of wars that no combatant nation could possibly win in the nuclear age. It can stem the diversion of economic and human resources to unproductive military expenditures. And it can demonstrate that it is possible to control technology when it is in everybody's long-term interest to do so.

It is not likely that we will see warfare abolished in our lifetimes, but we can at least hope that the human race can avoid crossing over the nuclear threshold. This should be no idle dream, for there are precedents for mutually advantageous limitations of military capacity. Through a mixture of ethical concerns, negotiated agreements, and a fear of retaliation, contending nations have exercised some restraint in the use of weaponry. During World War II poison gas was not used, and throughout the twentieth century prisoners have usually been treated in accordance with international conventions instead of being summarily massacred, as happened in many wars in the past. Nuclear weapons have created the possibility for unimaginable horrors, but at the same time they provide the best reason to solve our differences in a more intelligent manner than often has been the case for much of human history.

Questions for Discussion

1. In what ways have warfare and military preparation advanced technology? Can you think of any contemporary military technologies that might eventually have civilian applications? Do civilian "spinoffs" from the military sector partially justify today's tremendous expenditures on military technology?
2. Firearms were surpressed in Tokugawa Japan because it was in the interests of the ruling elite to do so. Are the political elites of modern nations similarly motivated to prevent the development and use of new weapons? Do individual nations differ from one another in this regard?
3. Is Mutually Assured Destruction (MAD) an inherently immoral policy? Can it continue to be the cornerstone of our military policy? Are there any alternatives other than the forging of effective arms control treaties?
4. India is a nuclear power. Pakistan, its historic enemy, has not yet built and tested nuclear weapons, although it has the capability to do so. Many Pakistanis are eager to proceed with a nuclear weapons program. What arguments would you use in order to convince them to hold back?
5. The history of arms-control agreements presents, at best, a mixed record; a few successful agreements are more than balanced by many failures. But nuclear weapons have been held in check up to now. Is there anything about these weapons that makes them better subjects for successful arms-control agreements? Is it reasonable to hope that we will succeed in controlling these weapons, or are we living on borrowed time?

Notes

1. Paul F. Walker, "Precision-guided Weapons," *Scientific American* 245, 2 (August 1981): 37–45.
2. See Jack S. Levy, "The Offensive/Defensive Balance of Military Technology and the Incidence of War," paper presented at the Annual Meeting of the International Studies Association, Mexico City, April 1983.
3. Lynn White, Jr., *Medieval Technology and Social Change* (New York: Oxford University Press, 1966), p. 28.
4. The narrative is based on Thomas Esper, "The Replacement of the Longbow by Firearms in the English Army," *Technology and Culture* 6, 3 (Summer 1965): 382–393.
5. Ibid., pp. 392–393.

6. Phillipe Contamine, *War in the Middle Ages* (Oxford: Basil Blackwell, 1984), p. 217.
7. John Francis Guilmartin, Jr., *Gunpowder and Galleys: Changing Technology and Mediterranean Warfare at Sea in the Sixteenth Century* (New York: Cambridge University Press, 1974), p. 152.
8. Richard A. Preston, Sydney F. Wise, and Herman O. Werner, *A History of Warfare and Its Interrelationships with Western Society* (New York: Frederick A. Praeger, 1956), p. 138.
9. Ibid., p. 137.
10. William H. McNeill, *The Pursuit of Power: Technology, Armed Force, and Society since A.D. 1000* (Chicago: University of Chicago Press, 1982), p. 83.
11. David A. Armstrong, *Bullets and Bureaucrats: The Machine Gun and the United States Army, 1861–1916* (Westport, Conn.: Greenwood Press, 1982).
12. Ibid., p. 173.
13. Armin Hall, *Tanks: An Illustrated History of Fighting Vehicles* (New York: Crescent Books, 1971), p. 28.
14. Witold Rybczynski, *Taming the Tiger: The Struggle to Control Technology* (New York: Viking/Penguin, 1985), p. 171.
15. Preston, Wise, and Werner, op. cit., p. 281.
16. Quoted in Hall, op. cit., p. 79.
17. Arch Whitehouse, *Tank: The Story of Their Battles and the Men Who Drove Them from the First World War to Korea* (Garden City, N.Y.: Doubleday, 1960), p. 127.
18. Preston, Wise, and Werner, op. cit., p. 281.
19. Walter Millis, *Arms and Men: A Study of American Military History* (New York: New American Library, 1956), p. 16.
20. B. H. Liddell Hart, *The Revolution in Warfare* (London: Faber and Faber, 1946), pp. 40–45.
21. Millis, op. cit., p. 48.
22. Michael Howard, *War in European History* (London: Oxford University Press, 1976), p. 99.
23. Theodore Ropp, *War in the Modern World* (New York: Collier Books, 1962), p. 111.
24. Preston, Wise, and Werner, op. cit., p. 320.
25. Seymour Melman, "Swords into Plowshares: Converting from Military to Civilian Production," *Technology Review* 89, 1 (January 1986): 64.
26. Stockholm International Peace Research Institute, *SIPRI Yearbook, 1984* (Stockholm, 1984) p. 170.
27. Robert S. Woodbury, "The 'American System' of Manufacturing," in Edwin T. Layton, Jr. (ed.), *Technology and Social Change in America* (New York: Harper & Row, 1973), pp. 47–63.
28. Maurice Pearton, *Diplomacy, War, and Technology since 1830* (Lawrence: University Press of Kansas, 1984), pp. 64–76.
29. Howard, op. cit., p. 116.
30. Bernard Brodie and Fawn Brodie, *From Crossbow to H-Bomb* (Bloomington: Indiana University Press, 1973), p. 10.
31. Ibid.
32. William Reid, *The Lore of Arms: A Concise History of Weaponry* (New York: Facts on File, 1984), p. 88.
33. Quoted in Daniel R. Headrick, *The Tools of Empire: Technology and European Imperialism in the Nineteenth Century* (New York: Oxford University Press, 1981), p. 86.
34. James E. Dougherty, *How to Think About Arms Control and Disarmament* (New York: Crane, Russak, 1973) pp. 43–44.

35. Quoted in Frederic J. Brown, *Chemical Warfare: A Study in Restraints* (Princeton, N.J.: Princeton University Press, 1968), p. 41.
36. Ibid., p. 40.
37. Noel Perrin, *Giving Up the Gun: Japan's Reversion to the Sword, 1543–1879* (Boston: David R. Godine, 1979), p. 19.
38. The narrative is based on Perrin, op. cit., pp. 33–45.
39. John H. Barton, *The Politics of Peace: An Evaluation of Arms Control* (Stanford, Calif.: Stanford University Press, 1981), p. 105.
40. The Harvard Nuclear Study Group, *Living with Nuclear Weapons* (Cambridge, Mass.: Harvard University Press, 1983), p. 34.
41. John Steinbruner, "Launch under Attack," *Scientific American* 251, 1 (January 1984): 38.
42. See Bernard T. Feld and Kosta Tsipis, "Land-based Intercontinental Ballistic Missiles," *Scientific American* 241, 5 (November 1979): 51–60.
43. Steinbruner, op. cit.
44. Herbert Scoville, Jr., "The SALT Negotiations," *Scientific American* 237, 2 (August 1977): 24.
45. The Harvard Nuclear Study Group, op. cit., p. 153.
46. Kosta Tsipis, "Cruise Missiles," *Scientific American* 236, 2 (February 1977): 20.
47. William Epstein, "The Proliferation of Nuclear Weapons," *Scientific American* 232, 4 (April 1975): 28.
48. Barton, op. cit., p. 208.
49. Richard L. Garwin, Kurt Gottfried, and Donald Hafner, "Antisatellite Weapons," *Scientific American* 250, 6 (June 1984): 55.
50. Barton, op. cit., pp. 220–232.

PART SIX

THE SHAPING AND CONTROL OF TECHNOLOGY

Chapter 15 Technology and Its Creators: Who's in Charge of Whom?

Technology is a human creation; indeed, one of the distinguishing characteristics of the human race is that its members consciously develop and use technologies to extend their natural capabilities. But although technology is the result of human efforts, individual humans usually have little control over the technologies that affect their lives. Crucial decisions about technology seem to be made anonymously. Some individuals and groups are making these decisions, but who are they, and how do they work? In this chapter we will look at the individuals who play a particularly important role in the creation of new technologies. The next chapter will then consider how the aggregation of these individuals into organizations affects the process of technological change. But before we do any of this, we should critically examine the basic assumption that, in the final analysis, technology is always subject to human control.

Technological Determinism

Nothing worthwhile in life comes without some costs attached. So it is with technology; while it has made our lives materially richer and expanded human power, the advance of technology has created many problems—environmental degradation, alienation, and the threat of nuclear annihilation, to name only the most obvious ones. And, most bothersome of all, there looms the possibility that technology is out of control. If this is so, what began more than a million years ago as a human creation has taken on a life of its own, with technology advancing according to its own inner dynamic, and unrestrained by social arrangements, culture, and thought.[1] The belief that technology acts as an independent force in our life is known as "technological determinism," and if it is true, we have become the servant of technology instead of its master.

There can be little question that technology exerts a great influence on social, political, and economic relationships. Everything from antibiotics to zippers has affected our lives to some degree; many of these have been explored in earlier portions of this book. With sufficient imagination it is possible to find a technological basis for an apparently unrelated social change, as the American sociologist W. F. Ogburn did in the 1930s when he attributed some of the increase in Black migration to Northern cities to the invention of the automatic stoker, which mechanically fed coal into the fireboxes of steam locomotives. These devices allowed the production of more powerful locomotives that could haul longer passenger trains. Longer trains meant longer platforms and longer distances to walk while carrying luggage, hence the need for more porters, an occupation usually held by Blacks who moved from the South in response to better job opportunities.[2] The reader is probably justified in thinking that this analysis seems rather farfetched. In fact, most of Ogburn's writings analyse the interaction of social and technological change in a more subtle manner. Still, this example shows what can happen when one is overly enthusiastic in seeking the technological roots of social change.

TECHNOLOGICAL ADVANCE AND CULTURAL LAG

While making technology the centerpiece of his theory of social change, Ogburn popularized the concept of "cultural lag"—the idea that habits, thoughts, values, and social arrangements often fail to change at the same speed as technological innovation. Technology moves ahead, but many other things lag behind. On the face of it, this seems a reasonable concept, and we should have no difficulty in coming up with a number of examples. To cite but one, the extension of modern medical technologies throughout the world has lowered the rate of infant mortality, but people continue to have large families due in part to traditional expectations that many children will not survive infancy. Only after several generations, when expectations and practices have "caught up" with the level of technology, will birth rates begin to fall.

Appealing as it is on the surface, the theory of cultural lag has its shortcomings. In the first place, technological changes are simply taken as givens, and no attempt is made to delineate how they came to be in the first place. To return to the above example, improvements in public health and medical technologies are themselves the result of social, cultural, and economic changes. Although many histories of technology convey the impression that inventions are carried forward largely by their own internal logic and are little influenced by external factors, this perspective leaves a great deal out of the story. As earlier chapters of this book have indicated, social, economic, political, and cultural arrangements strongly influence the course of technological change. It is therefore a mistake to think of technology as an independent source of social change.

Second, many attempts to demonstrate the connection between technological and social changes present us with the problem of measuring different phenomena. Technological changes can often be more easily measured than sociocultural changes, making it hard to determine the precise connection between the two. We can easily count the increase in the number of automobiles between 1900 and 1925, but can we measure the changes in sexual morality that the automobile supposedly produced during this period?

Third, at what point can it be said that a society "catches up" with a given technological innovation? New productive technologies have allowed an alteration of the traditional divisions between "men's work" and "women's work" but laws, customs, and social arrangements have not completely taken account of this fact. To be sure, there have been significant changes, but at what point can it be confidently said that a complete accommodation has occurred? Then too, some technological innovations have been followed by rapid adjustments, while in other cases these adjustments never seem to be made. Obviously, something other than technology itself is affecting the extent and speed at which particular cultural lags are eradicated.

Finally, there is a hidden value judgment contained in the concept of cultural lag. Implicit in it is the notion that technology is a progressive, dynamic element in human history, and that social and cultural arrangements are intransigent sources of obstruction. But isn't it possible that a cultural lag may demonstrate the danger or inappropriateness of a particular technological innovation? Once again it might be noted that just because something is technically possible, it does not necessarily follow that it should be done, or that people should submit to it and make the necessary alterations to their lives. We now have the capability to implant a fertilized ovum into a surrogate mother, who then carries and delivers the baby. Should we replace old-fashioned methods of reproduction with this "improved" technology?

CONVERGENCE THEORY

Closely associated with a belief in technological determinism is a set of hypotheses, orientations, and perspectives that goes under the title "convergence theory." According to this theory, the opportunities and demands presented by modern technology promote the convergence of all societies toward a single set of social patterns and individual behaviors. This occurs because (1) technological development requires a fixed series of steps and stages that all countries must follow, (2) technology imposes similar organizational constraints on a society, and (3) in response to technological imperatives, values and behavior patterns reflect the rational approach to life that was described in the first chapter of this book.[3]

Often implicit in this formulation is the expectation that technological advance in non-Western countries will result in cultural homogenization, as hitherto "backward" countries take on the cultural and social attributes of the Western world. After all, most advanced technologies have originated in Europe and the United States, so any nation making use of these technologies will presumably have to adopt Western institutions and culture if they are to make effective use of Western technologies. Societies converge, but they converge toward a place already occupied by the Western world. As one U.S. senator put it a generation ago, "We will lift Shanghai up and up, ever up, until it is just like Kansas City."

This was a prospect that many Chinese did not view with pleasure. Throughout most of their history the Chinese have been wary of foreign importations, and not even the adoption of Communist ideology altered this fundamental attitude. Especially when under the influence of Mao Zedong's ideology, many Chinese had misgivings about the adoption of foreign technologies. The underlying hostility toward the use of foreign technologies strongly asserted itself during the second half of the 1960s when China was racked by the internal struggles that went under the name of the Great Proletarian Cultural Revolution. Among the victims of the often bloody conflicts that occurred during this period were technical experts who were accused of being captivated by foreign technologies and opposed to the efforts of

China's workers and peasants to develop indigenous technologies. Opponents of the use of foreign technologies also feared that these technologies would be accompanied by unacceptable foreign cultural patterns. Imported technologies, it was feared, inevitably would be accompanied by "bourgeois" habits and "decadent" activities.[4]

China's concerns were not unique; throughout the Third World today can be found the apprehension that technologies developed in modern countries will result in the disruption of traditional social and cultural patterns. In some countries these apprehensions have motivated a return to traditional cultural patterns. This has been most notable in Iran, where the social and cultural disruption brought on by rapid economic modernization led to the resurgence of Islamic fundamentalism and the rule of the Ayatollah Khomeini. Even in places where the reaction has not been as strong, many have mourned the passing of traditional patterns of life and thought, and their replacement by a social order that rejects established ways in favor of the latest novelties, and where a cold rationality intrudes into all human relationships.

This has been the fear. But is it really true that technological modernization has to be obtained as a complete package, and that it destroys all existing cultural patterns that are not favorable to it? Although it has been widely accepted, convergence theory has not received a great deal of empirical support. For example, one study found that the values of individual workers were much more affected by their nationality than the kind of work they did; Brazilian oil refinery workers held values and beliefs that more closely resembled those held by farmers in their country than those held by their occupational counterparts in India or Ghana.[5] Then, too, there is the case of Japan, where traditional values and social patterns have shown a remarkable persistence in the face of rapid economic and technological change.

Although the evidence isn't complete, and probably never will be, it is not likely that technology acts as a universal solvent, dissolving all prior cultural and social patterns so that they conform to the presumed dictates of modern technology. The effects of specific technologies or of technological change in the aggregate are complex in the extreme, and it is simplistic to expect that all of the features of a society will be summarily transformed in accordance with technology's requirements. In this matter, it is useful to consider the words of David Landes: "Clearly there are no rigid compulsory relationships between a modern industrial economy and the entirety of its complex, multifaceted environment. Rather there is a wide range of links, direct and indirect, tight and loose, exclusive and partial, and each industrializing society develops its own combination of elements to fit its traditions, possibilities, and circumstances."[6]

Every society has its distinctive characteristics, and a particular technology will not necessarily produce identical social consequences for all that employ it. The actual consequences of a particular technology depend on why people have developed or adopted it in the first place.[7] Having invented gunpowder, the Chinese for many years employed it primarily for fireworks, while in late medieval Europe, gunpowder was immediately used for weaponry. The uses to which a technology is put depends on history, existing social arrangements, and the particular needs of the populace. It may even be the case, as George Daniels has argued, that "no single technological innovation—and no group of them taken together in isolation from nontechnological elements—ever changed the direction in which a society was going before the innovation. . . . Technology, in a word, is used to help people do better what they were already doing for other reasons, and what they are doing for other reasons determines the nature of their future technology."[8]

While this is an extreme view, it is certainly true that technological changes interact with other changes in a mutually influential manner. Social arrangements may themselves stimulate technological innovations that in turn produce social changes, leading to further technological developments.[9] It may be futile to try to determine where the process starts. Technological changes produce social changes, but at the same time technological change is a product of the society in which it takes place. Technologies exert a profound influence on human affairs, but they do not do so in a narrowly deterministic manner.

TECHNOLOGICAL CHANGE AND SOCIAL RELATIONSHIPS

In this discussion, we have been considering the interaction of technology and society in very general terms. In particular, there has been an unspoken assumption that out in the world there are unified things that we call "society" and "culture." This hardly seems realistic. Although social scientists have often assumed that societies are well-integrated constellations of mutually interdependent groups, roles, and statuses, an equally accurate representation of society shows it to be full of stresses, conflicts, and unequal distributions of power. These divisions may have important consequences for the pattern of technological change. Specific technologies may be devised, selected, and transmitted because they serve the interests of a particular group, possibly in opposition to the interests of another group. The development of technology expands human power, but this power is not wielded over only the nonhuman universe. As C. S. Lewis reminds us, "Man's power over nature is really the power of some men over others with nature as their instrument."[10] In considering the influence that technology in general or any single technology has over human affairs, it is therefore necessary to consider not only the technology and its presumed "imperatives," but also the key human agents of the technology, the organizations in which they operate, and how these influence the course of technological change. It is to this that we will next turn.

Experts and Expertise

At the end of the sixteenth century, the philosopher, essayist, and corrupt government official Francis Bacon formulated a famous maxim: "Knowledge is power." Since technology is ultimately based on knowledge, it stands to reason that the power to control technology will be wielded by those with the greatest technical knowledge. But is it so simple? Is the possession of the appropriate knowledge a sufficient basis for the exercise of control over the course of technological change?

During Bacon's time it might have been possible for a well-educated person to understand most of the basic principles underlying the science and technology of that era. The body of human knowledge was limited, and no intellectual barriers separated laypeople from scientists and technicians. Scientific inquiry was not the exclusive province of professional scientists, so amateurs could make significant contributions. Technological knowledge was more restricted, for many craft secrets were jealously guarded, but there were no intellectual reasons that these could not be comprehended by an outsider.

The situation today is vastly different. Knowledge has been growing at an accelerating rate of speed. Whereas a mere ten scientific journals existed in the middle of the eighteenth century, today there are at least 20,000 published throughout the world. Their number has doubled every fifteen years and increased by a factor of ten every fifty years, and the end is nowhere in sight. Part of this enormous increase is due to the proliferation of new scientific and technical specialities; today there are over 900 of them, and again there is no indication that this process has been arrested.[11] The enormous expansion of knowledge has fueled the great technological achievements of our time, but at the same time it has made all of us relative ignoramuses. Even well-trained scientists and technicians have to admit their own limitations when they leave their own field of specialization. In a world in which there is so much to know, nobody knows very much.

This ignorance can limit the ability of citizens and their representatives to control the course of technological change. Most technologies are built on a base of specialized knowledge; if most people are incapable of comprehending that knowledge, there looms the danger that the direction of technology will be left to the small cadre who are. When public choices depend on expert information, experts and not the electorate will supply and evaluate this information.[12] If this happens on a widespread scale, democracy becomes an illusion. Citizens can do nothing but accept the judgments of experts and hope that they are right.

But what does it mean to be "right"? Technologies cannot be judged purely according to their technical merit. Their social consequences have to be considered, and when this happens value judgments often have to be made. These judgments necessarily take the expert well outside the realm of his or her expertise. To illustrate, biologists might agree that a new medical technology is effective in allowing a couple to choose the sex of the child they intend to have, but is that the end of the matter? If the technology is used, it might result in a lopsided sex ratio (because boys are more desired than girls), with all kinds of consequences for society. And even when an issue seems to revolve solely around technical matters, hidden value judgments have a way of sneaking in. Sophisticated tools used by experts, such as cost-benefit analysis, often incorporate subjective judgments.[13] In carrying out cost-benefit analyses, one government agency, the Occupational Safety and Health Administration, reckons a human life to be worth $2 or $3 million, while another agency, the Federal Aviation Administration, puts it at $750,000. Which is the better figure? Are either of them reasonable? Can any monetary value be put on a human life? Is it valid to conduct cost-benefit analyses using such figures?

Public ignorance of technical matters may also be compounded by the desire of some experts to keep their knowledge secret. An expert can bolster his or her importance by being the sole source of important knowledge. The arcane nature of scientific and technological knowledge makes this inevitable to a certain degree, but experts have been known to draw a veil of mystery over their work in order to enhance their positions. The organizations in which they work have done the same thing. As Max Weber pointed out many decades ago, organizations commonly seek to expand their power by keeping their knowledge secret.[14]

In addition to undermining democratic participation, secrecy can also result in bad decisions because it promotes "groupthink." The only people deemed to be fully qualified experts are those who have access to secrets, and these secrets are accessible only to people within the same tight circle. As a result, everyone involved in a new technological project is an enthusiastic supporter, while potential opponents are excluded from the action. Major decisions can thus be made by a cadre of self-interested enthusiasts who are unchallenged by an informed opposition. The decisions to support the Concorde supersonic airliner and an ambitious atomic energy program in Britain occurred under precisely these circumstances.[15]

Most experts are employed by organizations that are likely to have a stake in an issue where expert opinion is sought. The free-standing expert is rare, so, for example, it can be very hard to find a petroleum engineer willing to give courtroom testimony that might be damaging to the oil industry. It is often the case that panels of experts who are brought together to judge a particular program have a strong interest in the continuation of that program. One such example is provided by a software systems analyst who was asked to serve on an advisory panel that was to make recommendations regarding the Strategic Defense Initiative's computer development program:[16]

> I have a project with the U.S. Navy that could profit from SDI funding and I suggested to the panel organizer that this might disqualify me. He assured me quite seriously that if I did not have such a conflict, they would not want me on the panel. He pointed out that the other panelists, employees of defense contractors and university professors dependent on Pentagon funds for their research, had similar conflicts. Citizens should think about such conflicts the next time they hear of a panel of "distinguished experts."

Even when an expert is not constrained by organizational ties, his or her independent judgment may be compromised. Although experts may believe in their intellectual independence, they can be consciously or unconsciously bought off. The enticements of power are all too real; like anyone else, experts can be tempted by the material and psychic rewards that go along with being close to centers of wealth and power. There are many potential rewards for not "rocking the boat," while being the bearer of bad news exposes one to substantial risks.[17]

Of course, even if some experts are not inclined to challenge technological policies, there are usually others who will. Ralph Nader, in calling attention to the shortcomings of contemporary automobiles, probably has done more to affect their design than any single engineer. Nuclear power has been under assault from a variety of quarters, and it is not likely that the issue will be resolved on the basis of purely technical criteria. Decisions to develop new military technologies can be intensely political. Technical considerations are important but not pre-eminent, for the decision to develop and deploy a new weapons system usually has important economic, political, and strategic ramifications.

When major policy decisions are at stake, experts will likely find that their technical knowledge cannot easily be converted into political influence. This can be seen in the usual fate of advisory committees of recognized experts that have been convened by the government in order to address a major scientific or technological issue. A great deal of effort is often expended by these committees, but their recommendations are likely to be ignored if they go against existing policies of the executive branch, expose the administration to political difficulties, or invite the wrath of special interest groups.[18] On occasion, presidents have buried committee reports if they challenge intended policies. For example, in the late 1960s a presidential advisory committee produced a report that was sharply critical of the proposed supersonic passenger airplane, a project strongly supported by the Nixon administration. Faced with this internal opposition, the White House simply kept the report confidential. Members of Congress were not allowed to see it for many months, until one Congressman was able to secure the release of some parts of it.[19]

This was not an isolated case. The use of confidentiality has prevented other reports from seeing the light of day. When this happens, the public at large may simply assume that the experts' advice has been followed or that it has been overridden for good reasons. As a result, the scientific advisory system may simply provide a facade that allows the president to legitimize his decisions about science and technology policy.[20]

In sum, experts are in a position to influence the course of technological change, but at the same time, they face many difficulties in converting their knowledge into power. Having the correct technical answers isn't enough; wielding power requires the mobilization of a constituency and the ability to create a broad base of support.[21] Even outside government, technological decisions have the same political character, with the same need for gathering support for projects and policies. This means that except for fairly trivial matters, issues are not settled on the basis of purely technical considerations.

To explore this issue further, we will consider the working environment of the people who are most closely involved with technological innovation: engineers. As we shall see, more is involved in their work than the exercise of purely technical skills.

ENGINEERS AND THE CONTROL OF TECHNOLOGY

In a modern economy the design and development of new technologies is largely the work of engineers and other technical specialists. It seems natural, then, to expect that a great deal of control over the direction of technological development should be exercised by members of these occupations. Moreover, since so much of our world is shaped by technology, technical experts might be expected to have a powerful influence on society as a whole.

This was the hope of a social theorist named Thorstein Veblen, who expounded his ideas during the first two decades of the twentieth century. According to Veblen, engineers and other technical experts were the "indispensible factor in the everyday work of carrying on the country's productive industry."[22] But at the same time, "they have nothing to say in the planning and direction of this industrial system, except as employees in the pay of financiers."[23] It was this contradiction between the technical acumen of the engineers and their subjugation to managerial control that Veblen saw as the greatest obstacle to continued economic progress. The only solution was a takeover of business enterprises by a "Soviet of Technicians" so that the economy would be continuously stimulated by technological advance instead of being retarded by the narrow financial interests of absentee owners and their delegated managers.[24]

Veblen's ideas provided inspiration for a technocratic political movement that flourished for a brief period during the 1930s, and then expired.[25] But the idea that technical expertise should be the proper basis of control did not die with it. Several decades later, a number of social scientists advanced the idea that Veblen's prescriptions had been achieved; the enterprises of advanced industrial economies were now under the control of engineers and other technologically oriented personnel. For sociologist Daniel Bell, the key personnel in modern corporations were the "corporate organizers," a "special breed, often engineers, whose self-conscious task was to build a new economic form."[26] A similar theme was sounded by economist John Kenneth Galbraith in his influential book, *The New Industrial State*. Galbraith argued that the owners of large business firms no longer ran them, as had been the case in the days of the traditional entrepreneur. Instead, the management of modern enterprises was now in the hands of the "technostructure": managers and engineers with the specialized knowledge and talent essential to planning and decision making.[27]

According to Galbraith's formulations, the members of the technostructure brought a new set of goals to corporations and other productive enterprises. Since they were not the main holders of company stock they would not be the primary beneficiaries of increased profits. Consequently, the new managerial elite was not primarily concerned with maximizing profits. Rather, the technostructure was primarily concerned with maximizing corporate growth and fostering the technological innovations that made this possible.[28] No longer was technological advance held in check by the narrow financial concerns of Veblen's absentee owners; the administration of productive enterprises by engineers and like-minded managers guaranteed the continual development of new products and production techniques. Equally important, the decisions of engineers and their intellectual kin determined the general trajectory of the economy and society.

THE WORK ENVIRONMENT OF ENGINEERS

If engineers have truly taken over corporate management, then at the very least we might expect them to be in control of their own work. As with other skilled occupational groups, engineers have wanted to be recognized as professionals. The work of a professional is based on the possession and application of knowledge that is found only within the ranks of the profession. This means that someone outside the profession cannot fairly pass judgment on the professional; only other professionals are able to do so. The essence of professionalism is autonomy.[29]

This has been evident in the long-established professions, such as medicine and law. Doctors and attorneys have been accustomed to the insulation of their careers from outside forces. In part, this has been due to the fact that, until fairly recently at least, they have been in business for themselves as independent practitioners. But this has not been the case with engineers. Except for a small number of private consultants, engineers have been the employees of an organization, usually some sort of business enterprise.

While serving as employees, engineers do enjoy a considerable amount of freedom as they go about their work. They are not subject to rigid supervision, and they have considerable discretion regarding how they plan their activities and schedule their work. They have to respond to the requests of management and deal with the problems presented to them, but they have wide latitude in determining how the job is done.[30]

At the same time, however, the work of engineers is closely constrained by the expectations of their employers. While they have the freedom to decide *how* a job is to be done, engineers have little influence over the choice of *what* is to be done. And, in truth, few engineers are concerned about their lack of influence in the setting of company policies or the purposes to which their talents are being put. The engineer's training and on-the-job socialization do not produce ambitions of this sort. The work of engineers exemplifies the rational search for the best way of attaining particular ends. But engineers rarely have the privilege of selecting these ends. Indeed, once they do, they no longer act as engineers.[31]

Many engineers cease to serve primarily as engineers when they move into managerial ranks. Engineering is a common first rung on the managerial ladder; nearly two-thirds of engineers in the U.S. become managers during at least part of their career, and during the present decade it is likely that half of U.S. corporations will be headed by managers with a previous background in engineering.[32] But the movement of engineers into management has the paradoxical result of further diluting the engineers' influence. For most engineers, success in one's career means attaining a managerial position, not being an increasingly proficient engineer.[33]

Engineering and management require different skills and orientations. As Samuel C. Florman summarizes, "The characteristics of a good manager—a feeling for people, politics and the bottom line—appear to conflict with those of a first-rate creative engineer—an aptitude for numbers, theorems, materials, and spatial relationships."[34] But while orientations and even fundamental values may conflict, with the prospect of a managerial career dangled in front of him, an engineer is not likely to challenge managerial values, even when they conflict with the engineer's professional values. Expert knowledge and a lifetime devotion to the acquisition and use of that knowledge are central to the professional's value system. In contrast, businessmen and managers value loyalty and personal initiative. The prospect of moving from engineering into management can affect a person's values and undercut his or her identity as an engineer.[35]

The problems of an engineer influencing company policy are illustrated by the dangers of "whistle-blowing." A whistle-blower is someone who detects some sort of wrongdoing within his or her organization and brings it to the attention of the public. An engineer can be put in this position when management makes a decision that is so technically unsound or otherwise defective that it poses a threat to the user or to society as a whole. Since an engineer knows the technical facts of a case better than anyone else, he or she is in the best position to reveal whatever problems have appeared.

Engineers may be willing to call attention to serious problems because of their conviction that "the facts speak for themselves."[36] In this they are sadly naive. Most engineers work in organizational settings with strong hierarchical structures. Decisions are made in accordance with vertical relationships; those at the upper levels can countermand those at the lower level, even if the latter have the "facts" in their favor. This is dramatically illustrated by the experiences of some engineers who worked for a major defense contractor. After a chief engineer designed an aircraft disc brake that was woefully deficient in stopping power, he enlisted the support of his superiors in falsifying crucial test data. When a lower-level engineer and a technical writer objected, they were told to keep their mouths shut and participate in the falsification. Subsequent testing by the Air Force revealed the brake's dangerous shortcomings. Deeply disturbed by the whole process, the dissenting engineer and the writer resigned from their positions. All of the conspirators retained their jobs, except for two who were promoted to higher positions.[37]

The hierarchical structure of organizations also allows the control of communication; employees are expected to "go through channels." Facts that pose a challenge to a superior can thus easily be suppressed. The only alternative is circumventing the hierarchy; a concerned employee can attempt to get the attention of officials at much higher levels of the organization or he or she can reveal malfeasance to the media or some other outside agency. These are risky courses of action. Top-level officials don't like disturbances of conventional channels of authority, and they are also likely to have doubts about an employee who has gone behind the boss's back. Revealing a problem to the media or an outside agency is an even worse breach of proper organizational behavior. Loyalty is much prized in organizations, and an employee who is perceived as being disloyal to his or her superior or to the organization as a whole is someone to be scorned.

The perils of whistle-blowing by an engineer were starkly revealed by the events following the explosion of the space shuttle *Challenger* shortly after launch. A group of engineers from Morton-Thiokol, the manufacturer of the shuttle's solid-fuel booster rocket, had warned the night before that cold weather could cause a failure of the O-rings that sealed the segments of the booster rockets. After the erosion of an O-ring resulted in the tragedy that many had feared, Roger Boisjoly, a specialist in seals who had worked on NASA projects for twenty years, used his testimony before the official government board of inquiry to narrate the technical and managerial failures that led to the tragedy. This soon led to ostracism and isolation at Morton-Thiokol. Eventually diagnosed as suffering from traumatic stress, Boisjoly left his secure and remunerative job, sold his house, and moved out of the community in which he had lived for many years. Although he earned some income by giving lectures on the causes of the disaster, Boisjoly continued to suffer considerable financial hardship and psychological strain.[38]

The battle of an engineer against his or her organization is likely to be a lonely one. Especially unfortunate is the fact that the professional engineering associations usually are not very helpful to the whistle-blower. These associations have often been dominated by businessmen who are more likely to be concerned with their industry than they are with the career concerns of individual engineers.[39] This is illustrated by the case of Ernest Fitzgerald, who was summarily fired after he had called attention to the huge cost overruns on the Lockeed C5A transport plane. His professional association, the American Institute of Industrial Engineers, spurned Fitzgerald's request to "investigate the professional and ethical questions involved." It refused on the grounds that it was a "technical organization," and not a "professional society." The inclusion of military contractors as members of the Institute made this a likely outcome.[40]

From this brief survey, it should be apparent that engineers do not exercise commanding influence over the course of technological change. Their work, while essential to technological development, is constrained by the organizational environment in which engineering operates. The crucial decisions are made by management, and even if engineers are often recruited into its ranks, when this happens they necessarily change their orientations. Managers must be primarily concerned with profit or some other organizational goal; technical virtuosity cannot be a goal in itself. If we are looking for a major source of control over technology, we therefore need to look beyond engineers to the organizations that employ them. This will be done in the next chapter.

Questions for Discussion

1. Which twentieth-century technology has produced more changes in our lives, television or the automobile? What have their social consequences been? Are these consequences solely due to these technologies, or have these technologies interacted with other sources of change?
2. Which technologies seem to confront the greatest amount of cultural lag? Why is this so? Are these lags necessarily bad, or do they serve some useful purposes?
3. All of the world's large steel mills and major airports use similar technologies, and as a result they seem to look about the same. Does this mean that the activities that take place within them are the same everywhere? How might cultural differences manifest themselves in these settings?
4. Some scientific and technological experts claim that nuclear power is safe and economical; other experts say just the opposite. How can a nonexpert choose between the two? What sort of procedures might be used in order to determine which group's claims are more valid?
5. In this chapter the assertion is made that engineers cease to act as engineers when they are involved in the selection of the goals to be pursued by their organization. Do you agree? How might the professional abilities of engineers make them effective participants in the goal-selection process? How might these abilities hinder them?

Notes

1. One of the most influential expositions of this idea is Jacques Ellul, *The Technological Society* (London: Jonathan Cape, 1965).
2. W. F. Ogburn, *Living with Machines* (Chicago: American Library Association, 1933), p. 13.
3. See Daniel J. Boorstin, *The Republic of Technology: Reflections on Our Future Community* (New York: Harper & Row, 1978), pp. 5–8, and Zbigniew Brzezinski and Samuel P. Huntington, *Political Power: USA/USSR* (New York: Viking, 1967), pp. 9–14.
4. Rudi Volti, *Technology, Politics, and Society in China* (Boulder, Colo.: Westview Press, 1982), pp. 64–67; Rennsselaer W. Lee III, "The Politics of Technology in Communist China," in Chalmers Johnson (ed.), *Ideology and Politics in Communist China* (Seattle: University of Washington Press, 1973), p. 301–325.
5. Frederick Fliegel et al., "Technology and Cultural Convergence: A Limited Empirical Test," *Journal of Cross-Cultural Psychology* 10, 1 (March 1979): 3–21.
6. David S. Landes, *The Unbound Prometheus: Technological Change and Industrial Development in Western Europe from 1750 to the Present* (Cambridge: Cambridge University Press, 1972), p. 545.
7. Robert S. Merrill, "The Role of Technology in Cultural Evolution," *Social Biology* 19, 3 (Spring 1972): 246.
8. George H. Daniels, "The Big Questions in the Historiography of American Technology," *Technology and Culture* 11, 1 (January 1970): 3, 8.
9. Edwin T. Layton, "Comment: The Interaction of Technology and Society," *Technology and Culture* 11, 1 (January 1970): 29.
10. Quoted in Ted Howard and Jeremy Rifkin, *Who Should Play God? The Artificial Creation of Life and What It Means for the Future of the Human Race* (New York: Dell, 1977), p. 8.
11. Daniel Bell, *The Coming of Post-Industrial Society: A Venture in Social Forecasting* (New York: Basic Books, 1973), pp. 178–179.
12. Duncan MacRae, Jr., "Science and the Formation of Policy in a Democracy," in Thomas J. Kuehn and Alan L. Porter (eds.), *Science, Technology, and National Policy* (Ithaca, N.Y.: Cornell University Press, 1981), p. 497.
13. David Bazelon, "Risk and Responsibility," in Ibid., pp. 360–361.
14. H. H. Gerth and C. Wright Mills (eds.), *From Max Weber: Essays in Sociology* (New York: Oxford University Press, 1958), p. 233.
15. David Collingridge, *The Social Control of Technology* (New York: St. Martin's Press, 1980), pp. 135–138.
16. David L. Parnas, "Why I Quit Star Wars," *Common Cause* 12, 3 (May–June 1986): 32, 34.
17. Mark R. Berg, "The Politics of Technology Assessment," in Kuehn and Porter, op. cit., p. 485.
18. Martin L. Perl, "The Scientific Advisory System: Some Observations," in Kuehn and Porter, op. cit., p. 267.
19. Ibid., p. 268.
20. Ibid., p. 269.
21. Berg, op. cit., p. 494.
22. Max Lerner (ed.), *The Portable Veblen* (New York: Viking, 1948), p. 440.
23. Ibid., p. 442.
24. Ibid., p. 463.
25. See William E. Akin, *Technocracy and the American Dream: The Technocrat Movement, 1900–1941* (Berkeley: University of California Press, 1977).

26. Daniel Bell, *The End of Ideology* (Glencoe, Ill.: The Free Press, 1967), p. 43.
27. John Kenneth Galbraith, *The New Industrial State*, 2d ed. (Boston: Houghton Mifflin, 1971), pp. 59–71.
28. Ibid., pp. 173–176.
29. See Terence J. Johnson, *Professions and Power* (London: Macmillan, 1972).
30. Robert Zussman, *Mechanics of the Middle Class: Work and Politics among American Engineers* (Berkeley: University of California Press, 1985), p. 58.
31. Ibid., pp. 122–123.
32. Bruce Krauskopf, "The Move from Engineering to Management," *Manufacturing Engineering* 90, 4 (April 1983): 95.
33. Zussman, op. cit., p. 151.
34. Samuel C. Florman, "Up and Down the Dual Ladder," *Technology Review* 88, 6 (August–September 1985): 12.
35. Edwin T. Layton, Jr., *The Revolt of the Engineers: Social Responsibility and the American Engineering Profession* (Cleveland: Case Western Reserve University Press, 1971), p. 8; Zussman, op. cit., p. 223.
36. Deena Weinstein, "Bureaucratic Opposition: Whistle-Blowing and Other Tactics," in Ron Westrum and Khalil Samaha (eds.), *Complex Organizations: Growth, Struggle, and Change* (Englewood Cliffs, N.J.: Prentice-Hall, 1984), p. 255.
37. Frank Vandiver, "Why Should My Conscience Bother Me?" in Robert Heilbroner et al., *In the Name of Profit* (New York: Doubleday, 1972), pp. 3–31.
38. Trudy E. Bell and Karl Esch, "The Fatal Flaw in Flight 51-L," *IEEE Spectrum* (February 1985).
39. Layton, *Revolt*, op. cit., passim.
40. Ralph Nader, Peter J. Petkas, and Kate Blackwell, *Whistle Blowing: The Report on the Conference on Professional Responsibility* (New York: Grossman, 1972), p. 52.

Chapter 16 Organizations and Technological Change

Organizations are a dominant part of modern life. Most of us are born in organizations, are educated in them, spend our working lives in them, and when we finally depart from this world one organization takes care of our mortal remains while another attempts to ease our passage into the hereafter. The pervasive influence of organizations has stimulated a considerable amount of research on their relationship to technological change. We will first consider how organizational structures are affected by technology. After that, we will turn things around and look at how organizational structures and processes can influence the course of technological change.

TECHNOLOGY AS A CAUSE OF ORGANIZATIONAL STRUCTURE

Most of the research on the interaction between organizations and technology has been concerned with how technology has affected organizational structures and processes. Many of these studies have come to the conclusion that an organization's structure—the extent to which authority is centralized, the amount of worker specialization, and the number and importance of formal rules—is strongly influenced by the principal technology that the organization uses.

One such study was conducted by Robert Blauner.[1] Blauner focused on a number of different industries that exemplified particular types of technologies and how they affect work. Printing, for example, was characterized by a set of activities reminiscent of traditional craft occupations that required a high degree of worker skill and involvement. Conducting his research at a time before the invention of computerized typesetting equipment, Blauner found that the absence of specialized machinery resulted in a work environment that allowed a great deal of worker autonomy. In contrast, textile mill operatives were little more than machine-minders. Their activities had to be closely attuned to the demands of the equipment, so there was little room for individual initiative. Assembly-line operations took mechanization to an even higher degree, and forced the worker to submit to a rigid scheduling of activities that was dictated by the inexorable movement of the line. Finally, continuous process industries, such as chemical manufacture or oil refining, made use of highly sophisticated technologies, but unlike textile production or automobile manufacture, they required considerable worker involvement and initiative. Workers had to carefully monitor production processes and take decisive action when the inevitable problems occurred. Their work could not be precisely scheduled or regulated, and they enjoyed a considerable amount of on-the-job-freedom.

Where Blauner's study focused on the relationship between the technology used by an industry and its effects on workers, Joan Woodward's inquiries centered on how different technologies affected organizational structures per se.[2] Of particular concern to her were such variables as the number of levels in the managerial hierarchy, the ratio of managers to other personnel, the number of people supervised by first-line managers, and the flexibility of the managerial system.[3] Woodward was not concerned with individual technologies, but with general manufacturing processes. Her mode of analysis made use of three broad categories: unit and small-batch production (such as shipbuilding or the manufacture of large transformers), large-batch and mass production (as exemplified by automobile manufacture), and process production (such as the production of chemical or petroleum products).

According to Woodward, each of these categories is distinguished by differences in technological complexity and the extent to which uncertainty can be reduced through the application of routine procedures. These differences in turn resulted in different types of organizational structures, each with their distinctive processes. Mass-production technologies required routinized processes, narrowly defined job duties, and a clear hierarchy of authority that kept a firm grip on things. In contrast, the two other methods of production—small-batch production and continuous process production—were much less rigid in their operations, and had a regular flow of worker-management communication that was not filtered according to hierarchical position. In similar fashion, the span of control—the number of workers under a single supervisor—was greatest in firms using mass-production technologies. Other kinds of organizational differences also seemed to reflect production technologies. Managerial specialization occurred more frequently in mass-production firms, and process production technologies seemed to require a higher ratio of managers to other personnel than was the case in organizations using different productive technologies. More relationships could be cited, but Woodward's essential point is that organizational forms must be matched to production technologies. In practical terms, a successful organization is one that uses structures and procedures that are appropriate to the productive technology employed.

A final example of an organizational theory guided by a consideration of technology is provided by the work of Charles Perrow. Perrow's theory is pitched at a more general level than those of Blauner and Woodward. Instead of considering particular types of productive technologies, Perrow considers two basic aspects of the work undertaken within an organization: the degree of variability of the raw materials processed by the organization, and the extent to which problem-solving procedures can be routinized. By raw materials, Perrow does not mean just inanimate objects; iron ore is a basic raw material for a steel mill, but for an employment agency people are its major raw material. In "processing" its raw materials, an organization will confront different degress of difficulty. Perrow's analysis of the problem-solving process centers on the number of exceptional cases that must be confronted and the extent to which the search procedures can be reduced to a routine. Different kinds of organizations are characterized by different combinations of these elements:[4]

Few exceptions and analyzable searches	routinized organizations—e.g., steel mills; custodial institutions—e.g., prisons
Few exceptions and unanalyzable searches	craft industries—e.g., custom-made products; socializing institutions—e.g., schools
Many exceptions and analyzable searches	engineering firms—e.g., heavy machinery; programmed learning schools
Many exceptions and unanalyzable searches	nonroutine manufacturing—e.g., aerospace; therapeutic institutions—e.g., elite psychiatric hospitals

Perrow's implicit definition of technology is quite general and abstract; technology is defined as a way of doing things, and Perrow is not concerned with the material artifacts that are part of the technology employed. What is important here is the nature of the raw material and methods needed to convert it into the organization's products. Above all, Perrow is concerned with organizational effectiveness. A major conclusion of his analysis is that the organizational structure should reflect the particular qualities of the raw materials being used and the technology used for their transformation. To be effective, an organization must use technologies appropriate to its tasks. His conclusions are thus similar to Woodward's: there is no such thing as an optimal organizational form that is universally valid. Everything depends on how things are produced and the materials from which they are made.

These studies, and others like them, have sensitized organizational theorists to the influence of technology on organizational structures and processes. Yet at the same time, few hard-and-fast conclusions can be asserted. Many empirical studies have found no relationship between an organization's technology and its structure.[5] To complicate matters further, apparent correspondences between organizational structures and the technologies employed may actually be the result of another factor, such as the size of the organization.[6] There have also been numerous problems of research strategy, most notably idiosyncratic definitions of technology that make each study unique and not easily comparable with others.

Even so, an important truth has emerged from these studies. As the idea took hold that an organization's structure was influenced by the technology it employed, it became apparent that one of the cherished notions of management, that there is "one best way" to organize things, is incorrect. What worked with one production technology did not work with another. Although the research results have been mixed, their general thrust seems to indicate that in order to be successful an organization needs to match its organizational structures and processes to the technologies it employs.

TECHNOLOGY AS A CONSEQUENCE OF ORGANIZATIONAL STRUCTURE

Up to now we have considered how technology affects organizations. Yet the reverse can also be true. Organizations can shape technological change through their ability to affect the supply and demand for a particular technology. We are a long way from Adam Smith's perfect market, where no individual firm is able to significantly influence the supply of a product or the demand for it. In regard to supply, an organization that accounts for a significant portion of an industry can strongly influence the technological development of the industry as a whole when it creates (or refrains from creating) new products. The choices of a few firms thus determine the kinds of cars we drive, the way steel is made, and the operating systems used by our personal computers. In similar fashion, when a few organizations are the major purchasers of particular products (as when a few aerospace firms buy highly specialized machine tools), their requirements will determine the nature of the product and the technology it embodies.

On what basis do organizations make technological choices? This is an important question, given the importance of organizational decisions for the course of technological change. Unfortunately, researchers have not paid much attention to this issue. Most of the studies centering on the connection between an organization's technology and its structure simply take technology as a given. In the studies summarized above, the technologies used by organizations appeared as independent entities. The implication of these studies is that organizations and their personnel react passively to the presumed dictates of technology.

It is possible, however, that the technologies employed by an organization are themselves the product of the organization's own structure and processes, and in particular, the configurations of power within the organization. Such a view would be in general accord with the perspective presented earlier in this book, that technologies can be, and often are, employed by individuals or groups in order to advance their own interests. Charles Perrow, whose theory was noted above, has recognized this, and in so doing has expressed some disenchantment with his earlier work in which he took technology to be an independent influence on an organization's structure. Perrow forcefully makes the point that after all is said and done, organizations are powerful tools for achieving goals. Much therefore depends on who controls the tools, and for what purposes.[7] A technology might be selected not because of its innate superiority, but because it meets the needs of the power holders within that organization.

Little systematic research has been conducted in this general area, and only a few studies of how technologies have been shaped by the key players of an organization can be cited. One of the most noteworthy is David Noble's study of the development of numerically controlled machine tools.[8] During the early 1950s the Servomechanism Laboratory at the Massachusetts Institute of Technology, with the financial suport of the U.S. Air Force, produced the first numerically controlled (NC) machine tools. Instead of being manipulated by skilled machinists, these devices were guided by programs stored on magnetic tape. NC technology increased productivity and allowed the precise machining of the complex shapes used in modern aircraft. At the same time, NC machine tools significantly reduced the amount of skilled labor required, and removed a great deal of decision-making power from the machinists. These consequences could be taken as an unavoidable by-product of technological advance; that is, the technology determined the skills, work patterns, and hierarchical relationships found in a machine shop using NC machine tools.

According to some critics, the goal of Numerical Control was the elimination of the skilled machinist. (Culver Pictures, Inc.)

The actual story is quite a bit more complicated. In the first place, the use of numerical control was not the only way to make machine tools run automatically. During the late 1940s a "record-playback" method for machine-tool guidance was developed. This technology recorded the motions of a machinist as he made a part. These motions were recorded on magnetic tape, which was then used to run the machine tool. The machinist, and not a programmer removed from the workplace, was thus the source of the machine's instructions. As it turned out, this was the greatest drawback of this technology. Numerical control was far more appealing to managers because it allowed them to gain more control over the work process. NC tools promised a reduction of the decision-making powers of workers, and vested it with managers, engineers, and programmers. In the words of one manager, "Look, with record-playback the control of the machine rests with the machinist—control of feeds, speeds, number of cuts, output; with NC there is a shift of control to management. Management is no longer dependent upon the operator and can thus optimize the use of their machines. With NC, control over the process is placed firmly in the hands of management—and why shouldn't we have it?"[9]

As things turned out, NC did not completely fulfill management's hopes for greater control. As we have seen in Chapter 9, even after more than three decades of development NC tools still have many shortcomings, and the fully automated factory is still a long way off. Human machinists are still necessary to keep production going. The programs governing the machine tool often have bugs; machines malfunction; and adjustments have to be made to compensate for tool wear, casting irregularities, and slight elasticity in the material. Machinists can also make important suggestions regarding how the design of a piece can be improved and how it can be more efficiently produced. The absence of regular feedback from the shop floor to the design room can result in considerable waste and inefficiency.[10]

These kinds of issues will take on a growing importance as computers and automated systems are increasingly incorporated into production processes. Modern information systems can be used to enhance the power and responsibilities of shop-floor workers. At the same time, however, they can produce the opposite effect by giving management a greater capacity to centralize operations and restrict the discretion of the workers. As one manager asserted, "The more we can control with the computer, the less dependence we will have on operators with in-depth knowledge. When there is a problem, we will have computer programs, logic, electronic sensors. We will be able to understand the problems."[11]

A certain degree of computer-based centralization may be justified in terms of its supposed contribution to coordination and efficiency. But this is hardly its only rationale. Many managers worry that computerized information systems can just as easily promote decentralization by making information more readily available to ordinary workers. A top executive from the same firm as the manager quoted above noted that:

> The classic managerial role has been that of the handler, manipulator, dealer, and withholder of information. An issue that the technology is forcing us to face involves the loss of managerial control. . . . Suddenly you are vesting in folks at the control station, folks who are interfacing with a tremendous technology power—power to see all the functions of the operation. That is kind of scary to some managers.[12]

Managers realize that the control of information has been an essential element of their authority. Consequently, in the words of one line manager, "Managers perceive workers who have information as a threat. They are afraid of not being the 'expert.' They are used to having everyone come to them and ask what to do."[13] This use of information was duly noted by a worker in the same firm: "[Managers] can't share information with us, because only by holding on to knowledge that we don't have can they maintain their superiority and their ability to order us around."[14]

In this setting, computer-based information systems are not neutral instruments, but are likely to be used in ways that are congenial to those occupying positions of authority within organizations. As with NC machine tools, a technology may be employed because managers see it as a way of extending their own power. Conversely, computerized information systems may be used to empower workers, enhance their responsibilities, and give them a claim to higher wages. There is no simple technological logic governing the use of these systems. The use of an existing technology may reflect the existing distribution of power in an organization, while the installation of a new technology may become an occasion for conflict over how it will be wielded in the future.

INTERORGANIZATIONAL RELATIONS AND TECHNOLOGICAL DEVELOPMENT

No business firm, no matter how large, develops all of the technologies that are embodied in its products. Every firm depends on a network of suppliers for the new materials, components, and even the ideas that are incorporated into its own products.[15] This is especially true of mature industries, where supplier firms are likely to be the principal source of substantial technological breakthroughs.[16] Without the efforts of outside suppliers to develop and sell innovative products it is likely that the technological level of many established industries' products would stagnate. Technological changes are therefore shaped not only by the structure and dynamics of individual organizations, but also by the pattern of relationships between organizations.

The automobile industry exemplifies the transfer of technology from supplier organizations to the automobilie manufacturers themselves. It also demonstrates that interorganizational relationships may act as obstacles to technological advance within the industry.[17] From its earliest years, the automobile industry has depended on thousands of outside suppliers for many of its components—everything from nuts and bolts to complete engines. These suppliers have provided some of the most important technological innovations for the automobile: disc brakes, radial tires, fuel injection, and electronic ignition, to name but a few. These innovations did not sell themselves, however. Before they appeared in automobiles they had to be bought by the automobile manufacturers (also known as "OEMs," or original equipment manufacturers).

Throughout most of the history of the industry, the relationship between suppliers and OEMs was that of independent buyers and sellers; cooperative approaches were generally absent. Suppliers were often left in the dark about the actual application of their products; a supplier might not even know which cars used its products or that they were about to be replaced by something totally different. The OEMs did not inform their suppliers about long-range intentions, and as a result suppliers could not effectively plan for the future.[18] Research done by suppliers usually took place without direct consultation with the OEMs, and the job of application came only after the component had been developed. There were few attempts to foster joint efforts between OEM and supplier in order to devise and develop new technologies.

The lack of supplier involvement in the design process was a major reason why American automobiles fared poorly in competition with those made in Japan, where there has been a greater inclination to incorporate suppliers in the design process at an early stage of a car's development.[19] However, faced with stringent competition from Japan, in recent years American car manufacturers have begun to intitiate joint activities with their suppliers. This is a logical and necessary step for an industry that needs to constantly upgrade the technological level of its products.

These cooperative activities stand in sharp contrast to traditional buyer-seller relationships, where both parties are only concerned with their own narrow interests. These relationships work well enough when standard commodities are bought and sold, and price is the paramount concern of the parties to the transaction. In contrast, the development and transfer of new technologies involve unique products and are fraught with many uncertainties. These uncertainties make it difficult to evaluate transactions on a precise cost-benefit basis. Moreover, marketplace exchanges by their nature do not foster the long-term vision necessary for the development and application of modern, sophisticated technologies. As the Japanese experience seems to indicate, the creation of effective working relationships between separate organizations is an essential element of technological and economic vitality.

ORGANIZATIONAL SIZE AND TECHNOLOGICAL INNOVATION

Technological innovation requires the commitment of personnel, money, and equipment if it is to be successful. It therefore seems logical to expect that large organizations, which are well endowed with these resources, should be the prime sources of technological changes in industry. This expectation has been strongly articulated by John Kenneth Galbraith: "a benign Providence . . . has made the modern industry of a few large firms an excellent instrument for inducing technical change. It is admirably equipped for financing technical development. Its organization provides strong incentives for undertaking development and for putting it to use."[20] For Galbraith and others who hold to this view, the day has long passed when independent inventors and small firms were the prime source of important new technologies. Their place has been taken by the well-financed laboratories and workshops of large corporations and government agencies.

On the face of it, this seems reasonable. It is hard to imagine how jumbo jets and mainframe computers could be produced in basement workshops. But the large-scale production of an innovation is not the same thing as its original conception and creation. In point of fact, many important inventions were first conceived and produced by small firms. Only when their technical feasibility had been demonstrated did they become products of large enterprises. Key patents on the turbojet were held by Frank Whittle, an Englishman who produced some of the world's first jet engines in a small workshop that was chronically short of funds and personnel. The first digital computer, the ENIAC, was put together at the University of Pennsylvania's Moore School of Electrical Engineering by John Mauchly and J. Presper Eckert with the assistance of ten engineers. These are not exceptional cases. University research teams, private inventors, and small firms seem to be responsible for a disproportionate share of inventive activity.[21] It is likely that half of today's new inventions are originated by small firms and individual inventors.[22]

There is no simple correlation between the size of firms and the extent of their research activity. Some large firms conduct very little technological research.[23] Many of course do, but there is little indication that within a given industry the largest firms pursue R&D more extensively than their smaller competitors.[24] And even if a correlation between size and inventive activity existed, it would tell us nothing about how effective the research efforts of large firms have been. In fact, it is likely that small firms do a better job of producing new technologies relative to the cost of their research expenditures.[25] If patents are taken as an indication of

R&D success, the very largest firms get more limited results for their expenditures than do smaller firms.[26] Moreover, a great deal of research conducted by large organizations with secure markets is not oriented toward the production of radically new things, but is directed at the creation of modest improvements.[27] Much of it is simply defensive and imitative.[28] Large firms that control a substantial share of their market are more likely to produce inventions that maintain the status quo, and avoid engaging in research that might radically alter the industry in which they already have a comfortable niche.

To be sure, invention is not the same thing as innovation. To actually bring an invention successfully into the marketplace may indeed require the resources of a large organization. As we have already seen, the development of new technologies can be a lengthy and expensive process. Large organizations are often better equipped to do development work, where scaling up and making relatively small but nonetheless crucial improvements are the key tasks. This may be their most important contribution to technological innovation.[29] Even so, gigantic size does not appear to be necessary or even beneficial. Large size may be necessary to muster the resources necessary for effective development work, but past a certain optimum size there appear to be no advantages to bigness.[30]

These generalizations of course do not apply to all industries. In some industries such as nuclear power, the bulk of both the research and development work has been done by large firms and government agencies. In other industries, such as computer software, a great deal has been done by small firms. About the safest statement that can be made is that technological advance requires the efforts of a wide variety of organizational types and sizes.[31] The initiation and development of some technologies require the massive efforts that only large organizations can provide. At the same time, many technologies have been—and will be—the result of the efforts of people working in small organizations or even by themselves.

ENTREPRENEURS AND ORGANIZATIONS

Organizations, be they large or small, make essential contributions to the development and use of technology. A solitary individual can only do so much; complex tasks, such as operating an airline or an oil refinery, require the efforts and skills of many people. Organizations coordinate these efforts and skills so that these tasks can be accomplished. This coordination is usually accomplished through the application of basic bureaucratic principles: specialization, formalization (the use of rules, regulations, and standard operating procedures), and hierarchical authority. Although bureaucracies are sometimes thought to be inherently inefficient, there is no reason that they should be. In fact, when it works as it should, a bureaucracy is like a well-oiled machine that faithfully does the job for which it was intended.

But bureaucratic organization has its limitations. Bureaucracies thrive on routine, working best in stable environments where nothing new comes along to disturb their machine-like operations. At the same time, these bureaucratic routines cannot be easily reconciled with the inherent uncertainty faced by someone attempting to invent something new. Furthermore, uncertainties surround not just the technology itself, but also all of the ancillary changes that may be required to use it effectively. All sorts of uncertainties may be involved in such matters as the acceptance of the technology by customers, the retraining of personnel, legal liabilities, performance of component suppliers, and changed government regulations.[32]

Bureaucracies rely on rules, regulations, and formal methods of management to get their employees to do what needs to be done. In most cases, activities and projects are closely scheduled and subject to careful financial scrutiny. But inventors need freedom, which puts them at odds with bureaucratic procedures. A large corporation that is organized according to bureaucratic principles can be a difficult environment for inventors. As Jack Kilby, the co-inventor of the integrated circuit, has noted, "There is a basic incompatibility of the inventor and the large corporation. Large companies have well-developed planning mechanisms which need to know at the beginning of a new project how much it will cost, how long it will take, and above all what it's going to do. None of these answers may be apparent to the inventor."[33]

As we have seen in Chapter 3, technological innovation is often the work of entrepreneurs. By definition, entrepreneurs are people who disrupt normal activities and procedures by vigorously pushing new ways of doing things. Promoting innovation is risky business, and routine-seeking bureaucratic organizations are generally disinclined to take risks. Not only is a bureaucracy's structure attuned to routine activities, its personnel usually seek to protect their positions and maintain their customary way of life. Under these circumstances, an entrepreneur's efforts to promote innovation within an organization may encounter a fair amount of resistance and even hostility.

The tension between the entrepreneur and an entrenched bureaucracy is well illustrated by an episode in the history of naval technology.[34] During the nineteenth century, gunnery was highly inaccurate; of the 9,500 shots fired from American naval vessels during the Spanish-American War, only 121 found their target. No attempt was made to follow a target; gunners used their cannon's elevating gear to give the shell the proper trajectory, fired after the roll of the ship brought the target into their sights, and hoped for the best. Then, during the last years of the nineteenth century, a British naval officer modified the elevating gear of his ship's guns so that they could be rapidly moved, allowing them to continuously track their target. He also modified the gun's telescopic sight so that it wouldn't poke the gunner in the eye when the gun recoiled.

The improvements in accuracy were spectacular, and before long an American officer named William S. Sims embarked on a personal mission to bring these improvements to American ships. Sims was the quintessential entrepreneur; although he hadn't invented the new system, he was utterly convinced of its value, and passionately devoted to seeing it installed. And he was willing to rattle a lot of cages in the course of promoting it. For its part, the Navy acted as a bureaucracy might be expected to act. At first, it ignored Sims' reports on the inadequacies of American gunnery technology and what could be done to improve it. The reports were simply filed away, where they provided a few good meals for the cockroaches that inhabited the file cabinets. When Sims became too insistent to be ignored, the Navy arranged a series of inappropriate tests that "proved" the inadequacy of the new system. Sims finally prevailed only after he succeeded in getting the attention of President Theodore Roosevelt, who installed him in a position that allowed him ultimately to become "the man who taught the Navy to shoot."

The case of Sims vs. the United States Navy is an admittedly extreme example of conflict between an entrepreneur and an entrenched bureaucracy. There are, after all, many examples of successful working relationships between innovative individuals and the organizations that employ them. Not every organization is structured along classic bureaucratic lines, and as we have already seen, an organization committed to innovation will likely have different structural characteristics than one pursuing routine activities. Still, there is no getting around the fact that organizations may find it hard to accommodate employees who take on the role of entrepreneur. These individuals can be arrogant, stubborn, and single-minded to the point of fanaticism. But at the same time, their vision and energy may be the most important element in their organization's success. For many organizations, technological dynamism has required continual efforts to find a balance between the needs of would-be entrepreneurs and established organizational structures and processes. We cannot discard bureaucratic organizations, but we need to find ways that allow them to continue doing what they do best, while at the same time preventing them from obstructing the individuals whose efforts are a major source of technological advance.

Complex societies could not exist without organizations, bureaucratic or otherwise. Organizations are no less important for the initiation, development, and application of most technologies. At the same time, however, inappropriate organizational structures can stifle technological advance. Equally important, the controllers of powerful organizations can select and cultivate technologies that serve only their own interests. This problem is not confined to business organizations; it can also manifest itself in the public sector. Accordingly, we will turn our attention to the biggest and potentially most powerful organization of all: the federal government.

Questions for Discussion

1. How would you organize work if you were the director of a research laboratory? How would you create a climate that would stimulate the creativity of your personnel, while at the same time preventing them from going off on unproductive tangents?
2. How might you design a research project that investigates the extent of managerial control over the selection of a firm's technologies? What sort of industries would be particularly suitable for such a project?
3. Many key industries are dominated by a handful of large firms. This tendency has been increasingly evident in recent years, as many separate firms have merged into larger ones. How might this trend affect technological innovation? Could it pose a threat to the overall development of the economy?
4. In 1986, Steven Jobs, the co-founder of Apple Computer, was eased out of the management post he held with the firm. Why do you suppose this happened? In

what ways did the evolution of the company affect Jobs's relationship to Apple Computer?

Notes

1. Robert Blauner, *Alienation and Freedom* (Chicago: University of Chicago Press, 1964).
2. Joan Woodward, *Industrial Organization: Theory and Practice* (London: Oxford University Press, 1965).
3. Richard H. Hall, *Organizations: Structure and Process*, 3d Edition (Englewood Cliffs, N.J.: Prentice-Hall, 1982), p. 64.
4. Adapted from W. Richard Scott, *Organizations: Rational, Natural, and Open Systems* (Englewood Cliffs, N.J.: Prentice-Hall, 1981), p. 37.
5. John R. Montanari, "Managerial Discretion: An Expanded Model of Organization Choices," *Academy of Management Review* 3, 2 (April 1978): 231–241.
6. Stephen P. Robbins, *Organization Theory: The Structure and Design of Organizations* (Englewood Cliffs, N.J.: Prentice-Hall, 1983), p. 136.
7. Charles Perrow, *Complex Organizations: A Critical Essay*, 2d ed. (Glenview, Ill.: Scott, Foresman, 1979), pp. 13–16.
8. David F. Noble, *Forces of Production: A Social History of Industrial Automation* (New York: Oxford University Press, 1986).
9. David F. Noble, "Social Choice in Machine Design: The Case of Automatically Controlled Machine Tools," in Andrew Zimbalist (ed.), *Case Studies in the Labor Process* (New York: Monthly Review Press, 1979), p. 34.
10. Harley Shaiken, *Work Transformed: Automation and Labor in the Computer Age* (New York: Holt, Rinehart and Winston, 1985), pp. 66–135.
11. Shoshana Zuboff, *In the Age of the Smart Machine: The Future of Work and Power* (New York: Basic Books, 1988), p. 268.
12. Ibid., p. 250.
13. Ibid., p. 252.
14. Ibid., p. 264.
15. James M. Utterback, "Innovation in Industry and the Diffusion of Technology," *Science* 183, 4125 (15 February 1974): 621.
16. James M. Utterback, "The Process of Technological Innovation within the Firm," *Academy of Management Journal* 14, 1 (March 1971); Donald A. Schon, *Technology and Change: The New Heraclitus* (New York: Dell, 1967), pp. 161ff.
17. This section is based on Rudi Volti, "Organizational Sets and Technology Transfer in the U.S. Automobile Industry," paper presented at the annual meeting of the Society for the History of Technology, Cambridge, Mass., November 1984.
18. David C. Smith, "Cooperating for Competition," *Ward's Auto World* (November 1981): 15.
19. Robert E. Cole, "Quality Control Practices in the Auto Industry: United States and Japan Compared," in Robert E. Cole (ed.), *The Japanese Automobile Industry: Model and Challenge for the Future* (Ann Arbor: The University of Michigan, Center for Japanese Studies, 1981), p. 92.
20. John Kenneth Galbraith, *American Capitalism: The Concept of Countervailing Power* (Boston: Houghton Mifflin, 1952), p. 86.
21. John Jewkes, David Sawers, and Richard Stillerman, *The Sources of Invention*, 2d ed. (New York: W.W. Norton, 1969).

22. Donald F. Hornig, "The Role of Government in Scientific Innovation," in James S. Coles (ed.), *Technological Innovation in the '80s* (Englewood Cliffs, N.J.: Prentice-Hall, 1984), p. 50.
23. Jewkes, Sawers, and Stillerman, op. cit., p. 124.
24. Oliver E. Williamson, *Markets and Hierarchies: Analysis and Antitrust Implications* (New York: The Free Press, 1975), pp. 181–182.
25. Christopher Freeman, *The Economics of Industrial Innovation*, 2d ed. (Cambridge, Mass.: The MIT Press, 1982), pp. 143–144.
26. F. M. Scherer, "Firm Size, Market Structure, Opportunity and the Output of Patented Inventions," *American Economic Review* 55 (December 1965): 1098–1125.
27. Williamson, op. cit., p. 187.
28. Freeman, op. cit., p. 176.
29. Ibid., p. 137.
30. Williamson, op. cit., p. 192.
31. Jewkes, Sawers, and Stillerman, op. cit., p. 168.
32. Freeman, op. cit., pp. 212–213.
33. T. R. Reid, *The Chip: How Two Americans Invented the Microchip and Launched a Revolution* (New York: Simon and Schuster, 1985), p. 191.
34. This account is based on Elting E. Morison, *Men, Machines, and Modern Times* (Cambridge, Mass.: The MIT Press, 1966), pp. 17–44.

Chapter 17 Government and the Control of Technology

The last chapter put forth the claim that organizations play a major role in determining the kind of technologies we get. To be sure, these organizations are not all-powerful. When business firms are involved, consumers can affect technological change by deciding whether or not to buy their products. As employees, we might be able to exert some influence through our acceptance or rejection of new workplace technologies. But not all of the major decisions regarding technological development are made in the private sector. To a large and perhaps increasing extent, government organizations determine the pattern of technological change. In a democratic society, this should expand the power of citizens to get the technologies they want. But do they? Does governmental involvement in the development of technology necessarily result in more beneficial technologies? Are democratic ideals realized in the operation of government policies and processes? These are some of the issues that will be addressed in the following survey of the government's role in shaping technological change.

GOVERNMENT ACTIONS AND THE SHAPING OF TECHNOLOGY

Government involvement in technological change is as old as the Republic. The United States of America emerged during an era when technological change was beginning to accelerate, and the framers of the Constitution believed that the government should encourage technological advance through the granting of patents and copyrights to those who had produced new things. Thus, Article 1, Section 8 of the Constitution stipulates that Congress is to "promote the Progress of Science and Useful Arts by securing for limited Times to Authors and Inventors the Exclusive Rights to their respective Writings and Discoveries."

The government's granting of patent rights provides a supportive legal framework for technological development. The government also is responsible for the provision of other types of infrastructure essential to the development and use of modern technologies. Since the ultimate basis of technological advance is the growth of knowledge, governments at all levels do much to indirectly sponsor the development of technology through their support of education. Government programs also provide the auxiliary goods and services that are essential elements of particular technologies. For example, high-speed air travel requires more than just jet aircraft; equally important are government-supported airports, air traffic controllers, weather forecasts, and safety inspections. To take another obvious example, the effective use of automobiles is dependent on a network of roads and highways, as well as driver training in the schools, the enactment and enforcement of traffic laws, and so on.

The case of the automobile brings up another reason for government involvement in technological matters. Every significant technological change brings with it an assortment of auxiliary consequences, some of them pernicious. The private automobile has generated many benefits, but it has also created a number of problems. These problems are not borne exclusively by individual car owners, and for this reason they are known as "negative externalities." That is, they are costs that are not directly taken into account when a person buys a car and puts it on the road: they are external to the market exchange between the buyer and the seller of the car. In operating a car, the owner incurs a number of personal costs: gas, insurance, repairs, and monthly payments that seem to go on forever. But at the same time, the operation of the car results in costs for the society as a whole: increased pollution, congestion, and so on. Since these costs are not included in the market transaction, governmental action is necessary. For example, the negative externality of automobile-generated pollution is commonly addressed by regulations that require the installation of pollution-control devices. Alternatively, the government could assess a pollution tax, thereby making it financially advantageous for an owner to install these devices.[1]

Some government policies and activities are not aimed directly at technological change, but they can strongly influence its scope and direction nonetheless. Tax policies can affect technological change by encouraging or discouraging capital investments that often embody new technologies. Capital investment can be stimulated to some degree by taxing the income derived from investments at a lower rate than other sources of income. As a result, individuals and firms are motivated to invest their money in new machines and other productive goods. The same thing could be done with in-house training programs that enhance workers' abilities to develop and use new technologies. Capital investment can also be encouraged by allowing the accelerated depreciation of capital equipment. This means that the cost of a machine or other piece of equipment with a useful life of, say, twenty years can be used to offset taxes during a much shorter period, say five years. As a result, a firm's immediate profits are increased. Since business executives tend to be primarily concerned with short-term performance, accelerated depreciation can stimulate capital investments that embody new technologies. An accountant would also note that an accelerated depreciation schedule increases what is known as the "present value" of a capital investment.[2]

The government can also encourage the development and use of specific technologies by offering tax credits to those willing to invest in them. For many years the federal tax code provided an important benefit for oil producers by granting a depletion allowance that was used to offset corporate taxes. One result was the expansion of petroleum-based energy technologies. Another approach was taken during the 1970s when the energy crises of that period generated a strong interest in alternative sources of energy. One governmental response was to offer generous tax advantages to people investing in wind-powered electrical generators, and large numbers of windmills were built in some areas.

As noted earlier, innovation does not always respect existing social and legal arrangements, and the courts may be called upon to resolve issues generated by technological change. To take one example, the movie industry claimed that people who used videocassette recorders to tape televised movies were guilty of copyright infringement. When the copyright laws were last amended in 1976, no mention was made of this emerging technology, and it was left to the Supreme Court to decide the issue. In a narrow 5–4 decision handed down in early 1984, the Court ruled that home recording constituted "fair use" under the existing copyright laws. Thus, what the *New York Times* called "a classic case of the law trying to keep pace with careering technology" was settled in favor of the makers of videocassette recorders and tapes, and as a result the movie industry has lost millions of dollars in potential revenues.[3]

In addition to providing a supportive environment for technological development, the federal government is directly involved in the generation of new technologies through its support of research and development. The scope of government involvement in R&D has grown massively in the postwar era. In 1939, the federal government spent only $50 million on the direct support of science and technology, most of which was expended by the Department of Agriculture.[4] In 1994, the federal government is expected to spend $70.4 billion, about half the nation's total expenditure on R&D.[5]

Government involvement in research has been heavily skewed in favor of defense and space exploration. The Apollo program to put men on the moon entailed expenditures of $24 billion.[6] In the 1984 fiscal year, a massive peacetime military buildup resulted in military R&D expenditures of $31.8 billion, nearly 70 percent of the total federal R&D budget.[7] The waning of the Cold War produced few immediate effects; the estimated military expenditure of R&D funds for 1994 is $41.6 billion, nearly 70 percent of federal outlays for Research and Development.[8] Much of this research lies outside the realm of public accountability; in 1987, a quarter of defense R&D funds was absorbed by classified programs that could not be scrutinized even by most members of Congress.[9] It is apparent that the defense sector absorbs vast R&D resources; even so, the funds allocated for other purposes are still considerable, and government financial support can be crucial to technological advance in many sectors.

BUT IS IT NECESSARY?

Why has government involvement in technological development increased to such a marked degree? In part it is due to the recognition that our prosperity is tied to our ability to advance technologically. Given the fierce competition of the international economy, the failure to maintain high technological levels can have severe consequences. Japan has recognized this fact, and its Ministry of International Trade and Industry (MITI) has aggressively sponsored the development of selected high-technology industries such as semiconductors, genetic engineering, robots, and composite materials. Not only has it provided research funds, MITI also has organized cooperative joint research efforts between firms in the same industry so that they can do collectively what they could not do individually.[10]

The government of the United States has been less explicit in its support of technological advance, but it still has played a major role. The development of computers and semiconductors during the early 1950s was extensively supported by federal funds for research and development. Equally important, government procurement of the resulting products created a market at a time when sales within the private sector did not seem promising.[11] In similar fashion, government procurement of aircraft engines and other components provided a strong stimulus for technological advances that spilled over into commercial aviation.[12] Military needs were of course the main source of these efforts. Accordingly, it can be argued that during the post-World War II era the United States has had a de facto policy of supporting technological development, but unlike Japan, much of it has been driven by military rather than commercial considerations.

At the same time, military requirements are not the sole justification for government support of technological advance. Although the market system has served as midwife to an abundance of innovative technologies, it cannot be expected to stimulate every type of technological innovation. Certain kinds of technological advances have little appeal to individual entrepreneurs, for they cannot be "packaged" for sale to individual consumers. Moreover, once these technologies are in place, they presumably benefit everyone, whether they pay for them or not. This is the so-called "free rider" problem. For example, some people may feel a strong need for a cleaner environment, but if they were to individually pay for a power plant's smokestack scrubbers, everyone in the community would benefit equally, even though most of them made no financial contributions to the installation of the scrubbers. Under these circumstances, the government has to take the lead by mandating the development and installation of pollution-control equipment or assessing a pollution tax that stimulated their installation. The costs would be borne by the operators of the power plant, who would likely pass them along to consumers in the form of higher utility bills. Alternatively, the government could subsidize the installation of scrubbers, and offset the subsidy by levying a tax.

Some activities produce benefits that are considerably greater than the returns that accrue to the individuals or firms that engage in them (economists characterize these as "positive externalities"). In some cases, the potential social benefits may be great, but the private rate of return may be so small that no private party is willing to undertake them. This is frequently the case with basic research that has no direct payoff but provides a foundation for subsequent technological applications. Firms in the private sector will also be disinclined to conduct research that is aimed at the extension of knowledge rather than a specific product or process. Ideas and information by themselves are not patentable, so there is no incentive for individuals and firms to seek pure knowledge in the hope that money can be made with it. The lack of direct payoffs leads to underinvestment by the private sector,[13] so government support is essential.

Moreover, once the knowledge is there, it can be made available to all at little additional cost, thereby increasing its benefits to society as a whole. Under these circumstances it makes sense for the government to sponsor basic research, paying for it through taxation.[14] This is precisely what has happened: two-thirds of "pure" scientific research is financed by the federal government.[15]

GOVERNMENT INSTITUTIONS FOR THE GUIDANCE OF TECHNOLOGY

A modern nation's economic and military security is now closely tied to its ability to generate and absorb new technologies. At the same time, simply coping with all of the consequences of technological change can require the leadership and organization that only government can provide. For these reasons, many governments have established special departments and ministries charged with the promotion, control, and governance of technology. But this has not been the case in the United States. There is no cabinet-level Department of Technology, only a collection of Congressional committees and government agencies that involve themselves in a myriad of separate issues and decisions. Some of these agencies reside within the executive branch as parts of cabinet-level departments (such as the Department of Energy's Nuclear Regulatory Commission), while others (such as the National Aeronautics and Space Administration and the National Science Foundation) are quasi-independent agencies directly under the president.

The executive branch is best situated to bring some degree of coherence to technology policy, and in fact the president and his staff have on occasion played significant roles in determining the course of a particular technology by influencing legislation, designating agency heads, impounding funds, and using the other tools available to the president. The importance of the executive branch in overseeing science and technology policy has been recognized by the establishment of the Office of Science and Technology Policy, whose director is in a favored position to influence presidential goals and policies, although his or her voice is only one among many.[16]

Direct presidential involvement in the shaping of technological policy is infrequent, although at times it has been highly significant. President Reagan's Strategic Defense Initiative (the "Star Wars" ballistic missile defense system) represented a major redirection of American defense policy, yet it began as an idiosyncratic effort. The president publicly announced the program only five days after informing his science adviser, while the joint chiefs of staff, the secretary of state and the secretary of defense got only two days notice. The chief scientist for the Pentagon learned of the proposal only nine hours before the speech that presented the program to the nation.[17]

With presidential involvement in the setting of technology policy largely confined to a few major policy decisions, the day-to-day shaping of technological policy tends to be the business of Congress. Here, the disjointed nature of American technology policy is apparent. There is no unified approach to the budgeting of funds for research and development: over a dozen appropriation subcommittees in each house of Congress take a hand in determining the budgets of the agencies they oversee.[18]

These committees and subcommittees often have to address issues directly connected to technological matters. Congressional involvement with technological issues was recognized in 1972 by the creation of the Office of Technology Assessment (OTA), which had as its purpose the provision of information to the Congress to help guide its decisions about technological policy, and to give it more leverage over the president in science and technology matters.[19]

OTA pursues a wide range of issues, ranging from African agriculture to space transportation systems. Its reports are often avidly read by scholars, congressmen and their staff, and interested members of the general public. At the same time, however, it is not certain that OTA has played a significant role in guiding technology policy. OTA's activities are overseen by a board of six Congressmen and six Senators that encompasses a broad range of political orientations. In its early years OTA tended to shy away from highly controversial topics, and it was only infrequently involved with the two largest categories in the federal R&D budget: defense and space.[20] More recently, it has tackled highly contoversial defense issues, most notably the Strategic Defense Initiative. Still, the conduct of the SDI study demonstrated the difficulty of remaining insulated from political pressures. Its classification review took nine months because of the large number of agencies that were concerned about the possibility of divulging military secrets. More significantly, three crucial chapters (which dealt with the survivability of the system should it be attacked) had to be deleted from the final report due to the Defense Department's objections. This decision was a source of considerable frustration to the Director of OTA, who has stated that "The three chapters have been thoroughly cleaned. They contain absolutely no surprises for the Soviets. You have to ask why they're denied to the American people."[21] Whatever the particular merits of witholding this information, the episode serves to demonstrate that technical issues affecting important groups will always be subject to political pressures. OTA has been subject to the criticism that it "does not always find the perfect balance between the goals of its political masters in Congress and the goals of objectivity and neutrality that are more common to the scientists and engineers who perform its tasks."[22] Given the contentiousness of many technological issues and the perpetual difficulty of separating facts from values, it is hard to imagine how such a balance could ever be achieved.

Much of the work of Congress in the setting of technology policy is done by its many committees and subcommittees, ranging from the House Armed Services Subcommittee to the Senate Agriculture Committee. But these units are not freestanding entities. They are closely associated with permanent governmental agencies that have as their chief concern the administration of a particular area of technology, such as NASA for the space program, or the Army, Navy, and Air Force for defense technologies. As we shall see, congressional committees often work in close partnership with these agencies.

PROCESSES

The most direct way that Congress determines the course of technological development is through passing laws that either forbid the use of a particular technology or regulate the way it is used. Indeed, the very notion that the government has a legitimate right to regulate private industry originated with the need to address a problem presented by a new technology—the explosion of steam boilers in the first half of the nineteenth century.[23] The Franklin Institute in Philadelphia was given the task of investigating the construction and operation of boilers. Its report provided a wealth of information about boilers and their components, as well as a set of recommendations regarding their operation and maintenance, many of which were subsequently written into law by Congress. Today, government regulations cover a vast number of technological concerns, everything from the siting of nuclear power plants to the length of trucks.

Along with the passage of laws and regulations, Congress affects the course of technological development through the budgetary process; that is, by deciding to pay for certain kinds of technologies and not others. In some cases, government financial support can be crucial. The supersonic transport plane (SST) literally never got off the ground because in 1971 Congress voted not to appropriate any more funds for its development. Conversely, some technologies, such as civilian nuclear power, have received a large amount of financial and legal support from the federal government that has been crucial to their development and use.

Governmental decisions that affect technological development do not take place in a political vacuum. It often happens that technologies are sponsored because they have substantial political support. Gaining government interest in addressing a technological issue is a crucial first step, and one that cannot be taken for granted. Here again, the piecemeal approach to technological change is evident. There is no clear and comprehensive approach to deciding what technological goals should (and should not) be pursued so that a governmental agenda can be set.[24] Instead of being the product of calm and careful planning, an issue is often put on the political agenda because of some "focusing event"; an oil embargo forces a consideration of the energy technologies we use; the launch of *Sputnik* by the Soviet Union demonstrates the apparent need for an accelerated U.S. space program; a serious accident reveals material and organizational defects in the operation of nuclear power plants.

A feeling of crisis and the need for decisive action also emerge when a social movement is directed at a particular technology. This is particularly likely to happen when a new technological possibility lies just over the horizon. Technologies that have already been established, whatever their unfortunate consequences, simply do not generate the kind of concerns that new ones do.[25] The automobile, along with giving us our prized freedom of mobility, also befouls our environment, encourages ugly sprawl, and kills nearly 40,000 Americans each year. Yet few people consider bringing it under tight governmental regulation or even abolishing it altogether.

To be sure, not all new technologies have produced much in the way of popular opposition. Such diverse technologies as the chlorination of water, the extensive use of medical x-rays, microwave ovens, and birth-control pills have received at best passing attention, and no sustained effort to subject them to governmental control.[26] In fact, it is not entirely clear why some technologies and not others become the focus of widespread concern. But it is certainly true that a technology that produces large-scale failures, such as nuclear reactor accidents, is more likely to evince opposition than one that produces a series of small-scale failures, even though in aggregate their consequences may be great, as has been the case with automobiles. And it is certainly true that media exposure can greatly heighten opposition to a technology with evident problems.[27] Again, this sort of exposure is more likely when the failure is large and spectacular rather than diffuse and important only in the aggregate.

Examples of widespread public debate over new technologies are infrequent; most innovations emerge with little public attention. But this does not mean that politics has been irrelevant. As noted earlier, many new technologies owe their development and diffusion to sponsorship by particular governmental agencies. Since there is no centralized approach to the governance of technology, many key decisions are made at the middle levels of government, that is, by government agencies and congressional committees. It is here that the decision is made to sponsor a particular technology and create the conditions necessary for its success. Technology policy is the sum of many separate policy actions undertaken at this level.

Under these circumstances, the ambitions of individual governmental agencies and their administrators can determine the course of technological development, for an agency has a strong interest in pushing technologies that are in accordance with its own goals, and having them adopted as national policy.[28] This is often achieved with little fanfare. Most decisions about technological policy do not generate much political heat, and a government agency may successfully promote a particular technology with little public discussion or debate.

The importance of government agency sponsorship can be seen in the history of atomic power in the United States. In the mid-1950s, the electrical utilities, the greatest potential users of nuclear power, had little reason to make use of the new technology. According to one utility company president, "We weren't anxious to get into nuclear power, and I don't think any other company in its right mind wanted to get into it either."[29] To overcome this reticence, the key government agency, the Atomic Energy Commission, engaged in a large-scale selling job. A demonstration plant was built at Shippingport, Pennsylvania, in 1954, and technical assistance was offered to the industry, along with subsidized prices for nuclear fuel.[30] This is not to say that the Atomic Energy Commission's actions were illegitimate or that civilian nuclear power owes its existence solely to government support, but it does show how national policy can be shaped by the interests of a particular government agency.

Since some government agencies are more powerful than others, their programs forge ahead, while other agencies' projects languish. As a result, potentially significant programs are not enacted, or if enacted they are chronically underfunded, while more dubious ones move ahead. But why are some government agencies more powerful than others? Some agencies are more influential than others because of the greater technical abilities of their staff and their greater interest in science and technology. But in many cases what really matters is who their clients are. Government agencies do not operate in splendid isolation; they are often closely tied to a particular set of clients. As we have seen with the Federal Communications Commission, even when the task of the agency is regulation, it may be "captured" by the industry that it is supposed to regulate. This happens because the agency needs the support of the industry it oversees in order to maintain its legitimacy and resources. An agency may even delegate to the industry some of the regulatory tasks that it is charged with performing. For example, much of the inspection work performed before the Federal Aviation Administration certifies the airworthiness of a new airplane is actually done by engineers in the employ of the plane's manufacturer.[31]

The close involvement of government agencies with powerful clients often results in the support of technologies that favor these clients. A great deal of government-sponsored agricultural research has focused on productivity improvements that require mechanization and generally high capital investments. Such research has benefited large, wealthy farmers most of all, and it has even contributed to the decline of small farms and the displacement of rural labor.[32] But large farmers provide the bedrock of political support for the Department of Agriculture, and as a result this agency has primarily sponsored technological developments that benefit its most important and powerful clientele. It has been argued that government-sponsored technological development in general has been strongly oriented to the support of centralized, capital-intensive industries; the result, according to some critics, has been unemployment, environmental damage, increased energy consumption, and the spread of monopoly.[33]

If an industry has a strong interest in a particular technology, it may exert a great deal of influence so Congress and the relevant government agencies will lend their support to that technology. In turn, members of Congress may find a particular technology to be politically attractive because it offers "pork barrel" benefits to their constituents. Some technologies, especially those in the military sector, have forged ahead because a project has contractors in a large number of congressional districts. Congressmen quite naturally are likely to support programs that provide jobs and other economic benefits for their states and districts, and are willing to support projects that benefit other Congressmen's constituents in return for support of projects that benefit their own.

The importance of client support can be seen in the career of one government official who had a general inclination to support technological change. As Assistant Secretary of Commerce for Science and Technology, he attempted to push through a number of innovative government programs for the development of new methods of housing construction. But instead of getting support from the building industry, he met with widespread resistance. In contrast, when he worked with the Weather Bureau, he found great enthusiasm for the independent development of weather satellites. As a result, the *Tiros* weather satellite was launched, while attempts to sponsor innovative building techniques languished.[34]

In sum, as with many other government policies, the direction of technological policy often results from the interaction of an "iron triangle" of congressional committees, government agencies, and the key groups whose interests are closely affected by governmental decisions. The decision to sponsor a new technology, along with the shape that it takes as it is developed, is often determined by these triangles. This means that the government usually doesn't sponsor the development of new technologies as part of a general effort to find solutions to public problems; rather, a technology is sponsored because a coalition can be formed from the members of the triangle, who see to it that the selected technologies develop in accordance with their own needs and interests.

It also happens that once under way, the development of a new technology builds up momentum. The personnel and agencies charged with its development have a strong interest in its continuation, as do potential users. As a result, a technological policy gets locked in place and options are closed off. In the absence of outside pressures, projects are destined to move ahead.[35] Even when a technology has manifest problems of considerable magnitude, the sponsors of the technology will tend to "throw good money after bad," for once the commitment has been made and the program is underway, it is far more difficult to terminate a program than to continue with it.

This may be the greatest defect of our system of governing technology. Technological change is an uncertain process, with many unforeseeable consequences. Efforts to forecast the exact consequences of a new technology so that policies can be constructed to deal with them are doomed to failure. The best that can be hoped is that technologies can be stopped or modified if this should prove necessary.[36] This is easier said than done. The effective governance of technology is faced with a fundamental paradox: change is easiest at the beginning, but this is when the need for it cannot be foreseen.[37] As Ernest Fitzgerald, the whistleblower we briefly met in Chapter 15, has said of military projects, ". . . there are only two phases to a major program. The first phase: 'It's too early to tell.' The second phase: 'It's too late to stop.'"[38]

In this way, technology appears to be out of control, as many technologies appear to be driven by their own momentum. But this is not really what is happening; seemingly out-of-control technologies move ahead because they suit the real interests of individuals and organizations. Were these to be taken away, there would be far less reason to continue with the development or deployment of the technology. Technologies do not simply stand or fall on their own merits; their success or failure can often be attributed to the political support that they receive. And this political support is often a reflection of the distribution of power both inside and outside the corridors of government. Under these circumstances, is there any hope that an unorganized citizenry can have any influence over the way technology develops? It is to this, our last topic, that we will next turn.

THE DEMOCRATIC CONTROL OF TECHNOLOGY

The decisions and activities of experts, enterprise managers, and government officials have powerfully affected the kinds of technologies that have been developed, where they have been applied, and how their costs have been met. Most of us have been little more than the consumers of technologies that have been developed by people and organizations largely unknown to us. This process has produced spectacular successes and spectacular problems. Technology has extended our lives, lightened some of our burdens, increased our wealth, and kept us amused. At the same time, the advance of technology has left us with some very unpleasant by-products. This is not the place to attempt to determine if the good has outweighed the bad, for so much depends on what we value. Which way of living is preferable: that of the !Kung Bushmen, with its relaxed work patterns, but material impoverishment and limited control of the environment, or our own, with its abundance of goods and knowledge, but frantic work schedules? In any event, for most of us the choice is moot; unless we decide to become nomadic hunter-gatherers or subsistence farmers, our lives will be lived in a world shaped by modern technology. At issue is not the acceptance or rejection of technology; what matters is how much individual influence we have over the choice of technologies that we use now and will use in the future.

As individuals, are we doomed to the passive acceptance of technological changes that have been decreed by others? Even if technological development has produced more good than bad, it has proceeded in a fashion that has put it beyond our individual control. It seems as though we have struck a bargain whereby we consume the fruits of technological advance in return for delegating to others the power to determine the technologies that shape the basic contours of our lives—everything from what we eat, to how we work, to the way we are entertained.

Most people seem to believe that this is a fair bargain. As we have seen in the first chapter of this book, there is not a great deal of anguish about our reliance on modern technology. Nor do we seem particularly distrustful of those who most strongly influence technological development; national opinion polls indicate that most people take the "establishment" side in technological controversies.[39]

Then too, it must be remembered that democratic participation has its costs. Above all, the expansion of the number of people engaging in technological decision-making can disrupt the orderly processes favored by managers, engineers, and bureaucrats. This follows from the simple arithmetic of participation; the more parties there are to a decision, the harder it is to decide. The result is often a stalemate—"my vote cancels yours." Everybody ends up with veto power, but no one is able to act in a positive fashion.[40] At best, the result is compromise, which may be a reasonable conclusion in some areas, but in matters that require decisive action, such as arms control or preventing the further deterioration of the ozone layer, a compromise decision may be worthless or even dangerous.

Does this mean that there is no middle ground between an apathetic acceptance of technological change decreed by others and a paralyzing stalemate? Can democratic participation be reconciled with technological advance? Before these questions can be answered, it is obviously necessary to have some workable definition of democracy. A vast amount of thinking and writing has gone into this issue, and we cannot presume to break any new ground here. But most of us could agree that a democratic process is one in which the people as a whole are able to participate in making meaningful choices about the things that affect their lives. The capitalist economic system does contain some democratic elements, for consumers can choose to buy or not buy particular products, and in so doing affect the course of technological change. There has been a "democratic" affirmation of videocasettes but not of videodiscs. And certainly consumer choice as much as government mandates dictated the shift to smaller cars in the 1970s. But these are choices made after the fact. Consumers can choose only between existing alternatives; they cannot determine which alternatives will be made available to them. This is a bit like voting in a referendum; the electorate can decide an issue, but it does not determine which issues are put on the ballot in the first place.

In similar fashion, democratic choice can be exercised by voting for or against candidates because of their stance on a key technological issue. There have been local elections that have hinged on a candidate's position on the fluoridation of water or on nuclear power. But this has not usually been the case. Elections are usually fought over a variety of issues; technological policy has been but one, and usually a minor one at that. And even where a technological issue is at the center of a campaign, the voters' voices are heard only after a technology is already in place, as with referenda on nuclear power. There is virtually no popular input into whether or not completely new technologies should be developed. In this sense, democratic controls are at best reactive.

Is it posssible to have democratic control of technology that goes beyond these after-the-fact votes? In some ways, the advance of technology has increased the potential for democratic participation in government. Most importantly, the electronic media have heightened our awareness of political issues. It has often been remarked, for example, that the widespread protests against the Vietnam war were stimulated by the public's viewing of the war on their television sets. Like no other medium before, television revealed to a vast audience the horrible realities of warfare. At the same time, however, exposure does not guarantee an enlightened response. As we have seen, the expansion of newspaper reading that was triggered by the invention of the steam-powered rotary press stimulated a great outpouring of jingoist sentiment and the fanning of imperialist ambitions in the United States. Finally, many critics have decried the present-day debasement of the political process by slickly produced advertisements for candidates and ballot initiatives. It thus seems that advances in communication technologies have had mixed consequences for the democratic process. The electronic media especially hold out the prospect for a more informed electorate, but this has often been overwhelmed by the negative consequences of television, just as happened with the rise of the popular press generations earlier.

Electronic media technology has also presented the possibility of direct democracy through a kind of electronic town hall. Instead of delegating decisions to elected officials, every citizen could be supplied with a computer terminal through which he or she could vote on the pressing issues of the day. Such a system is certainly technologically feasible, but is it a practical possibility? Could voters be expected to be sufficiently well informed to pass judgment on everything from levels of farm price supports to the wisdom of supplying arms to a revolutionary movement in some far-off land? And even if such a system were implemented, would it really result in a truly democratic order? It is obvious that people could not vote on every single issue, so some form of rationing would have to be instituted. Some agency would have to determine what got on the "ballot" in the first place, and this would be a source of power in itself. The ability to set governmental agendas—the determination of which issues make it into the political arena—is a crucial political decision, and this system would do nothing to democratize this process. All voters could do would be to make choices according to a pre-planned agenda.

Given the limitations of direct democracy, we necesarily have to rely on some form of representative democracy. Unfortunately, the established institutions of representative democracy in the United States have not always guided technological development in ways that serve the public interest. The key political actors have been especially responsive to large, well-organized bodies, such as industries, labor unions, and other special interests. All too often the common good is lost to the special pleadings of these groups. Compounding the problem is the fact that the political process is fragmented, which allows government agencies and officials to cut deals with special interests, while no one looks after the needs and concerns of the nation as a whole. Under these circumstances, government officials end up representing themselves and their powerful constituents rather than the people who voted them into office.

Does this mean that there is no hope that individual citizens can affect the course of technological change? Despite what has just been said, there is some room for optimism, and there are numerous occasions when democracy can assert itself in technological affairs. Democracy may work best when concerns are close to home, and the individual is able to work in a local context. When protests against particular technologies have occurred, as they have over the construction of nuclear power plants or the deployment of a new missile, they are often energized by local concerns. General issues are perceived as important because they are close to home.

Many pressing technological issues do not engage the interest of people distracted by their individual problems of raising a family, meeting mortgage payments, or studying for exams. Burdened with mundane daily problems, it is all too easy for us to deny that there is an urgent need to do something about nuclear proliferation or environmental damage. The perception that immediate, local concerns are being affected by technological change can therefore energize greater levels of public perception than is the case when the issues seem too large and remote to concern us.

It must be admitted that local politics is uninteresting to many people, as the usually low levels of voter turnout for local elections attest. Moreover, participation at the grass-roots level faces the same limits as other kinds of political participation; involvement can be time-consuming and stressful, and people would prefer to leave the tasks of governing to local officials. But at the same time, involvement in technological policies that have a direct impact on one's own community can reduce apathy and lethargy. And it might also be the case that expanding democratic participation and tying technological issues to local concerns would result in the emergence of technologies significantly different from those produced by our present system.

Finally, while the present system of governance has produced some technologies of very questionable value, we should also realize that no system of guiding technology will guarantee a utopia. No matter how advanced or benign our technologies, human life will always have its insecurities, disappointments, and tragedies. It can even be argued that it is dangerous to seek perfection in both our technological and political systems. There is something inhuman about a society that cannot tolerate the fact that life does not always go according to plan. Technological development has been characterized as an ongoing search for the "best way," and an inventor may on occasion produce an elegantly satisfactory solution to a technological problem. But most of life is not like that. We will always have to reconcile ourselves to the fact that even the most advanced and responsive technologies will not solve many of our basic problems. As we have seen, the spectacular successes of technological development should not blind us to the fact that some of the inherent difficulties of life are simply not amenable to technological solutions.

THE CHALLENGES OF THE FUTURE

In considering the years ahead, it is well to recall the waggish statement that it's dangerous to make predictions, especially about the future. The course of technological development has produced many surprises, and it will continue to do so. Twenty years ago few would have foreseen the emergence of personal computers with the power of existing mainframes. Yet at the same time it was confidently predicted that computers would soon be able to translate French into elegant English. We are still waiting for this to happen, and it is likely that we will wait for a long time. Forecasting is a highly inexact art, and although it can be safely said that our lives will be strongly affected by technological change, predictions beyond this general statement are fraught with hazards.

Still, the general contours of future technological developments can be discerned. Future developments in transportation and electronic communication will bring the world even closer together. High-speed personal computers with massive memory capacity will put enormous amounts of information at everyone's fingertips. New materials and microprocessors will result in the redesign of many familiar objects. The application of artificial intelligence could affect every aspect of work, play, and learning. And genetic engineering might allow nothing less than the redesign of human beings. Profound social changes will be necessary if we choose to make use of these technologies.

And yet, just because these things are possible does not mean that they should be done. We don't have to resign ourselves to the inexorable advance of technology while passively hoping that it produces more good than harm. Technology remains a human creation, and we have it in our power to determine its future course. Unfortunately, the power to make this determination is very unevenly distributed through society. A major reason for this unequal distribution of power is the unequal distribution of knowledge. A knowledgeable citizenry has always been essential to the functioning of democracy. In the words of Thomas Jefferson, "if we think [the people] not enlightened enough to exercise their control with a wholesome discretion, the remedy is not to take it from them, but to inform their discretion."[41] The expansion of citizens' knowledge is even more crucial than it was in the days of Jefferson, when most adults worked as farmers, and humans had never traveled faster than the speed of a horse. The relentless advance of technology has produced a dizzying pace of change that at times leaves us gasping. Even so, it bears repeating that technology is our own creation, and that its ultimate source is knowledge. But for technology to be truly beneficial, more than technical knowledge is required. Our challenge will be to develop and apply many different kinds of knowledge—ethical, philosophical, sociological, political, and economic—so that we can do a better job of defining our real needs and creating the technologies that serve them.

I hope that this book has given you some of the knowledge necessary to participate in shaping a future that will be strongly influenced by the technological decisions we make. But acquiring knowledge is not enough; it is also necessary to apply that knowledge. And that will be up to you.

Questions for Discussion

1. Do you think that the government should establish an agency similar to Japan's MITI as a way of stimulating technological advance? What obstacles would such an agency confront in the American political climate?
2. Very few elected officials in America have professional training in science or technology. Does this inhibit the effective governance of science and technology? Do you think that American government would be substantially different if most senators and congressmen previously worked as scientists or engineers instead of as lawyers, as is the case today?
3. Can you think of any potentially important technologies that have languished because they have lacked political support? How could they gain this support?
4. What is *your* definition of democracy? On the whole, has technological advance increased or decreased the amount of democratic participation in modern society?
5. Which emerging technologies will have the greatest impact on life during the opening decades of the twenty-first century? Will they have any unfortunate consequences? Should the government restrict or prevent any of them? Do you intend to exert any influence over these decisions? How will you do it?

Notes

1. Alan S. Blinder, *Hard Heads, Soft Hearts: Tough-minded Economics for a Just Society* (Reading, Mass.: Addison-Wesley, 1987), pp. 136–159.
2. Vincent H. Smith, *The Economics of Technology* (Stony Brook, N.Y.: The New Liberal Arts Program, 1990), pp. 12–16 and 27.
3. "Videotaping without Guilt," *New York Times* (22 January 1984), sec. 4:1.
4. W. Henry Lambright, *Governing Science and Technology* (New York: Oxford University Press, 1976), p. 17.
5. *Budget of the United States Government* (Washington, D.C.: United States Government Printing Office, 1993), p. 72.
6. Lambright, op. cit., p. 27.
7. Mary Acland-Hood, "Statistics on Military Research and Development Expenditure," in Stockholm International Peace Research Institute, *World Armaments and Disarmament* (London: Taylor and Francis, 1984), pp. 169–170.
8. *Budget of the United States Government*, op. cit.
9. Richard Barke, *Science, Technology, and Public Policy* (Washington, D.C.: CQ Press, 1986), p. 44.
10. Tom Redburn, "U.S., Japan Wage Battle of 'Chips'," *Los Angeles Times* (24 May 1983): part IV:1; Christopher Freeman, *Technology Policy and Economic Performance: Lessons from Japan* (London: Pinter, 1987), pp. 33–39.
11. Richard C. Levin, "The Semiconductor Industry," and Barbara Goody Katz and Almarin Phillips, "The Computer Industry," in Richard R. Nelson (ed.), *Government and Technical Progress* (New York: Pergamon Press, 1982).

12. David C. Mowery and Nathan Rosenberg, "The Commercial Aircraft Industry," in Ibid.
13. Christopher Freeman, *The Economics of Industrial Innovation*, 2d ed. (Cambridge, Mass.: The MIT Press, 1982), p. 168.
14. Harvey Averch, A *Strategic Analysis of Science and Technology Policy* (Baltimore: The Johns Hopkins University Press, 1985), p. 35.
15. Ibid., pp. 35–36.
16. The Office of Science and Technology Advisor to the President was established during the Eisenhower administration, abolished by Richard Nixon, and then revived by Gerald Ford. See Anne L. Hiskes and Richard P. Hiskes, *Science, Technology, and Policy Decisions* (Boulder, Colo.: Westview Press, 1986), pp. 46–50.
17. Barke, op. cit., p. 43.
18. Herbert Roback, "Congress and the Science Budget," in Thomas J. Kuehn and Alan L. Porter (eds.), *Science, Technology, and National Policy* (Ithaca, N.Y.: Cornell University Press, 1981), p. 300.
19. Lambright, op. cit., p. 24.
20. Barry M. Casper, "The Rhetoric and Reality of Congressional Technology Assessment," in Kuehn and Porter, op. cit., p. 336.
21. Laura van Dam and Robert Howard, "How John Gibbons Runs through Political Minefields: Life at the OTA," *Technology Review* 91, 7 (October 1988): 50.
22. Barke, op. cit., p. 31.
23. John G. Burke, "Bursting Boilers and Federal Power," *Technology and Culture* 7, 1 (Winter 1966): 1–23.
24. Lambright, op. cit., p. 204.
25. Allan Mazur, *The Dynamics of Technical Controversy* (Washington, D.C.: Communications Press, 1981), p. 97.
26. Ibid., p. 90.
27. Ibid., p. 111.
28. Lambright, op. cit., p. 202.
29. James W. Kuhn, *Scientific and Managerial Manpower in the Nuclear Industry* (New York: Columbia University Press, 1966), p. 115.
30. John F. Hogerton, "The Arrival of Nuclear Power," *Scientific American* 218, 2 (February 1968): 21–31.
31. Paul Eddy, Elaine Potter, and Bruce Page, "Destination Disaster: From the Tri-motor to the DC-10, the Risk of Flying," in Larry Hickman and Azizah Al-Hibri, *Technology and Human Affairs* (St. Louis: C.V. Mosby, 1981), pp. 591–592.
32. Lawrence Busch and William B. Lacy, *Science, Agriculture, and the Politics of Research* (Boulder, Colo.: Westview Press, 1983), p. 181.
33. Averch, op. cit., p. 67.
34. Lambright, op. cit., pp. 66–67, 91–93.
35. David Collingridge, *The Social Control of Technology* (New York: St. Martin's Press, 1980), p. 138.
36. Ibid., pp. 20–21.
37. Ibid., p. 1.
38. Quoted in Robert Bell, *Impure Science: Fraud, Compromise and Political Influence in Scientific Research* (New York: John Wiley and Sons, 1992), p. 80.
39. Mazur, op. cit., pp. 47–48.
40. See Jeffrey L. Pressman and Aaron Wildavsky, *Implementation*, 2d ed. (Berkeley: University of California Press, 1979), pp. 102–124.
41. Quoted in David Bazelon, "Risk and Responsibility," in Kuehn and Porter, op. cit., p. 359.

Suggestions for Further Reading

All of the topics introduced in this book merit more extended coverage. The following list is not a comprehensive bibliography, but is intended to provide sources that will allow the reader to pursue particular topics in greater depth. The number or numbers after each entry indicate the chapter or chapters of this book to which it most directly pertains.

Aitken, Hugh G. J. *The Continuous Wave: Technology and American Radio, 1900–1932* (Princeton, N.J.: Princeton University Press, 1985). 4, 5, 12

———. *Syntony and Spark: The Origins of Radio* (Princeton, N.J.: Princeton University Press, 1985). 4, 5, 12

Akin, William E. *Technocracy and the American Dream: The Technocrat Movement, 1900–1941* (Berkeley: University of California Press, 1977). 15

Armytage, W. H. G. *The Rise of the Technocrats: A Social History* (London: Routledge & Kegan Paul, 1965). 15

———. *A Social History of Engineering* (London: Faber & Faber, 1961). 15

Averch, Harvey. *A Strategic Analysis of Science and Technology Policy* (Baltimore: The Johns Hopkins University Press, 1985). 16, 17

Barke, Richard. *Science, Technology, and Public Policy* (Washington, D.C.: CQ Press, 1986). 17

Basalla, George. *The Evolution of Technology* (New York: Cambridge University Press, 1988). 1, 2, 3, 4, 5

Bernard, H. Russell, and Pertti Pelto (eds.). *Technology and Social Change* (Prospect Heights, Ill.: Waveland Press, 1987). 1, 5

Buchanan, R. A. *Technology and Social Progress* (London: Pergamon Press, 1965). 3, 9, 10

Burke, James. *Connections* (Boston: Little, Brown, 1978). 3, 10, 13, 14

Burke, John. *Technology and Change* (San Francisco: Boyd & Fraser, 1979). 1, 3, 4, 5

Burlingame, Roger. *Engines of Democracy: Inventions and Society in Mature America* (New York: Charles Scribner's Sons, 1940). 3, 9, 10

———. *March of the Iron Men: A Social History of Union through Innovation* (New York: Charles Scribner's Sons, 1938). 3, 9, 10

Cardwell, D. S. L. *Turning Points in Western Technology: A Study of Technology, Science and History* (New York: Science History Publications, 1972). 3, 4

Cipolla, Carlo M. *Guns, Sails, and Empires: Technological Innovation and the Early Phases of European Expansion, 1400–1700* (New York: Pantheon, 1966). 13, 14

Collingridge, David. *The Social Control of Technology* (New York: St. Martin's Press, 1980). 17

Constant, Edward W. *The Origins of the Turbojet Revolution* (Baltimore: The Johns Hopkins University Press, 1980). 3, 4, 5

Cooley, Mike. *Architect or Bee? The Human/Technology Relationship* (Boston: South End Press, 1982). 2, 15

Copp, Newton H. and Andrew W. Zanella. *Discovery, Innovation, and Risk: Case Studies in Science and Technology* (Cambridge, Mass.: MIT Press, 1993). 1, 2, 3, 4, 6, 7

Corn, Joseph J. (ed.). *Imagining Tomorrow: History, Technology, and the American Future* (Cambridge, Mass.: The MIT Press, 1986). 1, 2, 3

Cowan, Ruth Schwartz. *More Work for Mother: The Ironies of Household Technology from the Open Hearth to the Microwave* (New York: Basic Books, 1983). 2, 8, 10

Crawford, Stephen. *Technical Workers in Advanced Society: The Work, Careers, and Politics of French Engineers* (Cambridge: Cambridge University Press, 1989). 15

Cutcliffe, Stephen, and Robert Post (eds.). *In Context: History and the History of Technology* (Bethlehem, Pa.: Lehigh University Press, 1989). 1, 2, 3, 4, 5

Daumas, Maurice. *A History of Technology and Invention* (New York: Crown, 1969). 1, 3, 4, 5, 13, 14

de Camp, L. Sprague. *The Ancient Engineers* (Norwalk, Conn.: Brundy Library, 1966). 1, 3, 4, 5, 15

Dickson, David. *Alternative Technology and the Politics of Technological Change* (London: Fontana, 1974). 2, 5, 6, 8, 9, 10

Dutton, Diana B. *Worse than the Disease: Pitfalls of Medical Progress* (New York: Cambridge University Press, 1988). 7

Ellul, Jacques. *The Technological Society* (London: Jonathan Cape, 1965). 1, 2, 15

Etzioni, Amatai, and Richard Remp. *Technological Shortcuts to Social Change* (New York: Russell Sage Foundation, 1973). 2

Evans, Donald D., and Laurie Nogg Adler (eds.). *Appropriate Technology for Development: A Discussion and Case Histories* (Boulder, Colo.: Westview Press, 1979). 5

Ferkiss, Victor C. *Technological Man: The Myth and the Reality* (New York: George Braziller, 1969). 1, 2, 3, 9, 10, 17

Finch, James K. *The Story of Engineering* (Garden City, N.Y.: Doubleday, 1960). 1, 3, 4, 5, 15

Fischer, Claude S. *America Calling: A Social History of the Telephone to 1940* (Berkeley: University of California Press, 1992). 1, 3, 11

Forbes, R. J. *Man the Maker: A History of Technology and Engineering* (New York: Abelard-Schuman, 1958). 1, 3, 4, 5, 13, 15

Forester, Tom (ed.). *The Information Technology Revolution* (Cambridge, Mass.: The MIT Press, 1985). 1, 2, 3, 12

Freeman, Christopher. *The Economics of Industrial Innovation*, 2d ed. (Cambridge, Mass.: The MIT Press, 1982). 3, 4, 5

Haberer, Joseph. *Science and Technology Policy: Perspectives and Developments* (Lanham, Md.: University Press of America, 1985). 17

Gille, Bertrand. *Engineers of the Renaissance* (Cambridge, Mass.: The MIT Press, 1966). 3, 4, 5, 13, 14

Gimpel, Jean. *The Medieval Machine: The Industrial Revolution of the Middle Ages* (New York: Penguin, 1977). 1, 4, 5, 6, 13, 14

Headrick, Daniel R. *The Tools of Empire: Technology and European Imperialism in the Nineteenth Century* (New York: Oxford University Press, 1981). 2, 13, 14

Hindle, Brook, and Steven Lubar. *Engines of Change: The American Industrial Revolution, 1790–1860* (Washington, D.C.: The Smithsonian Institution Press, 1986). 2, 3, 5, 8, 10

Hirschhorn, Larry. *Beyond Mechanization: Work and Technology in a Post-Industrial Age* (Cambridge, Mass.: The MIT Press, 1984). 2, 4, 5, 9, 10

Hiskes, Anne L., and Richard P. Hiskes. *Science, Technology, and Policy Decisions* (Boulder, Colo.: Westview Press, 1986). 4, 15, 16, 17

Hodges, Henry. *Technology in the Ancient World* (New York: Alfred A. Knopf, 1970). 1, 3, 4, 5

Hounshell, David A. *From the American System to Mass Production, 1800–1932: The Development of Manufacturing Technology in the United States* (Baltimore: The Johns Hopkins University Press, 1984). 2, 3, 5, 8, 10

Howard, Ted, and Jeremy Rifkin. *Who Should Play God? The Artificial Creation of Life and What It Means for the Future of the Human Race* (New York: Dell, 1977). 2, 3, 4, 7, 15, 17

Hughes, Thomas P. *American Genesis: A Century of Invention and Technological Enthusiasm* (New York: Viking, 1989). 1, 2, 3, 4, 5

———. *Networks of Power: Electrification in Western Society, 1800–1930* (Baltimore: The Johns Hopkins University Press, 1983). 1, 3, 4, 5, 16, 17

Iannone, A. Pablo (ed.). *Contemporary Moral Controversies in Technology* (New York: Oxford University Press, 1987). 1, 2, 15, 17

Jewkes, John, David Sawers, and Richard Stillerman. *The Sources of Invention*, 2d ed. (New York: W.W. Norton, 1969). 3, 4, 5, 16

Johnston, Ron, and Philip Gummet (eds.). *Directing Technology* (New York: St. Martin's Press, 1979). 1, 3, 4, 5

Jones, Barry. *Sleepers, Wake! Technology and the Future of Work* (Melbourne: Oxford University Press, 1982). 2, 8, 9, 15, 16, 17

Kasson, John F. *Civilizing the Machine: Technology and Republican Values in America, 1776–1900* (New York: Penguin, 1977). 1, 2, 15, 17

Klemm, Friedrich. *A History of Western Technology* (Cambridge, Mass.: The MIT Press, 1964). 3, 4, 5, 6, 8

Kranzberg, Melvin, and Carroll W. Pursell (eds.). *Technology in Western Civilization* (New York: Oxford University Press, 1967). 1, 3, 4, 5, 6, 9, 10, 11, 12, 14, 16, 17

Kuehn, Thomas J., and Alan L. Porter (eds.). *Science, Technology, and National Policy* (Ithaca, N.Y.: Cornell University Press, 1981). 2, 3, 16, 17

Lambright, W. Henry. *Governing Science and Technology* (New York: Oxford University Press, 1976). 2, 3, 16, 17

Landes, David S. *The Unbound Prometheus: Technological Change and Industrial Development in Western Europe from 1750 to the Present* (Cambridge: Cambridge University Press, 1972). 1, 3, 4, 5, 8, 9, 11, 12

———. *Revolution in Time: Clocks and the Making of the Modern World* (Cambridge, Mass.: Harvard University Press, 1983). 8

Landels, J. G. *Engineering in the Ancient World* (Berkeley: University of California Press, 1978). 1, 4, 5, 6, 13, 14

Layton, Edwin T., Jr. (ed.). *Technology and Social Change in America* (New York: Harper & Row, 1973). 1, 2, 3, 4, 5, 15, 16, 17

Long, Franklin, and Alexandra Oleson (eds.). *Appropriate Technology and Social Values: A Critical Appraisal* (Cambridge, Mass.: Ballinger, 1980). 2, 5, 6, 17

Macksey, Kenneth. *Technology in War* (Englewood Cliffs, N.J.: Prentice-Hall, 1986). 13, 14

Mander, Jerry. *Four Arguments for the Elimination of Television* (New York: Quill, 1978). 12

Mansfield, Edwin. *Technological Change* (New York: W.W. Norton, 1971). 3, 4, 5

Marcus, Alan I., and Howard P. Segal. *Technology in America: A Brief History* (San Diego: Harcourt Brace Jovanovich, 1989). 3, 4, 5, 9, 10, 11, 12

Marx, Leo. *The Machine in the Garden: Technology and the Pastoral Ideal in America* (New York: Oxford University Press, 1964). 1, 2, 15

Mazur, Allan. *The Dynamics of Technical Controversy* (Washington, D.C.: Communications Press, 1981). 15, 17
McNeill, William H. *The Pursuit of Power: Technology, Armed Force, and Society since A.D. 1000* (Chicago: University of Chicago Press, 1982). 13, 14
Meehan, Richard L. *The Atom and the Fault: Experts, Earthquakes, and Nuclear Power* (Cambridge, Mass.: The MIT Press, 1984). 2, 16, 17
Merton, Robert. *Science, Technology, and Society in Seventeenth Century England* (New York: Howard Fertig, 1970). 3, 4
Mesthene, Emmanuel G. *Technological Change: Its Impact on Man and Society* (New York: New American Library, 1970). 1, 2, 3, 4, 5, 17
Mokyr, Joel. *The Lever of Riches: Technological Creativity and Economic Progress* (New York: Oxford University Press, 1990). 1, 2, 3, 4, 5, 11, 14, 15
Moore, Wilbert E. (ed.). *Technology and Social Change* (Chicago: Quadrangle, 1972). 2, 3, 4, 5
Morison, Elting E. *Men, Machines, and Modern Times* (Cambridge, Mass.: The MIT Press, 1966). 2, 3, 4, 5, 16, 17
———. *From Know-How to Nowhere: The Development of American Technology* (New York: New American Library, 1977). 2, 3, 4, 5, 16, 17
Mumford, Lewis. *The Pentagon of Power* (New York: Harcourt Brace Jovanovich, 1964). 1, 2, 3, 4, 5, 6, 13, 14, 15, 17
———. *Technics and Civilization* (New York: Harcourt, Brace & World, 1934). 1, 2, 3, 4, 5, 6, 8, 9, 10, 11, 12, 13, 14, 15
———. *Technics and Human Development* (New York: Harcourt Brace Jovanovich, 1967). 1, 2, 3, 4, 5, 6, 13, 14, 15, 17
Musson, A. E., and E. H. Robinson. *Science and Technology in the Industrial Revolution* (Toronto: University of Toronto Press, 1969). 3, 4, 5
Nelkin, Dorothy (ed.). *Controversy: The Politics of Technical Decisions* (Beverly Hills: Sage, 1979). 2, 16, 17
Noble, David F. *America by Design: Science, Technology, and the Rise of Corporate Capitalism* (New York: Oxford University Press, 1979). 2, 3, 4, 5, 15, 16, 17
———. *Forces of Production: A Social History of Industrial Automation* (New York: Oxford University Press, 1986). 2, 3, 4, 5, 15, 16, 17
Norman, Colin. *The God that Limps: Science and Technology in the Eighties* (New York: W.W. Norton, 1981). 5, 6, 8, 9, 10, 17
Pacey, Arnold. *The Culture of Technology* (Cambridge, Mass.: The MIT Press, 1983). 1, 4, 5
———. *The Maze of Ingenuity: Ideas and Idealism in the Development of Technology* (Cambridge, Mass.: The MIT Press, 1976). 1, 4, 5
———. *Technology in World Civilization* (Cambridge, Mass.: MIT Press, 1990). 1, 2, 3, 4, 5, 11, 13, 14
Parker, J. E. S. *The Economics of Innovation: The National and Multinational Enterprise in Technological Change*, 2d. ed. (London: Longman, 1978). 3, 4, 5, 16, 17
Perrucci, Robert, and Joel Gerstl. *The Engineers and the Social System* (New York: John Wiley & Sons, 1969). 15
Petroski, Henry. *To Engineer Is Human: The Role of Failure in Successful Design* (New York: St. Martin's Press, 1985). 1, 2, 3, 4, 15
———. *The Evolution of Useful Things* (New York: Alfred A. Knopf, 1992). 3, 5, 16
Postman, Neil. *Amusing Ourselves to Death: Public Discourse in the Age of Show Business* (New York: Viking/Penguin, 1985). 12

Pursell, Carroll W., Jr. (ed.). *Technology in America: A History of Individuals and Ideas* (Cambridge, Mass.: The MIT Press, 1981). 3, 4, 5
Rae, John, and Rudi Volti. *The Engineer in History* (New York: Peter Lang, 1993). 10, 15
Reid, T. R. *The Chip: How Two Americans Invented the Microchip and Launched a Revolution* (New York: Simon and Schuster, 1985). 3, 4, 5
Reinecke, Ian. *Electronic Illusions: A Skeptic's View of Our High-Tech Future* (Harmondsworth, England: Penguin, 1984). 2, 9, 10
Reich, Leonard S. *The Making of American Industrial Research: Science and Business at GE and Bell, 1876–1926* (Cambridge: Cambridge University Press, 1985). 3, 4, 5, 16
Reiser, Stanley Joel. *Medicine and the Reign of Technology* (New York: Cambridge University Press, 1978). 1, 2, 15
Richter, Maurice N., Jr. *Technology and Social Complexity* (Albany: State University of New York Press, 1982). 1, 2, 3, 4, 5, 15, 16, 17
Rogers, Everett M. *Diffusion of Innovations* (New York: The Free Press, 1962). 16
Ronan, Colin A. *The Shorter Science and Civilization in China: An Abridgement of Joseph Needham's Original Text*, 3 vols. (Cambridge: Cambridge University Press, 1978, 1981, 1986). 3, 4, 5, 13, 14
Rosenberg, Nathan. *Inside the Black Box: Technology and Economics* (Cambridge: Cambridge University Press, 1982). 3, 4, 5, 16
———. *Perspectives on Technology* (Armonk, N.Y.: M.E. Sharpe, 1985). 1, 3, 4, 5, 16
———. *Technology and American Economic Growth* (New York: Harper & Row, 1972). 3, 4, 5, 16, 17
Rothschild, Joan (ed.). *Machina ex Dea: Feminist Perspectives on Technology* (New York: Pergamon Press, 1983). 1, 2, 9, 10
———. *Women, Technology, and Innovation* (New York: Pergamon Press, 1982). 1, 2, 9, 10
Rybczynski, Witold. *Taming the Tiger: The Struggle to Control Technology* (New York: Viking/Penguin, 1985). 1, 2, 3, 4, 5, 13, 14
Sahal, Davendra. *Patterns of Technological Innovation* (Reading, Mass.: Addison-Wesley, 1981). 3, 4, 5
Schiffer, Michael Brian. *Technological Perspectives on Behavioral Change* (Tucson: University of Arizona Press, 1992). 1, 2, 3, 14
Schon, Donald A. *Technology and Change: The New Heraclitus* (New York: Dell, 1967). 3, 4, 5, 16, 17
Schroeer, Dietrich. *Science, Technology, and the Nuclear Arms Race* (New York: John Wiley & Sons, 1984). 13, 14
Segal, Howard P. *Technological Utopianism in American Culture, 1830–1940* (Chicago: University of Chicago Press, 1985). 2, 15
Shaiken, Harley. *Work Transformed: Automation and Labor in the Computer Age* (New York: Holt, Rinehart and Winston, 1985). 9, 10
Shapley, Deborah, and Rustum Roy. *Lost at the Frontier: U.S. Science and Technology Policy Adrift* (Philadelphia: ISI Press, 1985). 17
Singer, Charles et al. (eds.). *A History of Technology* (7 volumes) (Oxford: Clarenden Press, 1954–1978). 3, 4, 5, 9, 10, 11, 12, 13, 14
Smillie, Ian. *Mastering the Machine: Poverty, Aid and Technology* (Boulder, Colo.: Westview, 1991). 1, 2, 3, 4, 6
Spicer, Edward H. (ed.). *Human Problems in Technological Change: A Casebook* (New York: John Wiley & Sons, 1967). 1, 2, 5, 8
Stewart, Francis. *Technology and Underdevelopment* (Boulder, Colo.: Westview Press, 1977). 2, 5, 6

Teich, Albert H. (ed.) *Technology and the Future*, 5th ed. (New York: St. Martin's Press, 1990). 1, 2, 3, 4, 5, 6, 12, 15, 16, 17

Temple, Robert. *The Genius of China: 3,000 Years of Science, Discovery, and Invention* (New York: Simon and Schuster, 1989). 3, 5, 11, 13, 14

Toffler, Alvin. *Future Shock* (New York: Random House, 1970). 1, 2, 3, 9, 10, 15

Trescott, Martha Moore (ed.). *Dynamos and Virgins Revisited: Women and Technological Change in History* (Metuchen, N.J.: Scarecrow Press, 1979). 1, 2, 9, 10

Usher, Abbott Payson. *A History of Mechanical Inventions* (Cambridge, Mass.: Harvard University Press, 1954). 3, 4, 5, 7, 9, 10, 11

van Creveld, Martin. *Technology and War: From 2000 B.C. to the Present* (New York: The Free Press, 1989). 13, 14

Volti, Rudi. *Technology, Politics, and Society in China* (Boulder, Colo.: Westview Press, 1982). 4, 5, 15, 16, 17

Westrum, Ronald L. *Technologies and Society: The Shaping of People and Things* (Belmont, Calif.: Wadsworth, 1990). 1, 2, 3, 4, 5, 6, 10, 15, 16, 17

Whalley, Peter. *The Social Production of Technical Work: The Case of British Engineers* (Albany: State University of New York Press, 1986). 15

White, K. D. *Greek and Roman Technology* (Ithaca, N.Y.: Cornell University Press, 1984). 1, 4, 5, 6, 13, 14

White, Lynn, Jr. *Medieval Technology and Social Change* (New York: Oxford University Press, 1966). 5, 13, 14

Williams, Trevor I., and T. K. Derry. *A Short History of Technology* (Oxford: Oxford University Press, 1961). 1, 3, 4, 5, 6, 9, 10, 11, 12, 13, 14

Winner, Langdon. *Autonomous Technology: Technics-Out-of-Control as a Theme in Political Thought* (Cambridge, Mass.: The MIT Press, 1977). 1, 2, 15

Zimmerman, Jan. (ed.) *The Technological Woman: Interfacing with Tomorrow* (New York: Praeger, 1983). 1, 2, 9, 10

Zussman, Robert. *Mechanics of the Middle Class: Work and Politics among American Engineers* (Berkeley: University of California Press, 1985). 15

Index